Other Titles in This Series

59 **William Arveson, Thomas Branson, and Irving Segal, editors,** Quantization, nonlinear partial differential equations, and operator algebra (Massachusetts Institute of Technology, Cambridge, June 1994)

58 **Bill Jacob and Alex Rosenberg, editors,** K-theory and algebraic geometry: Connections with quadratic forms and division algebras (University of California, Santa Barbara, July 1992)

57 **Michael C. Cranston and Mark A. Pinsky, editors,** Stochastic analysis (Cornell University, Ithaca, July 1993)

56 **William J. Haboush and Brian J. Parshall, editors,** Algebraic groups and their generalizations (Pennsylvania State University, University Park, July 1991)

55 **Uwe Jannsen, Steven L. Kleiman, and Jean-Pierre Serre, editors,** Motives (University of Washington, Seattle, July/August 1991)

54 **Robert Greene and S. T. Yau, editors,** Differential geometry (University of California, Los Angeles, July 1990)

53 **James A. Carlson, C. Herbert Clemens, and David R. Morrison, editors,** Complex geometry and Lie theory (Sundance, Utah, May 1989)

52 **Eric Bedford, John P. D'Angelo, Robert E. Greene, and Steven G. Krantz, editors,** Several complex variables and complex geometry (University of California, Santa Cruz, July 1989)

51 **William B. Arveson and Ronald G. Douglas, editors,** Operator theory/operator algebras and applications (University of New Hampshire, July 1988)

50 **James Glimm, John Impagliazzo, and Isadore Singer, editors,** The legacy of John von Neumann (Hofstra University, Hempstead, New York, May/June 1988)

49 **Robert C. Gunning and Leon Ehrenpreis, editors,** Theta functions – Bowdoin 1987 (Bowdoin College, Brunswick, Maine, July 1987)

48 **R. O. Wells, Jr., editor,** The mathematical heritage of Hermann Weyl (Duke University, Durham, May 1987)

47 **Paul Fong, editor,** The Arcata conference on representations of finite groups (Humboldt State University, Arcata, California, July 1986)

46 **Spencer J. Bloch, editor,** Algebraic geometry – Bowdoin 1985 (Bowdoin College, Brunswick, Maine, July 1985)

45 **Felix E. Browder, editor,** Nonlinear functional analysis and its applications (University of California, Berkeley, July 1983)

44 **William K. Allard and Frederick J. Almgren, Jr., editors,** Geometric measure theory and the calculus of variations (Humboldt State University, Arcata, California, July/August 1984)

43 **François Trèves, editor,** Pseudodifferential operators and applications (University of Notre Dame, Notre Dame, Indiana, April 1984)

42 **Anil Nerode and Richard A. Shore, editors,** Recursion theory (Cornell University, Ithaca, New York, June/July 1982)

41 **Yum-Tong Siu, editor,** Complex analysis of several variables (Madison, Wisconsin, April 1982)

40 **Peter Orlik, editor,** Singularities (Humboldt State University, Arcata, California, July/August 1981)

39 **Felix E. Browder, editor,** The mathematical heritage of Henri Poincaré (Indiana University, Bloomington, April 1980)

38 **Richard V. Kadison, editor,** Operator algebras and applications (Queens University, Kingston, Ontario, July/August 1980)

37 **Bruce Cooperstein and Geoffrey Mason, editors,** The Santa Cruz conference on finite groups (University of California, Santa Cruz, June/July 1979)

(Continued in the back of this publication)

Quantization, Nonlinear Partial Differential Equations, and Operator Algebra

Proceedings of Symposia in
PURE MATHEMATICS

Volume 59

Quantization, Nonlinear Partial Differential Equations, and Operator Algebra

1994 John von Neumann Symposium on
Quantization and Nonlinear Wave Equations
June 7-11, 1994
Massachusetts Institute of Technology
Cambridge, Massachusetts

William Arveson
Thomas Branson
Irving Segal
Editors

American Mathematical Society
Providence, Rhode Island

PROCEEDINGS OF THE 1994 JOHN VON NEUMANN SYMPOSIUM
ON QUANTIZATION AND NONLINEAR WAVE EQUATIONS
HELD AT THE MASSACHUSETTS INSTITUTE OF TECHNOLOGY
CAMBRIDGE, MASSACHUSETTS
JUNE 7–11, 1994

with support from the National Science Foundation
Grant DMS-9400413

1991 *Mathematics Subject Classification.* Primary 81Txx, 81Rxx, 81Vxx, 35Lxx, 35Qxx, 47Bxx, 47Nxx, 58Cxx, 58Exx, 46Gxx, 46Lxx.

Library of Congress Cataloging-in-Publication Data

John von Neumann Symposium on Quantization and Nonlinear Wave Equations (1994 : Massachusetts Institute of Technology)

Quantization, nonlinear partial differential equations, and operator algebra : 1994 John von Neumann Symposium on Quantization and Nonlinear Wave Equations June 7–11, 1994, Massachusetts Institute of Technology, Cambridge, Massachusetts / William Arveson, Thomas Branson, Irving Segal, editors.

p. cm. — (Proceedings of symposia in pure mathematics, ISSN 0082-0717; v. 59)

Includes bibliographical references.

ISBN 0-8218-0381-6 (alk. paper)

1. Geometric quantization—Congresses. 2. Differential equations, Nonlinear—Congresses. 3. Differential equations, Partial—Congresses. 4. Operator algebras—Congresses. 5. Mathematical physics—Congresses. I. Von Neumann, John, 1903–1957. II. Arveson, William. III. Branson, Thomas, 1953– . IV. Segal, Irving Ezra. V. Title. VI. Series.

QC174.17.G46J67 1994

530.1′2′0151474—dc20 96-5187

CIP

∞ The paper used in this book is acid-free and falls within the guidelines
established to ensure permanence and durability.

♻ Printed on recycled paper.

10 9 8 7 6 5 4 3 2 1 00 99 98 97 96

Contents

Preface

John von Neumann was one of the greatest mathematical talents of the 20th Century, and a worthy successor to Poincare, Hilbert, and Weyl, in both intensity and breadth. His achievements are all the more impressive if one takes into consideration the short span of his life, and his very active national defense role during World War II and thereafter.

In addition to his personal research contributions he was immensely influential in general scientific terms, in a range extending from the abstraction of projective-like but continuous geometries to the practical application of high-speed computers. No single theme can adequately represent his broad scientific thrust. But his scientific impact as well as personal involvement was greatest in the work that provided mathematical coherence for quantum mechanics, and established the algebraic paradigm in modern analysis. The problem of developing an appropriate and rigorous formalism for quantum field theory was the chief motivation for his late monumental work on operator algebra.

This work remains unfinished, a long-standing challenge to advanced mathematical ideas and techniques, as well as fundamental for theoretical physics. In the meantime, in the past several decades, the global theory of nonlinear partial differential equations, the understanding and application of which was one of von Neumann's principal motivations in the development of computers, has made great strides. Nonlinear wave equations are what quantum field theory is all about, and thus a symposium bringing together these and related areas was a very fitting memorial to von Neumann's work and spirit.

Recent inroads in higher-dimensional nonlinear quantum field theory and in the global theory of relevant nonlinear wave equations have been accompanied by very interesting cognate developments. The latter include symplectic quantization theory on manifolds and in group representations, the operator algebraic implementation of quantum dynamics, as well as differential geometric, general relativistic, and purely algebraic aspects. "Quantization and Nonlinear Wave Equations" thus appeared highly appropriate as the theme of the first John von Neumann Symposium, a new series made possible by the establishment of a fund by Dr. and Mrs. Caroll V. Newsom in honor of his memory, intended to treat topics of emerging significance that are likely to underlie future mathematical developments.

The present volume is a microcosm of the recent seminal progress in the entire area. The work on quantization exemplifies both the Hilbert space and symplectic manifold approaches. That on nonlinear wave equations runs a gamut from microlocal theory through the Boltzmann equation to General Relativity. A variety of operator algebraic approaches to quantum dynamics, describing current developments and indicating key problems, are presented. Evolutionary aspects of what is essentially quantum probability, a new line of research branching off from the quantization problem, is exposed from different vantage points. Geometric aspects of partial differential operators are brought into relation to these matters. These topics are moreover interwoven in a coherent way based on the theme of the Symposium.

The Organizing committee for the Symposium, whose topic was selected by the AMS Committee on Summer Institutes and Special Symposia, consisted of Haim Brezis and Irving Segal, co-chairs, in addition to William Arveson, Robert Blattner, and Thomas Branson. The Symposium was supported in part by the National Science Foundation, and took place at the Massachusetts Institute of Technology from June 6–12, 1994. In addition to the papers presented here, invited lectures were given by Robert Blattner, Leonard Gross, Victor Guillemin, Roger Howe, Victor Kac, J. T. Stafford, and Zhengfang Zhou. Additional lectures were contributed by participants in the Symposium.

Irving Segal

Proceedings of Symposia in Pure Mathematics
Volume **59**, 1996

E_0-SEMIGROUPS IN QUANTUM FIELD THEORY

WILLIAM ARVESON

Department of Mathematics
University of California
Berkeley CA 94720, USA

8 July 1994

ABSTRACT. A lecture presented at the von Neumann symposium on Quantization and Nonlinear Wave Equations held at MIT in June, 1994. We describe the role of semigroups of endomorphisms of von Neumann algebras in algebraic formulations of quantum field theory, and present a summary of recent developments in the theory of E_0-semigroups.

1. Relation to symmetry groups of algebras of local observables.

In the algebraic formulation of quantum field theory, one works with a C^*-algebra $\mathcal{A}$ that is acted upon by a group of automorphisms representing the symmetries of spacetime (for example the Poincaré group), and which is also given a local structure (subalgebras of local observables). This local structure normally consists of associating a (von Neumann) subalgebra of $\mathcal{A}$ to every bounded region of spacetime in a coherent way, so that inclusions match up correctly, so that these subalgebras transform covariantly under the group action, so that the topologies fit together

1991 *Mathematics Subject Classification.* Primary 46L40; Secondary 81E05.
Key words and phrases. von Neumann algebras, semigroups, automorphism groups, quantum field theory.
This research was supported in part by NSF grant DMS92-43893

consistently, and which may satisfy additional conditions arising from the underlying physics. This C^*-algebra is not necessarily presented in a representation on a Hilbert space. However, given any state of $\mathcal{A}$ which restricts to a normal state on the local subalgebras, one can perform the usual GNS construction and obtain a representation of this algebra on a Hilbert space H.

Typically, the group of automorphisms (or large subgroups of it) leave the state invariant, and therefore one can naturally construct a unitary representation of the group on H which implements the action. Moreover, once one has a concrete representation of $\mathcal{A}$, one can associate von Neumann algebras with every (not necessarily bounded) open subset of spacetime by taking weak closures of appropriate unions.

For example, there is a von Neumann algebra $\mathcal{M}(C) \subseteq \mathcal{B}(H)$ associated with the forward light cone

$$C = \{(t,x) \in \mathbb{R} \times \mathbb{R}^3 : (x_1^2 + x_2^2 + x_3^2)^{1/2} < t\ \}.$$

If v is any vector of $\mathbb{R} \times \mathbb{R}^3$ which lies inside C then the one parameter group $\lambda \in \mathbb{R} \mapsto \lambda v$ defines a physical flow of time. This flow is represented by a one-parameter unitary group $\{U_\lambda : \lambda \in \mathbb{R}\ \}$ acting on H. Since $C + \lambda v \subseteq C$ for $\lambda \geq 0$, we may conclude from the covariance properties of the given structure that for *nonnegative* λ we have

$$U_\lambda \mathcal{M}(C) U_\lambda^* \subseteq \mathcal{M}(C). \tag{1.1}$$

Thus (1.1) defines a semigroup of isometric unit-preserving $*$-endomorphisms which acts on the von Neumann algebra $\mathcal{M}(C)$. In the simplest case where $\mathcal{M}(C)$ is a factor of type I_∞, we have a prototypical example of an E_0-semigroup $\{\alpha_\lambda : \lambda \geq 0\ \}$

$$\alpha_\lambda(A) = U_\lambda A U_\lambda^*, \qquad A \in \mathcal{M}(C), \quad \lambda \geq 0.$$

We will take up this situation in a more general setting. By an E_0-semigroup we mean a semigroup $\{\alpha_t : t \geq 0\ \}$ of self-adjoint normal endomorphisms of the algebra $\mathcal{B}(H)$ of all bounded operators on a Hilbert space H which preserves the identity ($\alpha_t(\mathbf{1}) = \mathbf{1}$ for every $t \geq 0$) and is such that

$$t \mapsto < \alpha_t(A)\xi, \eta >$$

is continuous for every $\xi, \eta \in H$ and every $A \in \mathcal{B}(H)$. We emphasize that for the results discussed in the sequel, it is essential that the Hilbert space H should be separable. Naturally, one may speak of E_0-semigroups that act on type I factors $\mathcal{M}$, and in this less concrete situation we require that the predual of $\mathcal{M}$ be separable. Two E_0-semigroups α (acting on $\mathcal{M}$) and β (acting on $\mathcal{N}$) are said to be *conjugate* if there is a $*$-isomorphism $\theta : \mathcal{M} \to \mathcal{N}$ such that $\theta \circ \alpha_t = \beta_t \circ \theta$ for every $t \geq 0$. One may also speak of E_0-semigroups that act on more general factors, but we shall not do so here.

The preceding discussion shows that one may obtain E_0-semigroups by starting with a strongly continuous one-parameter unitary group $U = \{U_t : t \in \mathbb{R}\}$ acting on a Hilbert space H, finding somehow a type I subfactor $\mathcal{M}_+ \subseteq \mathcal{B}(H)$ which is invariant in the sense that

$$U_t \mathcal{M}_+ U_t^* \subseteq \mathcal{M}_+$$

for nonnegative t and then defining an E_0-semigroup α on $\mathcal{M}_+$ by setting

$$\alpha_t(A) = U_t A U_t^*, \qquad t \geq 0, \quad A \in \mathcal{M}_+.$$

Notice that in this case there is a "complementary" E_0-semigroup. Indeed, the commutant $\mathcal{M}_- = \mathcal{M}'_+$ of $\mathcal{M}_+$ is also a type I factor, and is invariant under the automorphisms $A \mapsto U_t A U_t^*$ for *negative* t. Thus we can define a second E_0-semigroup β acting on $\mathcal{M}_-$ by way of

$$\beta_t(A) = U_t^* A U_t, \qquad t \geq 0, \quad A \in \mathcal{M}_-.$$

These remarks show that the E_0-semigroups that arise in this way from automorphism groups of a larger type I factor always occur in pairs. Several natural questions occur at this point. For example, does *every* abstract E_0-semigroup α which acts on a type I factor arise in this way from a one parameter group of automorphisms of a larger type I factor? If it does, then what is the relation between the "positive" and "negative" semigroups α and β? We will answer these and other questions presently. Let us first recall some known results pertaining to an analogous but much simpler situation.

Let $U = \{U_t : t \geq 0\}$ be a strongly continuous one parameter semigroup of isometries acting on a Hilbert space H. Then H can be decomposed into a direct sum $H_1 \oplus H_2$ giving rise to a corresponding decomposition

$$U = V \oplus W \tag{1.2}$$

of U into a direct sum of two semigroups of isometries with the property that V is *pure* in that

$$\bigcap_{t \geq 0} V_t H_1 = 0$$

and such that W_t is unitary for every $t \geq 0$. This decomposition is unique. Moreover, the pure summand V is unitarily equivalent to a direct sum of a countable number n of copies of the *shift* semigroup S, which acts on the Hilbert space $L^2(0, \infty)$ by way of

$$S_t f(x) = \begin{cases} f(x-t), & \text{for } x > t \\ 0, & \text{for } 0 < x \leq t. \end{cases}$$

The number n is also uniquely determined. Taken together, these results are often called the Wold decomposition. The summand H_2 is defined by

$$H_2 = \bigcap_{t \geq 0} U_t H,$$

and of course H_1 is the orthocomplement of H_2.

One of my first thoughts about E_0-semigroups was to speculate that there should be an effective decomposition resembling the Wold decomposition, and that this might lead to some kind of classification up to conjugacy. It quickly became clear that this idea was too naive. Since a discussion of this issue serves to make a significant point about E_0-semigroups, we offer the following comments.

Suppose that one is given an E_0-semigroup $\alpha = \{\alpha_t : t \geq 0\}$ which acts on a type I factor $\mathcal{M}$. We indicate the extent to which a decomposition resembling the Wold decomposition is valid for α. There are two natural von Neumann algebras associated with the action of α:

$$\mathcal{M}_\infty = \bigcap_{t \geq 0} \alpha_t(\mathcal{M}), \qquad \text{and}$$
$$\mathcal{M}^\alpha = \{A \in \mathcal{M} : \alpha_t(A) = A, t \geq 0\}.$$

$\mathcal{M}_\infty$ is the "tail" von Neumann algebra, and $\mathcal{M}^\alpha$ is the fixed algebra. We have $\mathcal{M}^\alpha \subseteq \mathcal{M}_\infty \subseteq \mathcal{M}$, and the action of α_t on $\mathcal{M}_\infty$ defines an automorphism of $\mathcal{M}_\infty$ for every $t \geq 0$. Thus the restriction of α to $\mathcal{M}_\infty$ defines a one-parameter *group* of automorphisms of $\mathcal{M}_\infty$. This W^*-dynamical system plays the role of the summand W in the Wold decomposition (1.2), and it is clearly a conjugacy invariant of the original E_0-semigroup α.

Let $\mathcal{N} = \mathcal{M}'_\infty$ denote the (relative) commutant of $\mathcal{M}_\infty$. Notice that since $\mathcal{M}$ is a factor of type I, we may treat relative commutants (in $\mathcal{M}$) as if they were true commutants. Since $\alpha_t(\mathcal{M}_\infty) = \mathcal{M}_\infty$ for every $t \geq 0$, it follows in a straightforward way that $\alpha_t(\mathcal{N}) \subseteq \mathcal{N}$, $t \geq 0$. Moreover, it is not hard to show that the restriction of α to $\mathcal{N}$ has trivial "tail" in the sense that

$$\bigcap_{t \geq 0} \alpha_t(\mathcal{N}) = \text{center of } \mathcal{N} = \text{center of } \mathcal{M}_\infty.$$

In the important case where $\mathcal{M}_\infty$ is a factor, then we have a decomposition resembling (1.2) to some extent, in that

$$\mathcal{M} = \mathcal{N} \vee \mathcal{M}_\infty \tag{1.3}$$

is generated by a pair of mutually commuting factors which are commutants of each other, that α determines an automorphism group on $\mathcal{M}_\infty$, that α leaves $\mathcal{N}$ invariant and in fact defines a "pure" semigroup on $\mathcal{N}$ in the sense that

$$\bigcap_{t \geq 0} \alpha_t(\mathcal{N}) = \mathbb{C}\mathbf{1}. \tag{1.4}$$

If $\mathcal{M}_\infty$ happens to be a factor of type I then (1.3) becomes a simple tensor product

$$\mathcal{M} = \mathcal{N} \otimes \mathcal{M}_\infty, \tag{1.5}$$

and if we define E_0-semigroups β on $\mathcal{N}$ and γ on $\mathcal{M}_\infty$ by restricting α in the obvious way then we have a decomposition

$$\alpha = \beta \otimes \gamma. \tag{1.6}$$

Formula (1.6) appears quite analogous to the Wold decomposition (1.2).

Unfortunately, $\mathcal{M}_\infty$ need *not* be a type I von Neumann algebra. Moreover, even if it were a type I factor, we do not know how to classify E_0-semigroups that satisfy (1.4). And if $\mathcal{M}_\infty$ is a factor not of type I, then almost nothing is known about semigroups satisfying (1.4). Finally, it is known that there exist E_0-semigroups whose fixed algebra $\mathcal{M}^\alpha$ is a factor of type II_∞ or of type III (we will have more to say about this in sections 7 and 8).

These remarks lead one to the conclusion that E_0-semigroups are much too complex to hope for a useful classification up to *conjugacy.*

2. Cocycle perturbations.

Let α be an E_0-semigroup acting on $\mathcal{B}(H)$. A *cocycle* for α is a *strongly continuous* family of unitary operators $U = \{U_t : t \geq 0\}$ satisfying

$$U_{s+t} = U_s\alpha_s(U_t), \qquad s, t \geq 0. \tag{2.1}$$

Notice that (2.1) implies that $U_0 = \mathbf{1}$. The condition also implies that the family of endomorphisms $\beta = \{\beta_t :\geq 0\}$ defined by

$$\beta_t(A) = U_t\alpha_t(A)U_t^*, \qquad t \geq 0 \tag{2.2}$$

is another E_0-semigroup acting on $\mathcal{B}(H)$. β is called a *cocycle perturbation* of α. Two E_0-semigroups are said to be *cocycle conjugate* if one of them is conjugate to a cocycle perturbation of the other. This is an important relation for the theory of E_0-semigroups, and we want to discuss the notion of cocycle perturbations in more detail.

Suppose that α and β are two E_0-semigroups which act on the same $\mathcal{B}(H)$, and that β is a cocycle perturbation of α as in (2.2):

$$\beta_t(A) = U_t\alpha_t(A)U_t^*, \qquad A \in \mathcal{B}(H), \quad t \geq 0.$$

Notice that the unitary operators $V_t = U_t^*$, $t \geq 0$ define a cocycle for β and of course we have

$$\alpha_t(A) = V_t\beta_t(A)V_t^*.$$

Thus, this notion of cocycle perturbation defines an equivalence relation on the set of all E_0-semigroups that act on a fixed $\mathcal{B}(H)$.

Note too that this relation is analogous to Connes' notion of exterior equivalence for one-parameter automorphism groups of a von Neumann algebra $\mathcal{M}$ [13]. In particular, Connes discovered that if one is given a pair of faithful normal weights of

$\mathcal{M}$, then the two associated modular automorphism groups are exterior equivalent, and this led to the establishment of an elegant classification theory for type III factors. Cocycle perturbations are of fundamental importance for the classification of E_0-semigroups as well, but for reasons that involve new phenomena that are unique to the theory of E_0-semigroups (see Theorem B below).

Remark 2.3. The cohomology of cocycle perturbations. Suppose first that α and β are *conjugate* E_0-semigroups which act on the same $\mathcal{B}(H)$, say $\theta \circ \beta_t = \alpha_t \circ \theta$ where θ is a $*$-automorphism of $\mathcal{B}(H)$. We may find a unitary operator W that implements θ in the sense that $\theta(A) = W^*AW$, $A \in \mathcal{B}(H)$. It follows that

$$\beta_t(A) = U_t\alpha_t(A)U_t^*, \qquad t \geq 0 \quad A \in \mathcal{B}(H),$$

where U is the α-cocycle defined by $U_t = W\alpha_t(W)^*$. An α-cocycle of this form is called *exact*. Conversely, if β is a perturbation of α by an exact cocycle, then α and β are conjugate. This remark explains why the problem of classifying E_0-semigroups to conjugacy is difficult: it is the same as the problem of computing a very subtle noncommutative cohomology group.

Remark 2.4. Perturbations of the infinitesimal generator. We want to point out a useful interpretation of cocycle conjugacy in terms of infinitesimal generators. Suppose that α acts on $\mathcal{B}(H)$, and let δ be the generator of α. For an operator A in the domain of δ, $\delta(A)$ is defined as the limit in the strong operator topology

$$\delta(A) = \lim_{t \to 0+} t^{-1}(\alpha_t(A) - A).$$

The domain of δ is a unital $*$-subalgebra $\mathcal{D}$ of $\mathcal{B}(H)$ and δ is a self-adjoint derivation of $\mathcal{D}$ into $\mathcal{B}(H)$. Let B be any bounded self adjoint operator in $\mathcal{B}(H)$ and put

$$\delta'(A) = \delta(A) + i(BA - AB).$$

δ' is another unbounded self-adjoint derivation having the same domain $\mathcal{D}$, and it can be shown that it is the generator of a second E_0-semigroup β which acts on $\mathcal{B}(H)$. In fact, β is a cocycle perturbation of α and the cocycle U relating β to α as in (2.2) can be defined as the global solution of the linear differential equation

$$\frac{d}{dt}U(t) = iU(t)\alpha_t(B), \qquad t \geq 0,$$

with the initial condition $U(0) = \mathbf{1}$. It is easy to see from the properties of this differential equation and the fact that B is bounded that the cocycle U is *norm* continuous:

$$\lim_{t \to t_0} \|U(t) - U(t_0)\| = 0,$$

for every $t_0 \geq 0$. Now in general, cocycles need not be norm continuous, and the generators of cocycle perturbations of α cannot be obtained by perturbing the generator of α by bounded derivations. Nevertheless, it is useful to think of the cocycle perturbations of an E_0-semigroup as having been obtained by perturbing its generator by an "unbounded" derivation. While this interpretation is merely a heuristic conceptual device, it can sometimes be made precise. In any event, one should consider the definition of cocycle perturbation as a precise formulation of this idea.

3. Numerical index.

The preceding discussion shows that one should look for cocycle conjugacy invariants. We now describe a numerical index invariant which is appropriately thought of as a quantized form of the Fredholm index of certain differential operators (or of the operator semigroups that they generate). It is defined as follows.

Let $\alpha = \{\alpha_t : t \geq 0\}$ be an E_0-semigroup acting on $\mathcal{B}(H)$. A *unit* for α is a strongly continuous semigroup of bounded operators $\{U_t : t \geq 0\}$ on H which intertwines α in the sense that

$$\alpha_t(A)U_t = U_t A, \qquad t \geq 0, \quad A \in \mathcal{B}(H),$$

and is not the zero semigroup $U_t = 0$. $\mathcal{U}_\alpha$ will denote the set of all units of α. Notice that if U and V belong to $\mathcal{U}_\alpha$ then for each $t \geq 0$ $V_t^* U_t$ is a bounded operator which, by 3.1, commutes with every bounded operator on H, and hence must be a scalar multiple of the identity

$$V_t^* U_t = f(t)\mathbf{1}, \qquad t \geq 0.$$

f is a continuous complex-valued funciton satisfying $f(0) = 1$, and one easily verifies that $f(s+t) = f(s)f(t)$ for all nonnegative s, t. Hence there is a unique complex number $c(U,V)$ satisfying

$$f(t) = e^{c(U,V)t},$$

for every $t \geq 0$. The bivariate function

$$c : \mathcal{U}_\alpha \times \mathcal{U}_\alpha \to \mathbb{C}$$

is called the *covariance function* of α. The covariance function is easily seen to be conditionally positive definite. Thus one may use the pair $(\mathcal{U}_\alpha, c)$ in a natural way to construct a Hilbert space H_α. It can be shown that H_α is *separable.* We define

$$d_*(\alpha) = \dim H_\alpha.$$

Thus the possible values of $d_*(\alpha)$ are $0, 1, 2, \ldots, \aleph_0$. Significantly, there are E_0-semigroups α for which $\mathcal{U}_\alpha = \emptyset$ [18], and in this case it is convenient for the arithmetic of the index to define $d_*(\alpha) = 2^{\aleph_0}$. See [2] for more detail.

The basic property of this index is that it behaves well under the formation of tensor products [6]. If α and β are two E_0-semigroups acting, respectively, on $\mathcal{B}(H)$ and $\mathcal{B}(K)$ then it is easy to see that there is a unique E_0-semigroup $\alpha \otimes \beta$, acting on $\mathcal{B}(H \otimes K)$, which satisfies

$$(\alpha \otimes \beta)_t : A \otimes B \mapsto \alpha_t(A) \otimes \beta_t(B)$$

for every $t \geq 0, A \in \mathcal{B}(H), B \in \mathcal{B}(K)$.

Theorem A.

$$d_*(\alpha \otimes \beta) = d_*(\alpha) + d_*(\beta).$$

Notice that $d_*(\alpha)$ is invariant under cocycle conjugacy, or what is in substance the same, that $d_*(\alpha)$ is stable under cocycle perturbations of α. That is easily seen as follows. Let $W = \{W_t : t \geq 0\}$ be any cocycle for α and consider the associated perturbation β

$$\beta_t(A) = W_t \alpha_t(A) W_t^*, \qquad t \geq 0, \quad A \in \mathcal{B}(H).$$

One may check that if $U = \{U_t : t \geq 0\}$ is a unit for α and if we define

$$\tilde{U}_t = W_t U_t, \qquad t \geq 0,$$

then $\tilde{U}$ is a unit for β. Moreover, $U \mapsto \tilde{U}$ is a bijection of $\mathcal{U}_\alpha$ onto $\mathcal{U}_\beta$ which carries one covariance function into the other. Thus the Hilbert spaces H_α and H_β are isomorphic and it follows that $d_*(\alpha) = d_*(\beta)$.

This definition of numerical index is equivalent to the one given in [3], and differs significantly from Powers' earlier definition of numerical index [18]. In the latter the index was defined as the multiplicity of a certain representation of a C^*-algebra associated with the infinitesimal generator of α. Since the representation depended on making a particular choice of a unit, it was not clear that this index was unambiguously defined. Later, Powers and Robinson [22] gave another definition of index which was obviously well-defined, but which took values in an abstract set of equivalence classes rather than in the nonnegative integers. Recently, Powers and Price [23] have shown that $d_*(\alpha)$ actually agrees with Powers' "infinitesimal" definition in all cases. In particular, it is now clear that Powers' original definition of the numerical index was unambiguous.

4. Continuous tensor product systems.

We now want to emphasize a fundamental relationship between E_0-semigroups and continuous tensor products of Hilbert spaces. Indeed, one could argue that up to cocycle conjugacy, the theory of E_0-semigroups *is* the theory of continuous tensor products of Hilbert spaces.

Heuristically, a *product system* is a measurable family of Hilbert spaces $E = \{E_t : t > 0\}$ which behaves as if each E_t were a continuous tensor product

$$E_t = \bigotimes_{0<s<t} H_s, \qquad H_s = H$$

of copies of a single separable Hilbert space H. While this heuristic picture is often useful, one must be careful not to push it too far. Indeed, we will see that this picture is basically correct for the simpler examples of product systems, but that there are other examples with the remarkable property that the "germ" H fails to exist. Rather than reiterate the details of the precise definition here, we illustrate the essentials of the structure of product systems in the discrete case, where the positive real line is replaced with the discrete set $\mathbb{N} = \{1, 2, \dots\}$ of positive integers. Then we will indicate briefly how to change the axoims to pass from $\mathbb{N}$ to $\mathbb{R}^+$. Full details can be found in [2].

Let H be a separable Hilbert space. For every $n = 1, 2 \dots$ let $E(n)$ be the full tensor product of n copies of H:

$$E(n) = \underbrace{H \otimes H \otimes \cdots \otimes H}_{n \text{ times}}.$$

We may organize these spaces into a family of Hilbert spaces $p : E \to \mathbb{N}$ over $\mathbb{N}$ by setting

$$E = \{(t, \xi) : t \in \mathbb{N}, \xi \in E(t)\ \},$$

with projection $p(t, \xi) = t$. We introduce an associative multiplication on the structure E by making use of the tensor product

$$(s, \xi) \cdot (t, \eta) = (s + t, \xi \otimes \eta),$$

$\xi \in E(s)$, $\eta \in E(t)$. This multiplication is bilinear on fibers, and has the two additional properties

$$E(s + t) = \overline{\text{span}}\ E(s) \cdot E(t), \qquad s, t \in \mathbb{N} \tag{4.1}$$

$$< ux, vy > = < u, v >< x, y >, \quad \forall u, v \in E(s), \quad x, y \in E(t). \tag{4.2}$$

Notice that the Hilbert space associated with the sections of $p : E \to \mathbb{N}$ is the direct sum

$$\sum_{t \in \mathbb{N}} E(t) = \sum_{n=1}^{\infty} H^{\otimes n},$$

essentially the full Fock space over the one-particle space H.

A *unit* is a section $n \in \mathbb{N} \mapsto u_n \in E(n)$ satisfying

$$u_{m+n} = u_m u_n,$$

and which is not the zero section. The most general unit has the form

$$u_n = \underbrace{x \otimes x \otimes \cdots \otimes x}_{n \text{ times}}$$

$n \geq 1$, where x is a nonzero element of the one-particle space H.

Fix $n = 1, 2 \ldots$ and let u be a vector in E_n. u is called *decomposable* if for every $k = 1, 2, \ldots, n - 1$ there are vectors $v_k \in E(k)$, $w_k \in E(n - k)$ such that

$$u = v_k w_k.$$

Notice that the most general decomposable vector in $E(n)$ is an elementary tensor of the form

$$u = x_1 \otimes x_2 \otimes \cdots \otimes x_n,$$

where $x_k \in H$ for $k = 1, 2, \ldots, n$.

A product system is a similar structure, except that it is associated with the space of positive reals rather than $\mathbb{N}$. More precisely, a *product system* is a family of separable Hilbert spaces over the semi-infinite interval $(0, \infty)$

$$p : E \to (0, \infty)$$

which admits an associative multiplication that is bilinear on fiber spaces and has properties analogous to properties corresponding to (4.1) and (4.2). Additionally,

E is endowed with a *standard Borel structure* which is compatible in the natural way with the other structures of E, and with the further property that there should be a separable Hilbert space H such that

$$E \cong (0,\infty) \times H, \tag{4.3}$$

where $\cong$ denotes an isomorphism of measurable families of Hilbert spaces. Condition (4.3) is nontrivial, and is the property in this category that corresponds to local triviality of Hermitian vector bundles. There is a natural notion of isomorphism for the category of product systems, which we will not write down explicitly here (see [2]). We may also define *units* of E and *decomposable* vectors of the fiber spaces E_t in a way analogous to the above. For example, a unit is a measurable cross section

$$u : t \in (0,\infty) \mapsto u(t) \in E_t$$

satisfying $u(s+t) = u(s)u(t)$ for all $s,t > 0$, and which is not the zero section.

One might expect that it should be possible to write down a comprehensive list of (continuous) product systems as we have done above for their discrete analogues. In that case there is, up to isomorphism, exactly one "discrete" product system for every integer $d = 1,2,\dots,\aleph_0$. d can be taken to be the dimension of the one-particle space E_1. In the continuous case, however, nothing like that is true. While there is a family of "natural" examples parameterized by the values $d = 1,2,\dots,\aleph_0$, there are many others as well. The problem of classifying general product systems is an unsolved problem which, as we will see, is of central importance in the theory of E_0-semigroups.

Every E_0-semigroup $\alpha = \{\alpha_t : t \geq 0\}$ gives rise to a product system E_α in the following way. Suppose α acts on $\mathcal{B}(H)$. For every $t > 0$, let $E_\alpha(t)$ be the intertwining space

$$E_\alpha(t) = \{T \in \mathcal{B}(H) : \alpha_t(A)T = TA \quad \forall A \in \mathcal{B}(H)\}.$$

$E_\alpha(t)$ is obviously a complex vector space, and in fact it is a Hilbert space. Indeed, if $S,T \in E_\alpha(t)$ then because of the intertwining property it follows that T^*S commutes with every operator $A \in \mathcal{B}(H)$. Hence there is a unique complex number $< S,T >$ such that

$$T^*S = < S,T > \mathbf{1}.$$

$<,>$ is an inner product on $E_\alpha(t)$ which makes it into a Hilbert space. Thus we have a family of Hilbert spaces $p : E_\alpha \to (0,\infty)$ defined by

$$E_\alpha = \{(t,\xi) : t > 0, \quad \xi \in E_\alpha(t)\}$$

where $p(t,\xi) = t$. If we use operator multiplication to define multiplication in E_α by

$$(s,S)\cdot(t,T) = (s+t, ST),$$

then it is not hard to establish the properties (4.1) and (4.2). E_α inherits a natural standard Borel structure as a subspace of $(0,\infty) \times \mathcal{B}(H)$, where $\mathcal{B}(H)$ is endowed with the Borel structure generated by its weak* topology. Finally, it is a nontrivial fact that property (4.3) is valid as well [2]. Thus E_α is a product system. The importance of product systems in this context derives from the following result from [2].

Theorem B. *α and β are cocycle conjugate iff E_α and E_β are isomorphic product systems.*

In order to illustrate what lies behind Theorem B, let us consider the case in which α and β both act on $\mathcal{B}(H)$ and β is a cocycle perturbation of α:

$$\beta_t(A) = U_t\alpha_t(A)U_t^*, \qquad t \geq 0, \quad A \in \mathcal{B}(H),$$

where $U = \{U_t : t \geq 0\}$ is an α-cocycle. In this case it is quite easy to exhibit an isomorphism of product systems $\theta : E_\alpha \to E_\beta$. Indeed, if we fix $t > 0$ and choose $T \in E_\alpha(t)$, then one may verify directly that U_tT belongs to $E_\beta(t)$. Thus we can define a unitary operator $\theta_t : E_\alpha(t) \to E_\beta(t)$ by $\theta_t(T) = U_tT$; θ is defined as the total map. θ is a (measurable) bijection which is unitary on fibers, hence it is an isomorphism of families of Hilbert spaces. The cocycle property implies that for $s, t > 0$ and $S \in E_\alpha(s), T \in E_\alpha(t)$ we have

$$\begin{aligned}\theta_s(S)\theta_t(T) &= U_sSU_tT = U_s\alpha_s(U_t)ST \\ &= U_{s+t}ST = \theta_{s+t}(ST).\end{aligned}$$

Thus θ preserves multiplication, and hence it is an isomorphism of product systems.

To prove the converse direction (still assuming that α and β act on the same $\mathcal{B}(H)$), one basically has to start with an isomorphism $\theta : E_\alpha \to E_\beta$ and show that θ is associated with a unitary α-cocycle U as above. This is technically more difficult, but the basic idea is similar (see [2], Theorem 3.18).

Theorem B implies that the problem of classifying E_0-semigroups up to cocycle conjugacy is equivalent to that of classifying *certain* product systems...namely, those product systems that can be associated with an E_0-semigroup as above. It is a basic result in our approach to E_0-semigroups that *every* product system arises in this way.

Theorem C. *For every product system E there is an E_0-semigroup α such that E is isomorphic to E_α.*

The proof of Theorem C is very indirect [5], and makes essential use of the spectral C^*-algebras discussed in sections 7 and 8.

We want to point out that there is a general notion of *dimension* of an abstract product system that generalizes the numerical index of E_0-semigroups. This dimension function takes values in the set $\{0, 1, 2, \ldots, \aleph_0, 2^{\aleph_0}\}$, and corresponding to Theorem A it obeys

$$\dim(E \otimes F) = \dim E + \dim F$$

where $\otimes$ denotes the natural tensor product in the category of product systems. The relation of d_* to dim is given by the expected formula

$$d_*(\alpha) = \dim(E_\alpha).$$

It follows that, in order to classify E_0-semigroups up to cocycle conjugacy, one should seek to determine the structure of product systems. In particular, we may consider the set Σ of all isomorphism classes of product systems. The class of a product system E will be denoted $[E]$. There is a natural "addition" in Σ, defined by the natural tensor product operation

$$[E] + [F] = [E \otimes F],$$

which makes Σ into an abelian semigroup. There is a neutral element, which arises from the *trivial* product system Z. Z is defined as the trivial family of one-dimensional Hilbert spaces

$$Z = \{(t, z) : t > 0, z \in \mathbb{C}\ \}$$

where the inner product in $\mathbb{C}$ is the usual one $< z, w >= z\bar{w}$. The multiplication in Z is given by

$$(s, z) \cdot (t, w) = (s + t, zw).$$

It can be seen that if α is an E_0-semigroup which is trivial in the sense that each α_t is an automorphism, then E_α is isomorphic to Z. Moreover, it is also a fact that there are no nontrivial line bundles in Σ: every product system E with one-dimensional fiber spaces $E_t, t > 0$, is isomorphic to Z [6]. One can verify directly that for every $[E] \in \Sigma$ one has

$$[E \otimes Z] = [Z \otimes E] = [E].$$

Therefore $[Z]$ functions as an additive zero for Σ.

There is also a natural involution in Σ, defined by $[E] \mapsto [E^o]$, where E^o is the product system *opposite* to E. The structure of E^o is identical to that of E except that multiplication is reversed. With this involution, Σ becomes an abelian involutive semigroup with a zero element. *The problem of classifying E_0-semigroups up to cocycle conjugacy becomes the probem of determining the structure of the involutive semigroup Σ.*

At this point, we are not even certain of the *cardinality* of Σ! It is expected that Σ is uncountable, but this has not been proved. Notice for example that by the more general version of Theorem A alluded to above, the dimension function defines a homomorphism of Σ into the additive semigroup of extended nonnegative integers $\{0, 1, 2, \dots, \aleph_0, 2^{\aleph_0}\}$. Little is known about the quotient structure in Σ defined by this homomorphism. For example, by recent work of Powers in which a new family of E_0-semigroups is constructed [21], we now know that for each $k = 1, 2, \dots, \aleph_0$ there are infinitely many elements x of Σ that satisfy

$$\dim(x) = k.$$

But it is still not known if there are distinct elements $x, y \in \Sigma$ satisfying $\dim(x) = \dim(y) = 0$. Equivalently, is there a nontrivial E_0-semigroup α with the property that there is a nonzero unit $U = \{U_t : t \geq 0\ \}$ and such that every other unit V is related to U by a relation of the form

$$V_t = e^{i\lambda t} U_t, \qquad t \geq 0$$

where λ is a complex number?

Finally, let us return to some questions raised in section 1 concerning the problem of extending E_0-semigroups to automorphism groups acting on a larger type I factor. Suppose that we are given a *pair* of E_0-semigroup s α, β acting respectively on $\mathcal{B}(H)$ and $\mathcal{B}(K)$. We are interested in obtaining conditions on the pair α, β which imply that there is a one-parameter group of unitaries $W = \{W_t :\in \mathbb{R}\}$ acting on the tensor product $H \otimes K$ whose associated automorphism group $\gamma_t(C) = W_t C W_t^*$ satisfies

$$\gamma_t(A \otimes \mathbf{1}) = \alpha_t(A) \otimes \mathbf{1}, \qquad \text{for } t \geq 0$$
$$\gamma_t(\mathbf{1} \otimes B) = \mathbf{1} \otimes \beta_{-t}(B), \qquad \text{for } t \leq 0.$$

In case such a group exists, then α and β are said to be *paired.* This relation was introduced by Powers and Robinson in [22] as an intermediate step in their definition of another index. The Powers-Robinson index is an equivalence relation defined in the class of all E_0-semigroups; α and β are said to have the *same index* if there is a third E_0-semigroup σ with the property that α can be paired with σ and σ can be paired with β.

Using the theory of product systems one can determine the precise nature of this pairing, and thus give a more concrete form to the Powers-Robinson index. The details are as follows. It is not hard to show that α and β are paired iff their product systems E_α and E_β are *anti* isomorphic. Moreover, with any particular anti isomorphism $\theta : E_\alpha \to E_\beta$ one can write down a *specific* one parameter unitary group W acting on $H \otimes K$ which simultaneously extends α and β in the above sense (the details can be found in [2, pp 27–28]).

Carrying this one step further, we can answer a question posed in section 1 which asks how to describe the possible ways of extending an E_0-semigroup to a larger type I factor. More precisely, starting with a particular E_0-semigroup α we seek to describe all possible ways of finding a one-parameter unitary group $W = \{W_t : t \in \mathbb{R}\}$ acting on some other Hilbert space K and a type I subfactor $\mathcal{M} \subseteq \mathcal{B}(K)$ with the property that

$$W_t \mathcal{M} W_t^* \subseteq \mathcal{M}, \qquad \text{for } t \geq 0$$

and such that α is conjugate to the restriction of $\mathrm{ad}W_t$, $t \geq 0$ to $\mathcal{M}$. The preceding remarks show that one should begin by considering the product system E_α^o opposite to E_α. Notice that by Theorem C, there exist E_0-semigroups whose product systems are isomorphic to E_α^o. Moreover, the set of all possible extensions of α is described by the set of all anti isomorphisms of E_α to E_α^o. In turn, these are obtained by composing the natural anti isomorphism of E_α to E_α^o with an arbitrary automorphism of E_α itself. Thus the set of all possible extensions of α is parameterized by the group of all automorphisms of the product system E_α. This group is computed explicitly for the "standard" examples in ther last section of [2]; its structure in the case of general product systems remains mysterious.

In particular, it follows from these remarks that two E_0-semigroups have the same index in the sense of Powers-Robinson iff their product systems determine the same element of the semigroup Σ. Thus, this discussion also gives a somewhat more concrete description of the Powers-Robinson index: it is now identified with this Σ-valued index map

$$\alpha \to [E_\alpha] \in \Sigma.$$

5. CCR **flows.**

We have already remarked that there is a sequence of *standard* E_0-semigroups, and corresponding to them a sequence of *standard* product systems. These have been described in detail in [2] and [7]. The purpose of this section is to give a brief description of these standard examples and to describe recent results on the problem of characterizing E_0-semigroups that are cocycle conjugate to a standard one.

It is useful to think of this construction as a functor related to second quantization; that interpretation makes explicit the precise sense in which the index $d_*(\alpha)$ of an E_0-semigroup is a quantized form of the Fredholm index of certain differential operators. A fuller discussion of these issues can be found in [7].

Consider the category $\mathcal{S}$ whose objects are semigroups of isometries $U = \{U_t : t \geq 0\}$ each of which acts on a separable Hilbert space H_U. $\hom(U, V)$ consists of unitary operators $W : H_U \to H_V$ which intertwine U and V:

$$WU_t = V_t W, \qquad t \geq 0.$$

This category admits a natural direct sum operation $\oplus$, in which $U \oplus V$ is the semigroup of isometries on $H_U \oplus H_V$ defined by

$$(U \oplus V)_t = U_t \oplus V_t, \qquad t \geq 0.$$

Every semigroup of isometries U decomposes uniquely into a direct sum

$$U = V \oplus W \tag{5.1}$$

where W is a semigroup of unitary operators and where V is *pure* in the sense that

$$\bigcap_{t>0} V_t H_V = 0.$$

Moreover, every pure semigroup of isometries is isomorphic to a direct sum of a countable number d of copies of the simple *shift* semigroup S which acts on $L^2(0, \infty)$ by way of

$$S_t f(x) = \begin{cases} f(x-t), & \text{for } x > t \\ 0, & \text{for } 0 < x \leq t. \end{cases}$$

The number d of copies of S is uniquely determined by V, and is called the *index* of V. We remark that there are other ways to define the index of V, involving the deficiency spaces of the infinitesimal generator of V. But the definition we have given is the quickest. All of this information about the decomposition (5.1) is often referred to as the *Wold decomposition.*

The index obeys the expected law of addition

$$\operatorname{ind}(U_1 \oplus U_2) = \operatorname{ind}(U_1) + \operatorname{ind}(U_2),$$

and it can take on any of the values $0, 1, 2, \ldots, \aleph_0$. Notice that there is also a tensor product operation defined on $\mathcal{S}$, but it has terrible arithmetic properties with respect to the index. For example, if U_1 and U_2 have index 1 then $U_1 \otimes U_2$ has index $\aleph_0$. Thus we must consider $\mathcal{S}$ as a category with a single operation $\oplus$.

Let $\mathcal{E}$ be the category whose objects are E_0-semigroups and whose maps are conjugacies. Thus, if α and β are E_0-semigroups acting respectively on $\mathcal{B}(H_\alpha)$ and $\mathcal{B}(H_\beta)$, then $\hom(\alpha, \beta)$ consists of $*$-isomorphisms $\theta : \mathcal{B}(H_\alpha) \to \mathcal{B}(H_\beta)$ satisfying

$$\theta(\alpha_t(A)) = \beta_t(\theta(A))$$

for every $t \geq 0$, $A \in \mathcal{B}(H_\alpha)$. The natural operation in $\mathcal{E}$ is the tensor product of E_0-semigroups that has already been defined in section 3.

One might well ask if there is a direct sum operation in $\mathcal{E}$. Assuming that there were such an operation, one would expect $\alpha \oplus \beta$ to be an E_0-semigroup acting on $\mathcal{B}(H_\alpha \oplus H_\beta)$. By replacing β with a conjugate copy of itself if necessary, we may assume that $H_\alpha = H_\beta = H$. Then $\mathcal{B}(H \oplus H)$ consists of 2×2 matrices over $\mathcal{B}(H)$. One would expect *at least* that $\alpha \oplus \beta$ should restrict to α and β on the appropriate summands, that is $\alpha \oplus \beta$ should also have the property

$$(\alpha \oplus \beta)_t : \begin{pmatrix} A & 0 \\ 0 & B \end{pmatrix} \to \begin{pmatrix} \alpha_t(A) & 0 \\ 0 & \beta_t(B) \end{pmatrix}, \tag{5.2}$$

for all $t \geq 0$ and all $A, B \in \mathcal{B}(H)$. However, if there were an E_0-semigroup on $\mathcal{B}(H \oplus H)$ which satisfied (5.2), then one could show that α and β must in fact be cocycle perturbations of each other. This argument is a variation of Connes' elegant observation about exterior equivalence of modular automorphism groups (the idea can be found in Lemma 8.11.2 of [17]). In particular, if α and β are not cocycle conjugate then it is impossible to make a reasonable definition of $\alpha \oplus \beta$. One might summarize this state of affairs as follows: *the only appropriate operation in the category $\mathcal{S}$ is the direct sum and the only possible operation in the category $\mathcal{E}$ is the tensor product.*

Finally, the index is defined on the objects of $\mathcal{E}$ and because of Theorem A we have

$$d_*(\alpha \otimes \beta) = d_*(\alpha) + d_*(\beta).$$

We now describe a construction which can be considered a functor from $\mathcal{S}$ to $\mathcal{E}$. This is (Boson) quantization, and it is also a form of exponentiation in that direct sums carry over to tensor products. The details are as follows.

Let H be a Hilbert space. We will write H^n for the symmetric tensor product of n copies of H if $n \geq 1$; H^0 is defined as $\mathbb{C}$. Let

$$e^H = \sum_{n=0}^{\infty} H^n$$

be the symmetric Fock space over H. The natural exponential map $\exp : H \to e^H$ is defined by

$$\exp(\xi) = \sum_{n=0}^{\infty} \frac{1}{\sqrt{n!}} \xi^{\otimes n}.$$

e^H is spanned by the vectors of the form $\exp(\xi)$, $\xi \in H$, and we have

$$< \exp(\xi), \exp(\eta) >= e^{<\xi,\eta>}.$$

For every $\xi \in H$ there is a unique unitary operator $W(\xi)$ on e^H which satisfies

$$W(\xi)\exp(\eta) = e^{-\frac{1}{2}\|\xi\|^2 - <\eta,\xi>} exp(\eta + \xi).$$

W is strongly continuous, obeys Weyl's form of the canonical commutation relations

$$W(\xi)W(\eta) = e^{i\mathrm{Im}<\xi,\eta>}W(\xi + \eta),$$

and $W(H)$ is an irreducible set of operators on e^H.

Now let $U \in \mathcal{S}$. We define an E_0-semigroup α^U on $\mathcal{B}(e^H)$ as follows. Fix $t \geq 0$. Because of the irreducibility of W, it is not hard to verify that there is a unique normal endomorphism α_t^U of $\mathcal{B}(e^H)$ satisfying

$$\alpha_t^U(W(\xi)) = W(U_t\xi), \qquad \xi \in H_U.$$

$\alpha^U = \{\alpha_t^U : t \geq 0\}$ is an E_0-semigroup. Because of the natural identification

$$e^{H_1 \oplus H_2} = e^{H_1} \otimes e^{H_2}$$

we have a natural identificaiton

$$\alpha^{U \oplus V} = \alpha^U \otimes \alpha^V \tag{5.3}$$

for every $U, V \in \mathcal{S}$. With these observations in hand, one can establish the functoriality of the map $U \to \alpha^U$ (see [7] for more detail).

Let $U \in \mathcal{S}$. Applying (5.3) and the Wold decomposition (5.1), we find that α^U decomposes into a tensor product

$$\alpha^U = \beta \otimes \gamma$$

where γ is a trivial E_0-semigroup (i.e., each γ_t is an automorphism) and where β is conjugate to an E_0-semigroup of the form $\alpha^{d\cdot S}$ where $d \cdot S$ is a direct sum of d copies of the simple shift semigroup S acting on $L^2(0,\infty)$. An E_0-semigroup such as β is called a *CCR flow of index d.* This terminology is justified by the following index theorem [2],[7].

Theorem D.

$$d_*(\alpha^U) = \mathrm{index}(U).$$

The proof of Theorem D involves some work: one must find *all* the units of α^U and compute the associated covariance function in order to calculate the dimension of its associated Hilbert space. It follows immediately from Theorem B that if α is any E_0-semigroup and γ is a *trivial* E_0-semigroup, then $\alpha\otimes\gamma$ is cocycle conjugate to α (actually, it is not hard to prove the latter directly) . If we collect this observation together with Theorem D then we are led to the conclusion that for any $U, V \in \mathcal{S}$,

α^U is cocycle conjugate to α^V iff α^U and α^V have the same index. Thus the numerical index is a complete cocycle conjugacy invariant for these examples.

We also point out that there is a corresponding construction of "standard" examples of product systems which parallels what we have done for E_0-semigroups, and we refer the reader to [2] for the details.

Finally, one may use the Fermionic Fock space to construct standard examples of E_0-semigroups having index $1, 2\dots, \aleph_0$ in a way that is roughly parallel to what we have done above (though we point out that the construction of the corresponding product systems is not so explicit in the Fermionic setting). The E_0-semigroups obtained from either the Bosonic or the Fermionic construction turn out to be conjugate, provided that their numerical index is the same.

6. Classification results.

In this section we will describe two characterizations of product sytems which are isomorphic to one of the standard product systems. These results provide a classification of E_0-semigroups which have sufficiently many units, or which have enough decomposable operators.

Let $\alpha = \{\alpha_t : t \geq 0\}$ be an E_0-semigroup acting on $\mathcal{B}(H)$, and fix $t > 0$. Consider the set $\mathcal{S}_t$ of all operators which can be decomposed into a finite product of the form

$$T = U_1(t_1)U_2(t_2)\dots U_n(t_n), \tag{6.1}$$

where the $U_1, U_2, \dots, U_n$ are units for α, where $t_1, t_2, \dots, t_n$ are positive real numbers summing to t, and where n is an arbitrary positive integer. Because $t_1 + t_2 + \dots + t_n = t$ it is clear that $\mathcal{S}_t \subseteq E_\alpha(t)$ for every $t > 0$. We say that α is *completely spatial* if there is a $t > 0$ such that

$$H = \overline{\operatorname{span}}\{T\xi : T \in \mathcal{S}_t, \xi \in H\}. \tag{6.2}$$

It is easy to see that if (6.2) is true for some particular positive t, then it is true for every positive t.

It was proved in ([2] section 7) that every completely spatial E_0-semigroup is cocycle conjugate to a CCR flow. It follows that completely spatial E_0-semigroups are classified by their numerical index. This is proved at the level of product systems, using Theorem B.

That result has recently been extended significantly [9]. The extended version does not assume the existence of units, and is formulated as follows. Fix $t > 0$. An operator $T \in E_\alpha(t)$ is called *decomposable* if, for every $0 < s < t$ there are operators $A_s \in E_\alpha(s)$, $B_s \in E_\alpha(t-s)$ such that

$$T = A_s B_s.$$

We write $\mathcal{D}_t$ for the set of all decomposable operators in $E_\alpha(t)$. Any operator of the form (6.1) is decomposable because of the semigroups property of each U_k, and therefore $\mathcal{D}_t$ contains $\mathcal{S}_t$. In fact, it is not hard to see that the following conditions are equivalent

$$E_\alpha(t) = \overline{\operatorname{span}}\,\mathcal{D}_t \tag{6.3.1}$$

$$H = \overline{\operatorname{span}}\{T\xi : T \in \mathcal{D}_t, \xi \in H\}, \tag{6.3.2}$$

and that if the conditions (6.3) are satisfied for some particular t then they are satisfied for every positive t. We remark that one uses the Hilbert space topology on $E_\alpha(t)$ in (6.3.1); hence it is apparent that the conditions (6.3) depend only on the structure of the product system E_α associated with α. The main result of [9] is

Theorem E. *Every E_0-semigroup satisfying the conditions (6.3) is cocycle conjugate to a CCR flow.*

Utilizing an ingenious construction in [19], Powers showed that there are E_0-semigroups which possess no units whatsoever. In a recent paper [21] he also proved that there are E_0-semigroups which have units but which are *not* cocycle conjugate to a CCR flow. It follows that there are product systems which a) have no units, and others which b) have units but not enough units to generate the product system.

We believe that it should be possible to give a more direct consturction of product systems with these properties. Unfortunately, we do not yet know how to do this. There are examples of product systems that arise naturally in probability theory (see [2], pp 14–16). Some of these examples do not *appear* to contain enough units. However, in all such cases we have studied we eventually found many units that were not initially obvious. What *is* immediately obvious in these probabilistic examples is that the product systems are generated by their decomposable vectors. Theorem E tells us that such product systems must be standard ones. In particular, any attempt to construct nonstandard product systems from "decomposable" sets must fail.

7. Spectral invariant.

Theorem C above asserts that every abstract product system is associated with an E_0-semigroup. This result is analogous to the fact that every locally compact group G has a faithful unitary representation on a Hilbert space. The proof of the latter assertion about groups follows from an analysis of the properties of the group C^*-algebra $C^*(G)$, together with the Gelfand-Neumark theorem. In that result, $C^*(G)$ functions as the "spectrum" of the group G.

In this section we show how, starting with a product system E, one can construct a *spectral* C^*-algebra $C^*(E)$. Results like Theorem C are obtained by exploiting the properties of $C^*(E)$. More generally, $C^*(E)$ provides a "topological" invariant that is important for understanding the nature of product systems and their associated E_0-semigroups.

Let $p : E \to (0,\infty)$ be a product system, and let us write $E(t) = p^{-1}(t)$ for the Hilbert space over $t > 0$. We form the Hilbert space of L^2 sections

$$L^2(E) = \int_{(0,\infty)}^{\oplus} E(t)\,dt\ .$$

The inner product in $L^2(E)$ is the natural one

$$<\xi,\eta>=\int_0^\infty <\xi(t),\eta(t)>\ dt\ .$$

Let $f \in L^1(E)$ be an integrable section. Using the multiplication in E we see that for every $\xi \in L^2(E)$ and every $0 < t < x$,

$$f(t)\xi(x-t) \in E(x),$$

and hence we can define a measurable section $f * \xi$ by

$$f * \xi(x) = \int_0^x f(t)\xi(x-t)\,dt\ .$$

For fixed $f \in L^1(E)$, left convolution by f, $\xi \mapsto f * \xi$, defines a bounded linear operator on $L^2(E)$ of norm at most $\|f\|_1$. This operator is denoted l_f. A straightforward computation shows that for any two functions $f, g \in L^1(E)$, there are functions $h_1, h_2 \in L^1(E)$ such that

$$l_f^* l_g = l_{h_1} + l_{h_2}^*.$$

It follows that the linear span of all products of the form $l_f l_g^*$ is a self-adjoint subalgebra of $\mathcal{B}(L^2(E))$. $C^*(E)$ is defined as the norm-closure of this algebra

$$C^*(E) = \overline{\operatorname{span}}\{l_f l_g^* : f, g \in L^1(E)\ \}. \tag{7.1}$$

The funcamental property of $C^*(E)$ is that its representations correspond to all possible E_0-semigroups α for which E_α is isomorphic to E. This is a key result in the theory and we want to state it precisely. It is convenient to slightly generalize the notion of E_0-semigroup. By an e_0-semigroup we mean a semigroup $\alpha = \{\alpha_t : t \geq 0\}$ of normal $*$-endomorphisms of $\mathcal{B}(H)$ that satisfies all of the conditions of an E_0-semigroup except that $\alpha_t(\mathbf{1})$ is not required to be $\mathbf{1}$. Thus, for an e_0-semigroup α,

$$P_t = \alpha_t(\mathbf{1})$$

defines a strongly continuous family of projections which decreases as $t \to \infty$. The limit

$$P_\infty = \lim_{t\to\infty} \alpha_t(\mathbf{1})$$

is an α-invariant projection which induces a decomposition of the underlying Hilbert space

$$H = H_\infty \oplus H_0,$$

where H_∞ and H_0 are, respectively, the ranges of the projections P_∞ and $\mathbf{1} - P_\infty$. The restriction of α to $\mathcal{B}(H_\infty)$ is an E_0-semigroup, and the restriction α^0 of α to $\mathcal{B}(H_0)$ is an e_0-semigroup whose limiting projection is zero:

$$\lim_{t\to\infty} \alpha_t^0(\mathbf{1}) = 0. \tag{7.2}$$

To this extent the study of e_0-semigroups reduces tc the study of E_0-semigroups and the extreme case of e_0-semigroups satisfying (7.2).

We need to relate the representations of $C^*(E)$ more directly to E. By a *representation* of E we mean a measurable mapping $\phi : E \to \mathcal{B}(H)$ which restricts to a linear map on each fiber $E(t), t > 0$, which preserves multiplication in that $\phi(u)\phi(v) = \phi(uv)$ for all $u, v \in E$, and which obeys the following partial "commutation relation" on each fiber:

$$\phi(v)^*\phi(u) = < u, v > \mathbf{1}, \qquad u, v \in E(t). \tag{7.3}$$

If $\phi : E \to \mathcal{B}(H)$ is an arbitrary representation then we can define an e_0-semigroup $\alpha = \{\alpha_t : t > 0\}$ which acts on $\mathcal{B}(H)$ as follows. For each positive t, choose an orthonormal basis $\{e_1(t), e_2(t), \dots\}$ for $E(t)$ and put

$$\alpha_t(A) = \sum_{n=1}^{\infty} \phi(e_n(t))A\phi(e_n(t))^*, \tag{7.4}$$

$A \in \mathcal{B}(H)$. We define $\alpha_0(A) = A$. The left side is independent of the particular choice of basis $\{e_n(t)\}$, and it is true (though nontrivial) that α is an e_0-semigroup whose canonical product system is isomorphic to E. Indeed,

$$E_\alpha(t) = \{\phi(v) : v \in E(t)\}$$

and ϕ itself implements the stated isomorphism of E and $E\alpha$. These things are proved in [3].

More significantly, any representation $\phi : E \to \mathcal{B}(H)$ determines a unique representation $\pi : C^*(E) \to \mathcal{B}(H)$ by way of

$$\pi(l_f l_g^*) = \phi(f)\phi(g)^*, \qquad f, g \in L^1(E) \tag{7.5}$$

where for $f \in L^1(E)$, $\phi(f)$ denotes the operator integral

$$\phi(f) = \int_0^\infty \phi(f(x))\,dx.$$

The key property of $C^*(E)$ is that this association $\phi \to \pi$ is in fact a bijection [3]:

Theorem F. *The nondegenerate separable representations of $C^*(E)$ correspond bijectively with all e_0-semigroups α for which E_α is isomorphic to E.*

Because of Theorem F, we are let to examine the structure of $C^*(E)$, and attempt to describe its state space in terms that are as explicit as possible. The remainder of this paper is devoted to a discussion of progress on these issues.

The principal result of [3] is that $C^*(E)$ is a simple C^*-algebra in most (and perhaps all) cases. More precisely,

Theorem G. *For every product system E, $C^*(E)$ is a unitless nuclear C^*-algebra. If E possesses a nonzero unit, then $C^*(E)$ has no closed nontrivial ideals.*

In particular, the C^*-algebras $C^*(E_n), n = 1, 2, \dots, \aleph_0$ associated with the standard examples E_n (i.e., the product systems of the CCR flows) are all simple. These C^*-algebras are most like continuous versions of the Cuntz algebras $\mathcal{O}_n$, $n = 2, 3, \dots, \infty$ (see [3], [10]).

We still do not know if $C^*(E)$ is simple in the cases where $\mathcal{U}_E = \emptyset$. However, we do have the following information in general. Consider the one-parameter unitary group W defined on $L^2(E)$ by

$$W_t\xi(x) = e^{itx}\xi(x).$$

The generator N of W,

$$W_t = e^{itN}$$

replaces the *number operator* on ordinary Fock space. N has Lebesgue spectrum distributed throughout $[0, \infty)$ with infinite multiplicity. The associated one parameter group of automorphisms leaves $C^*(E)$ invariant,

$$W_t C^*(E) W_t^* = C^*(E), \qquad t \in \mathbb{R}$$

and thus induces a natural one parameter group of automorphisms $\gamma = \{\gamma_t : t \in \mathbb{R}\}$ of $C^*(E)$. γ is called the *gauge group*. The best general result in this direction is the following, from which Theorem G is easily deduced (see [3]).

Theorem G1. *There are no closed proper ideals J in $C^*(E)$ which are gauge invariant in the sense that $\gamma_t(J) = J$, for every $t \in \mathbb{R}$.*

Theorem F tells us that in order to specify an e_0-semigroup whose product system is isomorphic to E, it is enough to specify a state of $C^*(E)$. However, there remains a significant question: how does one know when a representation gives rise to an E_0-semigroup rather than, say, merely an e_0-semigroup? In order to discuss this, let us say that a representation π of $C^*(E)$ is *essential* if it gives rise to an E_0-semigroup, and *singular* if it gives rise to an e_0-semigroup satisfying 7.2. Similarly, a state of $C^*(E)$ (i.e., a nonzero positive linear functional on $C^*(E)$ with no condition on its norm) is called essential or singular according as the representation it defines via the GNS construction has the corresponding property.

The state space $\mathcal{P}$ of $C^*(E)$ is a norm-closed cone, and it is known that this cone decomposes into a direct of order ideals

$$\mathcal{P} = \mathcal{E} \oplus \mathcal{S}$$

where $\mathcal{E}$ (resp. $\mathcal{S}$) denotes the set of essential (resp. singular) states [4]. A detailed description of this decomposition and the singular summand $\mathcal{S}$ is given in [4]. Here, we want to concentrate on the description of the essential summand. In view of the precdding discussion, the assertion that every product system E is associated with an E_0-semigroup becomes the assertion that $C^*(E)$ has (nonzero) essential states...i.e., that $\mathcal{E} \neq \emptyset$. Our main result along these lines is the following result from [5] which will be discussed further in the following section.

Theorem H. *For every product system E, there is an essential state of $C^*(E)$ whose E_0-semigroup α is ergodic in the sense that*

$$\{A \in \mathcal{B}(H_\alpha) : \alpha_t(A) = A, \quad \forall t \geq 0\ \} = \mathbb{C}\mathbf{1}.$$

Because of Theorem F we have $E_\alpha \cong E$. This result is analogous to the fact that a locally compact group has irreducible unitary representations. Indeed, the relation that exists between a representation $\pi : C^*(E) \to \mathcal{B}(H)$ and its associated e_0-semigroup $\alpha = \{\alpha_t : t \geq 0\}$ acting on $\mathcal{B}(H)$ is expressed in (7.4) and (7.5). From the nature of this relation it follows that the fixed algebra of α,

$$\{A \in \mathcal{B}(H) : \alpha_t(A) = A \quad \forall t \geq 0\}$$

is precisely the commutant of $\pi(C^*(E)$. Thus, *the proof of Theorem G amounts to showing that $C^*(E)$ has nonzero essential pure states.* Such states will be discussed in the following section.

There are other consequences one can obtain along similar lines. For example, one knows that the C^*-algebra$C^*(E)$ is not GCR, and hence it has representations which generate factors of type II_∞ or III. There are even essential states with these properties, and hence we may conclude that every E_0-semigroup is cocycle conjugate to an E_0-semigroup whose fixed agebra is a factor of type II or III.

8. States in the regular representation.

$C^*(E)$ is defined as a C^*-algebra of operators on the Hilbert space $L^2(E)$, and this gives rise to a representation $\lambda : E \to \mathcal{B}(L^2(E))$. For $v \in E(t)$, $\lambda(v)$ is defined as

$$\lambda(v)\xi(x) = \begin{cases} v \cdot \xi(x-t), & \text{if } x > t \\ 0, & \text{if } 0 < t < x. \end{cases}$$

Notice that for $f \in L^1(E)$ we have

$$l_f = \int_0^\infty \lambda(f(x))\, dx.$$

λ is called the (left) regular representation. The associated e_0-semigroup α is singular, since $\alpha_t(\mathbf{1})$ is the projection onto to the subspace $\{\xi \in L^2(E) : \xi(x) = 0, \text{for } 0 < x \leq t\}$ and these subspaces decrease to 0 as $t \to \infty$. Actually, we will be more concerned with the e_0-semigroup generated by the right regular anti representation $\rho : E \to \mathcal{B}(L^2(E))$, where for $v \in E(t)$, $\rho(v)$ is defined as the operator

$$\rho(v)\xi(x) = \begin{cases} \xi(x-t) \cdot v, & x > t \\ 0, & 0 < x \leq t. \end{cases}$$

Notice that ρ reverses multiplication in the sense that $\rho(uv) = \rho(v)\rho(u)$, for $v, v \in E$. Nevertheless, we can use ρ to define a second e_0-semigroup $\beta = \{\beta_t : t \geq 0\}$ by way of

$$\beta_t(A) = \sum_{n=1}^\infty \rho(e_n(t))A\rho(e_n(t))^*, \quad A \in \mathcal{B}(L^2(E))$$

for $t > 0$, $\{e_1(t), e_2(t), \dots\}$ being any orthonormal basis for $E(t)$, and where β_0 is defined as the identity endomorphism. It is true (and nontrivial) that β is an e_0-semigroup [3],[5].

The generator of β is defined as the limit in the strong operator topology

$$\delta(A) = \lim_{t\to 0+} \frac{1}{t}(A - \beta_t(A)),$$

A ranging over the set $\mathcal{D}(\delta)$ of all operators for which the indicated limit exists. It is convenient here to use a different sign in the definition of δ than that of section 2. $\mathcal{D}(\delta)$ is a unital $*$-subalgebra of $\mathcal{B}(L^2(E))$ and δ is an unbounded self adjoint derivation from $\mathcal{D}(\delta)$ into $\mathcal{B}(L^2(E))$. We will first give an alternate description of $C^*(E)$ in terms of δ. This description of $C^*(E)$ is of key importance.

For every $t \geq 0$, let P_t denote the projection onto the subspace

$$\{\xi \in L^2(E) : \xi(x) = 0 \quad \text{a.e., for } 0 < x \leq t\ \}.$$

Notice that $\alpha_t(\mathbf{1}) = \beta_t(\mathbf{1}) = P_t$, and that $\{P_t : t \geq 0\ \}$ is a strongly continuous family of projections that increases from $\mathbf{0}$ to $\mathbf{1}$ as t moves from 0 to ∞. An operator $A \in \mathcal{B}(L^2(E))$ is said to have *bounded support* if there is a $t > 0$ such that $A = P_tAP_t$, and we write

$$\mathcal{B}_0 = \bigcup_{t>0} P_t\, \mathcal{B}(L^2(E))\, P_t \tag{8.1}$$

for the $*$-algebra of all operators of bounded support.

Significantly, every operator in $\mathcal{B}_0$ belongs to the range of δ. Indeed, if $A = P_tAP_t$, then the integral defined by the strong limit

$$I(A) = \lim_{T\to\infty} \int_0^T \beta_s(A)\, ds$$

exists and obeys

$$\|I(A)\| \leq t, \tag{8.2.1}$$

$$\delta(I(A)) = A \tag{8.2.2}$$

(see [5], Theorem 2.2). Indeed, the restriction of $I(\cdot)$ to $P_t\mathcal{B}(L^2(E))P_t$ is a normal completely positive linear map for every $t > 0$.

We will also write

$$H_0 = \bigcup_{t>0} P_tL^2(E)$$

for the linear space of all vectors $\xi \in L^2(E)$ which have bounded support, and

$$\mathcal{K}_0 = \mathcal{B}_0 \cap \mathcal{K}$$

for the $*$-algebra of all compact operators of bounded support.

The following result characterizes $C^*(E)$ in terms of the derivation δ.

Theorem I. *The set $\mathcal{A}$ of all operators A in the domain of δ satisfying $\delta(A) \in \mathcal{K}_0$ is a $*$-algebra whose norm closure is $C^*(E)$.*

Recall that $C^*(E)$ is spanned by operators of the form $l(f)l(g)^*$ where f and g are arbitrary integrable sections of E. Notice that if ξ, η are elements of $L^2(E)$ which have bounded support, then we may consider ξ and η as elements of $L^1(E)$, and operators of the form $l(\xi)l(\eta)^*$ also span $C^*(E)$. Such an operator belongs to $\mathcal{A}$ and the following formula implies that $\delta(l(\xi)l(\eta)^*$ is a rank-one operator in $\mathcal{K}_0$,

$$\delta(l(\xi)l(\eta)^*) = \xi \otimes \overline{\eta}, \qquad \xi, \eta \in H_0, \tag{8.3}$$

$\xi \otimes \overline{\eta}$ denoting the operator

$$\zeta \in L^2(E) \mapsto < \zeta, \eta > \xi.$$

The proof of (8.3) can be found in ([5], p. 288).

Theorem I and formula (8.3) open the way to a very explicit description of the state space of $C^*(E)$, which we now describe. Let ω be a linear functional defined on the algebra $\mathcal{B}_0$ of all operators having bounded support. ω is called a *locally normal weight* if, for every $t > 0$, the restriction of ω to $P_t\mathcal{B}(L^2(E))P_t$ is a normal positive linear functional. Locally normal weights are generalizations of normal weights (more precisely, of noncommutative Radon measures). Indeed, if

$$\omega : \mathcal{B}(L^2(E))^+ \to [0, +\infty]$$

is a normal weight satisfying $\omega(P_t) < +\infty$ for every $t > 0$, then the restriction of ω to $\mathcal{B}_0$ is a locally normal weight. We emphasize, however, that *not every locally normal weight can be extended to a normal weight of* $\mathcal{B}(L^2(E))$ (see [5], appendix A for an example). Thus, locally normal weights are more general than normal weights.

Notice that $\beta_t(\mathcal{B}_0) \subseteq \mathcal{B}_0$ for every $t > 0$. Thus we can make the following

Definition 8.4. *A locally normal weight ω is called decreasing if for every $t \geq 0$ we have*

$$\omega(\beta_t(A^*A)) \leq \omega(A^*A), \qquad A \in \mathcal{B}_0.$$

ω is called invariant if equality holds for every $t > 0$ and every $A \in \mathcal{B}_0$.

We will call such an ω simply a *decreasing weight.* Let $\mathcal{W}$ denote the cone of all decreasing weights satisfying the growth condition

$$\sup_{t>0} \frac{1}{t}\omega(P_t) < \infty.$$

Noting that $P_t = \mathbf{1} - \beta_t(\mathbf{1})$, it is clear that an invariant weight ω satisfies

$$\omega(\beta_s(\mathbf{1}) - \beta_{s+t}(\mathbf{1})) = \omega(\mathbf{1} - \beta_t(\mathbf{1}))$$

for all $s, t > 0$, and from this it follows that there is a constant $c \geq 0$ such that

$$\omega(P_t) = c \cdot t.$$

In particular, *every invariant weight belongs to* $\mathcal{W}$.

We can now describe the state space of $C^*(E)$ (see [5], Theorems 4.15 and 5.7). Every locally normal weight ω defines a linear functional $d\omega$ on $\mathcal{A}$ by way of

$$d\omega(A) = \omega(\delta(A)).$$

Notice that $d\omega$ is the "derivative" of ω in the direction of the flow of the e_0-semigroup β. We need to know when $d\omega$ is a positive linear functional which has finite norm. The characterization is as follows.

Theorem J. *For every decreasing weight $\omega \in \mathcal{W}$, $d\omega$ is a positive linear functional on $\mathcal{A}$ having norm*

$$\|d\omega\| = \sup_{t>0} \frac{1}{t}\omega(P_t).$$

$\omega \to d\omega$ is an affine order isomorphism of the cone $\mathcal{W}$ onto the state space of $C^*(E)$, which maps the subcone of invariant weights onto the cone of essential states.

With Theorem J in hand, it is easy to show that $C^*(E)$ must have essential states. A straightforward construction allows one to write down an invariant weight ω on $\mathcal{B}_0$ which is normalized so that

$$\omega(P_t) = t, \qquad \forall t > 0$$

(see Theorem 5.9 of [5]). It follows from Theorem J that $d\omega$ is an essential state of $C^*(E)$ satisfying $\|d\omega\| = 1$.

Corollary. *For every abstract product system E, there is an E_0-semigroup α for which E_α is isomorphic to E.*

As we have pointed out previously, with a little care, one can arrange that α is ergodic in the sense that

$$\alpha_t(A) = A, \quad \forall t \geq 0 \implies A = \text{scalar}.$$

The details can be found in [5].

There are numerous interesting unsolved problems concerning the spectral C^*-algebras $C^*(E)$. For example, if E_n is the standard product system of dimension n, $n = 1, 2, \ldots, \aleph_0$, then $C^*(E)$ is known t be a "continuous time" analogue of the Cuntz algebra $\mathcal{O}_{n+1}$ (see [10]). However, we do not yet know if these C^*-algebras $C^*(E_n)$ are mutually non-isomorphic for different values of n. Cerainly this is the case for the $\mathcal{O}_n$. In fact, Cuntz showed that $\mathcal{O}_m$ is not isomorphic to $\mathcal{O}_n$ essentially by calculating the K-theory of these C^*-algebras [15],[16]. But while the K-theory of the C^*-algebras $C^*(E_n)$ has not been computed, there is some evidence that K-theory may not be capable of distinguishing between them.

Finally, we want to point out the remarkable fact that, like $\mathcal{O}_\infty$, every spectral C^*-algebra $C^*(E)$ has an unbounded trace τ. This is easily seen using the description of $C^*(E)$ given in Theorem I. Let $\mathcal{A}_1$ denote the set of all operators A in the domain of δ such that $\delta(A)$ is a *trace class* operator of bounded support. $\mathcal{A}_1$ is a self-adjoint ideal in $\mathcal{A}$ which is clearly norm-dense in $C^*(E)$. We can define a linear functional τ on $\mathcal{A}_1$ by way of

$$\tau(A) = \text{trace}(\delta(A)).$$

Notice that $\tau(AB) = \tau(BA)$, since

$$\begin{aligned}\text{trace}(\delta(AB)) &= \text{trace}(A\delta(B)) + (\delta(A)B) = \text{trace}(\delta(B)A) + \text{trace}(B\delta(A)) \\ &= \text{trace}(\delta(BA)).\end{aligned}$$

It is possible to show that τ is *not* a positive trace in that there exist operators $A \in \mathcal{A}_1$ satisfying $\tau(A^*A) < 0$. Such traces are uncommon in operator theory, but notice that the Wodzicki residue provides another example of an unbounded non-positive trace on an algebra of pseudo-differential operators [14]. As yet, the role of this trace in the theory of E_0-semigroups remains mysterious.

References

1. Araki, H. and Woods, E. J., *Complete Boolean algebras of type I factors*, Publ. RIMS (Kyoto University) **2, ser. A, no. 2** (1966), 157–242.
2. Arveson, W., *Continuous analogues of Fock space*, Memoirs Amer. Math. Soc. **80 no. 3** (1989).
3. ______, *Continuous analogues of Fock space II: the spectral C^*-algebra*, J. Funct. Anal. **90** (1990), 138–205.
4. ______, *Continuous analogues of Fock space III: singular states*, J. Oper. Th. **22** (1989), 165–205.
5. ______, *Continuous analogues of Fock space IV: essential states*, Acta Math. **164** (1990), 265–300.
6. ______, *An addition formula for the index of semigroups of endormorphisms of $\mathcal{B}(H)$*, Pac. J. Math. **137** (1989), 19–36.
7. ______, *Quantizing the Fredholm index*, Operator Theory: Proceedings of the 1988 GPOTS-Wabash conference (Conway, J. B. and Morrel, B. B., ed.), Pitman research notes in mathematics series, Longman, 1990.
8. ______, *The spectral C^*-algebra of an E_0-semigoup*, Operator Theory Operator Algebras and applications, Proc. Symp. Pure Math. (Arveson, W. and Douglas, R. G., ed.), vol. 51, part I, 1990, pp. 1–15.
9. ______, *Decomposable E_0-semigroups*, (in preparation).
10. ______, *C^*-algebras associated with sets of semigroups of isometries*, Int. J. Math. **2, no. 3** (1991).
11. Arveson, W. and Kishimoto, A., *A note on extensions of semigroups of $*$-endomorphisms*, Proc. A. M. S. **116, no 3** (1992), 769–774.
12. Bratteli, O. and Robinson, D. W., *Operator algebras and quantum statistical mechanics I, II*, Springer-Verlag, 1989.
13. Connes, A., *Une classification des facteurs de type III*, Ann. Scient. Ecole Norm. Sup. **6 , fasc. 2 ser. 4e** (1973), 133–253.
14. ______, *Non Commutative Geometry*, Academic Press (to appear).
15. Cuntz, J., *Simple C^*-algebras generated by isometries*, Comm. Math. Phys. **57** (1977), 173–185.
16. ______, *K-theory for certain C^*-algebras*, Ann. Math. **113** (1981), 181–197.
17. Pedersen, G. K., *C^*-algebras and their automorphism groups*, Academic Press, 1979.
18. Powers, R. T., *An index theory for semigroups of endomorphisms of $\mathcal{B}(H)$ and type II factors*, Can. J. Math. **40** (1988), 86–114.
19. ______A non-spatial continuous semigroup os $*$-endomorphisms of $\mathcal{B}(H)$, Publ. RIMS (Kyoto University) **23** (1987), 1053–1069.
20. ______On the structure of continuous spatial semigroups of $*$-endomorphisms of $\mathcal{B}(H)$, Int. J. Math. **2, no 3** (1991), 323–360.
21. ______, *New examples of continuous spatial semigroups of endomorphisms of $\mathcal{B}(H)$*, (preprint 1994).
22. Powers, R. T. and Robinson, D., *An index for continuous semigroups of $*$-endomorphisms of $\mathcal{B}(H)$*, J. Funct. Anal. **84** (1989), 85–96.
23. Powers, R. T. and Price, G, *Continuous spatial semigroups of $*$-endomorphisms of $\mathcal{B}(H)$*, Trans. A. M. S. (to appear).

Proceedings of Symposia in Pure Mathematics
Volume **59**, 1996

Nonlinear Phenomena in the Spectral Theory of Geometric Linear Differential Operators

Thomas Branson

ABSTRACT. The extremal problem for the functional determinant of a natural linear elliptic operator a on Riemannian manifold is studied. Viewing the determinant as a function of the Riemannian metric, we encounter nonlinear geometric analytic phenomena: sharp inequalities comparing nonlinear functionals of the metric and its derivatives. The derivation and use of such inequalities in new situations, especially essentially tensor-valued inequalities, leads back to linear theory and the classification of conformally covariant differential operators.

0. Introduction

A central object of study in Geometric Analysis is the space $\mathcal{G}(M)$ of Riemannian metrics on a smooth compact manifold M. The most revealing data in this study are the spectra of differential operators A_g which are functorially, or naturally, associated to the metric g; for example, the Laplacian. The diffeomorphism group $\text{Diffeo}(M)$ is the gauge transformation group in this setting; the spectrum of a natural A_g will be unaffected by diffeomorphisms and thus gauge invariant. The multiplicative group $C^\infty_+(M)$ of smooth positive real functions e^ω on M also acts on $\mathcal{G}(M)$ in a natural way, by conformal change $g \mapsto e^{2\omega}g$. The space $\mathcal{G}(M)/\text{Diffeo}(M)$ can thus be broken down into several parts (and hopefully reassembled in the long run): the quotient $\mathcal{G}(M)/(\text{Diffeo}(M) \ltimes C^\infty_+(M))$, the conformal classes $C^\infty_+(M)g$, and the intersection of the two group actions: conformal changes that are actually implemented by diffeomorphisms, the conformal transformation group $\mathbf{C}(M,g)$.

The functional determinant $\det A_g$ is a spectral invariant that is apparently quite revealing of the geometry of g. Originally of interest in quantum field theory, where it provides a regularization of the functional integral, the functional determinant has recently become the object of intense study in connection with String Theory in Physics, and the isospectral problem in Mathematics. These two pursuits illustrate complementary approaches to an understanding of the space of metrics: (1) one can do an extremal problem, to try to get a representative of a conformal class which is somehow "uniform"; and (2) one can use spectral invariants to bound the metric, showing that there is a unique or nearly unique metric (modulo gauge)

1991 *Mathematics Subject Classification.* Primary 58C40; Secondary 53A30.
Partially supported by NSF grant INT-9114401.

with a given A-spectrum. In approach (2), the hope is that spec A_g will be a separating function, or coordinate, on the space of metrics. In either approach, when looking for functionals that can control or be controlled by the functional determinant, one immediately runs into the most delicate nonlinear geometric analytic aspects of the Riemannian manifold (M, g): the borderline Sobolev embeddings $L^2_\nu \hookrightarrow L^{2m/(m-2\nu)}$, where $m = \dim M$, their best constants and extremals, and (especially) the exponential class embedding $L^2_{m/2} \hookrightarrow e^L$ at the edge of the borderline. Thus, as in General Relativity, when one tries to incorporate a study of the states of space into the study of a linear field equation, nonlinearities, and in fact the most delicate possible ones, introduce themselves.

It is perhaps no surprise that when there is a uniform metric g_0 with a large (i.e., noncompact) conformal group $G = \mathbf{C}(M, g_0)$, the representation theory of G enters the theory. Since invariant nonlinearities are present, one is comparing invariant Hilbert space norms, the everyday object of study in Representation Theory, with invariant Banach space norms that are outside the province of Representation Theory as it is currently constituted. As a result, there is an interplay of several subjects which fall under the general rubric of Harmonic Analysis:

Spectral theory of differential operators;
Sharp inequalities of Sobolev embedding and Moser-Trudinger type;
Conformal geometry;
Noncompact Representation Theory;
Compact Representation Theory.

In this paper we concentrate on the extremal problem for the functional determinant on the even-dimensional spheres S^m, in the conformal class of the standard (round) metric g_0, and the many subsidiary problems from subjects in the above list that necessarily enter. Here the conformal transformation group, isomorphic to $\mathrm{SO}(m+1, 1)$, is the largest possible. (The edge of the Sobolev borderline, i.e. the transition from Lebesgue class to exponential class target in the Sobolev embedding, is strikingly reflected here as the endpoint of the complementary series of representations.) The two-dimensional theory, in the case where the operator A is the Laplacian Δ, was handled by Onofri [**O**] and by Osgood, Phillips, and Sarnak [**OPS1-2**]. Here the results have very strong implications: the space $\mathcal{G}(S^2)/(\mathrm{Diffeo}(S^2) \ltimes C^\infty_+(S^2))$ is a single point, so the conformal class of g_0 is really all there is. Our main focus here is higher dimensions, where the space of metrics is infinite dimensional even after quotient by the groups of gauge transformations and conformal changes. In particular, there is a complete treatment of the four-dimensional case due to Branson, Chang, Ørsted, and Yang [**BØ3, BCY**] which we describe here. Specifically, it is shown that for L the conformal Laplacian and $\nabla\!\!\!/$ the Dirac operator, $\det L_g$ and $\det \nabla\!\!\!/^2_g$ have their extremals exactly on the orbit $\mathbf{C}(S^4, g_0) \cdot g_0$ (under the normalizing assumption $\mathrm{vol}\, g = \mathrm{vol}\, g_0$). This is done by showing that $\log((\det L_g)/(\det L_{g_0}))$, and the analogous quantity for $\nabla\!\!\!/^2$, are definite linear combinations of the sharp forms of the quantities describing the embeddings $L^2_2 \hookrightarrow e^L$ and $L^2_1 \hookrightarrow L^4$. We carry this method to dimension six, and show that again, the extremals are on the orbit of g_0 under the conformal transformation group. Here the log-determinant formulas are definite linear combinations of several quantities describing, among other things, the $L^2_3 \hookrightarrow e^L$ and $L^2_1 \hookrightarrow L^3$ embedding, and a tensor-valued $L^2_1 \hookrightarrow L^3$ embedding. *Apropos* this and higher-dimensional problems, we present a new result classifying second-order conformally covariant operators; the precise form of these is needed to get the sharp form of tensor-spinor valued Sobolev embedding inequalities.

1. Conformal covariants

Let $\mathcal{R}_m$ be the category of smooth, m-dimensional Riemannian manifolds (M, g); let $\mathcal{R}_m^{\text{or}}$ be the subcategory consisting of oriented manifolds (M, g, E), E being the volume element; and and let $\mathcal{R}_m^{\text{spin}}$ be the smaller subcategory of manifolds (M, g, E, γ) with spin structure, γ being the fundamental tensor-spinor. The structure group of Riemannian geometry in $\mathcal{R}_m$ (resp. $\mathcal{R}_m^{\text{or}}$, $\mathcal{R}_m^{\text{spin}}$) is $\mathrm{O}(m)$ (resp. $\mathrm{SO}(m)$, $\mathrm{Spin}(m)$); we use H as a common abbreviation for these structure groups. To study conformal geometry, we enlarge the structure group to $CH = H \times \mathbb{R}_+$. The finite-dimensional representations of CH are the $(V(\lambda), \lambda^r)$, where $(V(\lambda), \lambda)$ is a finite dimensional representation of H, and $\lambda^r(h, \alpha) = \alpha^r \lambda(h)$ for $h \in H$, $\alpha \in \mathbb{R}_+$. If $\mathcal{F}_H$ and $\mathcal{F}_{CH}$) are the bundles of H-frames and CH-frames respectively, the associated bundle construction gives natural vector bundles

$$\mathbb{V}(\lambda) = \mathcal{F}_H \times_\lambda V(\lambda), \qquad \mathbb{V}^r(\lambda) = \mathcal{F}_{CH} \times_{\lambda^r} V(\lambda)$$

with structure groups H and CH. r is the *conformal weight* of the CH-bundle $\mathbb{V}^r(\lambda)$. If I^r is the bundle of scalar (r/m)-densities, $\mathbb{V}^r(\lambda)$ is naturally CH-isomorphic to $I^r \otimes \mathbb{V}^0(\lambda)$.

A *tensor-spinor bundle* is a natural subbundle of the $\otimes$-algebra generated by the tangent, cotangent, spinor, and cospinor bundles TM, T^*M, ΣM, $\Sigma^* M$. Since the defining representation of $\mathrm{O}(m)$ and the spin representations of $\mathrm{Spin}(m)$ are faithful, each $\mathbb{V}(\lambda)$ can be realized as a tensor-spinor bundle. If V is a natural subbundle of $(TM)^{\otimes p} \otimes (T^*M)^{\otimes q} \otimes (\Sigma M)^{\otimes r} \otimes (\Sigma^* M)^{\otimes s}$, then V naturally carries the conformal weight $q - p$. Note that g, E and γ are sections of tensor-spinor bundles ($\Sigma M \otimes \Sigma^* M \otimes TM$ in the case of γ). The symmetric Riemannian connection gives a connection on $\mathcal{F}_H$, and thus provides a covariant derivative ∇ for each $\mathbb{V}(\lambda)$; note that ∇g, ∇E, and $\nabla \gamma$ vanish.

DEFINITION 1.1. *A conformal covariant is a CH-equivariant differential operator $D : \mathbb{V}^r(\lambda) \to \mathbb{V}^s(\sigma)$ for some λ, σ, r, s, which is functorial and polynomial in g if $H = \mathrm{O}(m)$; in (g, E) if $H = \mathrm{SO}(m)$; or in (g, E, γ) if $H = \mathrm{Spin}(m)$, and satisfies a conformal covariance law*

$$\omega \in C^\infty(M), \quad \bar{g} = e^{2\omega} g, \quad \bar{E} = e^{m\omega} E, \quad \bar{\gamma} = e^{-\omega}\gamma \quad \Longrightarrow \quad \bar{D} = e^{-b\omega} D \mu(e^{a\omega}),$$

where a and b are real numbers, and $\mu(e^{a\omega})$ is multiplication by $e^{a\omega}$.

REMARK 1.2. The scalings of E and γ are chosen to be consistent with that of g; in the case of γ, this reflects the Clifford relations. Given a differential CH-operator $D : \mathbb{V}^r(\lambda) \to \mathbb{V}^s(\sigma)$, we can "forget" the conformal weights to get a differential H-operator $D : \mathbb{V}(\lambda) \to \mathbb{V}(\sigma)$, then assign new conformal weights to get a CH-operator $D : \mathbb{V}^{r'}(\lambda) \to \mathbb{V}^{s'}(\sigma)$. (Calling all of these operators D is, of course, an abuse of notation.) If $D : \mathbb{V}^r(\lambda) \to \mathbb{V}^s(\sigma)$ is a conformal covariant of bidegree (a, b), then $D : \mathbb{V}^{r'}(\lambda) \to \mathbb{V}^{s'}(\sigma)$ will be a conformal covariant of bidegree $(a - r' + r, b - s' + s)$; in particular, $D : \mathbb{V}^{r+a}(\lambda) \to \mathbb{V}^{s+b}(\sigma)$ is conformally invariant. We call $(r + a, s + b)$ the *reduced conformal bidegree* of D; it is independent of the particular realization of the domain and target H-bundles as tensor-spinor bundles.

REMARK 1.3. It is sufficient to check the infinitesimal form

$$\left.\frac{d}{d\varepsilon}\right|_{\varepsilon=0} D_{\varepsilon\omega} = -(b-a)\omega D_0 + a[D_0\,, \mu(\omega)] \tag{1.1}$$

of the covariance relation in Definition 1.1, in order to establish conformal covariance. Here $D_{\varepsilon\omega}$ is the evaluation of D in the metric $g_{\varepsilon\omega} = e^{2\varepsilon\omega} g_0$. For if (1.1) holds generally, it holds with $g_{\varepsilon_0\omega}$ in place of g_0, whence

$$\left.\frac{d}{d\varepsilon}\right|_{\varepsilon=\varepsilon_0} D_{\varepsilon\omega} = -(b-a)\omega D_{\varepsilon_0\omega} + a[D_{\varepsilon_0\omega}\,, \mu(\omega)].$$

But then

$$\frac{d}{d\varepsilon}\left(e^{b\varepsilon\omega} D_{\varepsilon\omega}\mu(e^{-a\varepsilon\omega})\right) = 0,$$

as desired.

REMARK 1.4. Irreducible representations of $\mathrm{Spin}(m)$ (and thus irreducible $\mathrm{Spin}(m)$-bundles) are parameterized by *dominant weights* $(\lambda_1, \dots, \lambda_\ell) \in \mathbb{Z}^\ell \cup (\frac{1}{2} + \mathbb{Z})^\ell$, $\ell = [m/2]$, with

$$\begin{aligned} &\lambda_1 \geq \dots \geq \lambda_\ell\,, && m \text{ odd}, \\ &\lambda_1 \geq \dots \geq \lambda_{\ell-1} \geq |\lambda_\ell|\,, && m \text{ even}. \end{aligned} \tag{1.2}$$

The dominant weight λ is the highest weight in the corresponding representation. The representations factoring through $\mathrm{SO}(m)$ are exactly those with $\lambda \in \mathbb{Z}^\ell$. We shall sometimes abuse notation by intentionally confusing an irreducible representation λ with its highest weight $(\lambda_1, \dots, \lambda_\ell)$. When writing dominant weights, we shall sometimes omit terminal strings of zeros, and denote, for example, a string of k ones as 1_k. Thus, on oriented manifolds, the differential form bundles $\Lambda^k M \cong_{\mathrm{SO}(m)} \Lambda^{m-k} M$ are $\mathbb{V}(1_k)$ for $0 \leq k < m/2$, and the middle-form bundle $\Lambda^{m/2} M$ for even m is $\mathbb{V}(1_{m/2}) \oplus \mathbb{V}(1_{(m-2)/2}\,, -1)$; the summands are the ± 1 or $\pm\sqrt{-1}$ eigenbundles of the Hodge $\star$ operator. The spinor bundle ΣM is $\mathbb{V}((\frac{1}{2})_{(m-1)/2})$ if m is odd, and $\mathbb{V}((\frac{1}{2})_{m/2}) \oplus \mathbb{V}((\frac{1}{2})_{(m-2)/2}\,, -\frac{1}{2})$ if m is even; here the summands are the positive and negative spinors $\Sigma_+ M$, $\Sigma_- M$.

EXAMPLES 1.5. The *conformal Laplacian* is $L = \Delta + (m-2)\tau/4(m-1)$, where Δ is the Laplacian and τ is the scalar curvature. L is a conformal covariant of reduced bidegree $((m-2)/2, (m+2)/2)$ on scalar functions [**Y**]. The *Dirac operator* $\nabla\!\!\!/ = \gamma^\alpha \nabla_\alpha$ (with the usual invariant index notation) is a conformal covariant of reduced bidegree $((m-1)/2, (m+1)/2)$ on the spinor bundle [**Kosm**].

EXAMPLES 1.6. In 1976, Fegan [**F**] showed that the *gradients* defined in [**SW**] are conformal covariants. Specifically, let $\mathbb{V}(\lambda)$ be an irreducible $\mathrm{Spin}(m)$-bundle, and let $\mathbb{V}(\sigma)$ be an irreducible constituent of $T^*M \otimes \mathbb{V}(\lambda) \cong_{\mathrm{Spin}(m)} \mathbb{V}(1) \otimes \mathbb{V}(\lambda)$. By [**F**], each $\mathbb{V}(\sigma)$ that appears does so with multiplicity 1. Thus there is a differential operator

$$\mathbb{V}(\lambda) \xrightarrow{\nabla} T^*M \otimes \mathbb{V}(\lambda) \xrightarrow{\mathrm{Proj}_\sigma} \mathbb{V}(\sigma), \tag{1.3}$$

which we call $G_{\lambda\sigma}$. $\mathbb{V}(\sigma)$ appears if and only if σ is a dominant weight and

$$\begin{aligned} &\sigma = \lambda \pm e_a\,, && \text{some } a \in \{1, \dots, \ell\}, \\ &\underline{\text{or}} && m \text{ is odd},\ \lambda_\ell \neq 0,\ \sigma = \lambda, \end{aligned} \tag{1.4}$$

where e_a is the ath standard basis vector in $\mathbb{R}^\ell$. [**F**] shows that $G_{\lambda\sigma}$ has reduced conformal bidegree $(\mathfrak{m}, \mathfrak{m}+1)$, where

$$\mathfrak{m} = \mathfrak{m}_{\lambda\sigma} = \tfrac{1}{2}(m - 1 + \langle \lambda + \sigma + 2\rho, \lambda - \sigma \rangle),$$

the inner product is the standard one in $\mathbb{R}^\ell$, and $2\rho := (m-2, m-4, \dots, m-2\ell)$ is the sum of the positive $\mathfrak{so}(m)$ roots. In odd dimensions, the Dirac operator is a gradient; in even dimensions, there are two Dirac operators $\Sigma_+ M \leftrightarrow \Sigma_- M$ interchanging the bundles of positive and negative spinors; these are gradients. The exterior derivative d and its formal adjoint δ are gradients, except when middle-forms are involved for even m; in this case, one must restrict or project to an eigenbundle of $\star$.

EXAMPLES 1.7. In 1981, the present author discovered conformal covariants $D_{2,k}$ of reduced bidegree $((m-2)/2, (m+2)/2)$ on differential forms of all orders k when $m > 2$ [**Br1**]. These have the form $(m-2k+2)\delta d + (m-2k-2)d\delta + Z_k$, where Z_k is a certain 0th-order action of the Ricci curvature. This generalizes the conformal Laplacian (the case $k = 0$) and, in even dimensions, the *Maxwell operator* δd on $(m-2)/2$-forms. On middle-forms in the even-dimensional oriented case, $D_{2,m/2}$ interchanges the two eigenbundles of the Hodge $\star$.

EXAMPLES 1.8. In 1983, Paneitz discovered a 4th-order conformal covariant P_4 acting on scalar functions, with leading term Δ^2 and lower-order terms involving the Ricci tensor [**Pa**]. Wünsch [**W**] found a similar 6th-order covariant P_6 provided $m \neq 4$. Graham, Jenne, Mason and Sparling [**GJMS**] generalized these results by constructing, in principle, covariants $P_{2\ell}$ of even order 2ℓ on scalar functions, with leading term Δ^ℓ, provided m is odd or $m \geq 2\ell$. The reduced bidegree of $P_{2\ell}$ is $((m-2\ell)/2, (m+2\ell)/2)$.

EXAMPLES 1.9. In [**W**], Wünsch also found second-order covariants acting on the bundles $\mathrm{TFS}^p M$ of trace-free symmetric p-tensors; their reduced bidgree is $((m-2)/2, (m+2)/2)$. In [**J**], Jenne found covariants acting between the $\mathrm{TFS}^p M$ for different p.

2. Functional determinants

In this section, we work in the category of compact Riemannian manifolds (M, g) and use zeta function and dimensional regularization to get formulas for functional determinant quotients $(\det \bar{A}_\omega)/(\det A_0)$, where A is a suitable differential operator, A_0 is A evaluated in a metric g_0, and A_ω is A evaluated in a conformally related metric $g_\omega = e^{2\omega} g_0$. In fact, we shall use this notation consistently: if S is some geometric object, and a choice of background metric g_0 is understood, S_ω will be the evaluation of S in $e^{2\omega} g_0$. We first need to say what we mean by "suitable differential operator".

ANALYTIC ASSUMPTIONS 2.1. *A is a differential operator of positive order on sections of a tensor-spinor bundle $\mathbb{V}$. A is formally self-adjoint and has positive definite leading symbol.*

REMARK 2.2. It is automatic from the assumption on the leading symbol that such A have even order; we fix 2ℓ as a notation for the order of A.

Under the analytic assumptions, we have a *heat expansion* [**G3**]: if f is an indeterminate element of $C^\infty(M)$,

$$\mathrm{Tr}_{L^2} fe^{-tA} \sim \sum_{n=0}^{\infty} t^{(n-m)/2\ell} \int_M fU_n[A]dv, \qquad t \downarrow 0,$$

where $U_n[A]$ is locally computable from the symbol of A in local coordinates. Parity considerations show that $U_n[A] = 0$ for n odd. The analytic assumptions also guarantee a real eigenvalue spectrum $\{\lambda_j\}$ for A, with Weyl asymptotics $\lambda_j \sim \mathrm{const} \cdot j^{2\ell/m}$ as $j \uparrow \infty$. In particular, there are only finitely many nonpositive λ_j. Let $a_n[A] = \int_M U_n[A]dv$. If $q[A]$ is the multiplicity of 0 as an eigenvalue,

$$\begin{aligned}\sum_{\lambda_j \neq 0} e^{-t\lambda_j} &= -q[A] + 2\sum_{\lambda_j<0} \sinh(t\lambda_j) + \sum_{n=0}^{N} a_n[A] t^{\frac{n-m}{2\ell}} + O(t^{(N-m+2)/2\ell}) \\ &=: \sum_{n=0}^{N} \tilde{a}_n[A] t^{\frac{n-m}{2\ell}} + O(t^{(N-m+1)/2\ell}).\end{aligned}$$

($\tilde{a}_n[A]$ might not vanish for some odd n, but this can only happen for odd m.) The Mellin transform yields a meromorphic continuation of the *zeta function*

$$\zeta_A(s) := \sum_{\lambda_j \neq 0} |\lambda_j|^{-s}$$

to $\mathbb{C}$:

$$\begin{aligned}\Gamma(s)\zeta_A(s) = \sum_{n=0}^{N} \left(s - \frac{m-n}{2\ell}\right)^{-1} \tilde{a}_n[A] + \int_0^1 t^{s-1} O(t^{(N-m+1)/2\ell})dt \\ + \int_1^\infty t^{s-1} \sum_{\lambda_j \neq 0} e^{-t|\lambda_j|}dt.\end{aligned}$$

$\zeta_A(s)$ is thus regular at $s = 0$; we define the *functional determinant* of A by

$$\det A := (-1)^{\#\{\lambda_j<0\}} e^{-\zeta_A'(0)}.$$

NATURALITY ASSUMPTIONS 2.3. *A is constructed functorially and polynomially from the metric g, its inverse $g^\sharp$, the covariant derivative ∇, and the Riemann curvature R, with coeffiecients that depend rationally on the dimension m. (In particular, A might be undefined for m in a finite set $\mathcal{M}_A$.) Under uniform dilations of the metric, A has homogeneity degree 2ℓ (equal to its order): $\bar{g} = \alpha^2 g$, $0 < \alpha \in \mathbb{R} \Rightarrow \bar{A} = \alpha^{-2\ell} A$.*

REMARK 2.4. We do not allow A to depend on orientation or the fundamental tensor-spinor γ. (Note that by the Lichnerowicz formula, $\nabla\!\!\!/^2$ does not depend on γ.) The rational coefficient assumption makes sense because spaces of Riemannian invariants have natural m-stable bases for large m; this is a consequence of Gilkey's Theorem [**G1**] and straightforward generalizations. By [**BGP**], there are also m-stable bases in the category of conformally flat manifolds; this will be of interest to us later.

REMARK 2.5. It is an important point that the functional determinant is not invariant under uniform dilation of the metric. In the notation just above, $\zeta_A(0)$ is dilation invariant, and

$$\det \bar{A} = \alpha^{-2\ell\zeta_A(0)/m} \det A.$$

Thus

$$\mathcal{D}(A,g) = (\operatorname{vol} g)^{2\ell\zeta_A(0)/m} \det A$$

is a scale-invariant normalization of the determinant. Like the determinant, it is a spectral invariant, since the volume is a positive multiple of $a_0[A]$. We can thus do the extremal problems for the determinant as a function of the metric if we normalize the volume to 1, or, equivalently, treat the functional $\mathcal{D}(A,g)$.

CONFORMAL ASSUMPTIONS 2.6. *A is a positive integer power D^h of a differential operator D which satisfies the above naturality assumptions, and in addition is allowed to depend on γ. h is independent of m, and D is conformally covariant of bidegree $(a(m), b(m))$, where $a(m)$ and $b(m)$ are rational functions.*

REMARK 2.7. A need not be conformally covariant, and D need not have positive leading symbol; for example, let $D = \nabla\!\!\!/$ and $h = 2$. The rationality of the conformal bidegree in some particular tensor-spinor realization is equivalent to the rationality of the reduced conformal bidegree.

CONFORMAL INDEX THEOREM 2.8 [**BØ1**]. *Let M be an m-dimensional compact manifold, and suppose that A satisfies 2.1, 2.3, and 2.6, with $m \notin \mathcal{M}_A$. Then the quantities $q[A]$, $\#\{\lambda_j \neq 0\}$, $a_m[A]$, and $\zeta_A(0)$ are constant on each conformal class in M. If $\omega \in C^\infty(M)$ and $g_{\varepsilon\omega} = e^{2\varepsilon\omega} g_0$ is a conformal curve of metrics,*

$$(d/d\varepsilon)|_{\varepsilon=0} a_n[A_{\varepsilon\omega}] = (m-n)\int_M \omega(U_n[A]dv)_0\,. \tag{2.1}$$

GENERALIZED POLYAKOV FORMULA 2.9 [**BØ2**]. *Suppose the assumptions of Theorem 2.8 are satisfied, and that the null space of A vanishes on (M,g). Then*

$$(d/d\varepsilon)|_{\varepsilon=0} \zeta'_{A_{\varepsilon\omega}}(0) = 2\ell \int_M \omega(U_m[A]dv)_0\,. \tag{2.2}$$

Thus the determinant supplies the *conformal primitive* for the function $U_m[A]$ which (2.1) fails to provide. Suppose now we want to calculate determinant quotients

$$-\log \frac{\det A_\omega}{\det A_0} = \zeta'_{A_\omega}(0) - \zeta'_{A_0}(0). \tag{2.3}$$

We could integrate the Polyakov variational formula along a conformal curve, using explicit knowledge of $U_m[A]$; this is the strategy employed in low dimensions in [**Po1-2, O, OPS1, BØ3**]. Another possibility is to analytically continue (2.1) in the dimension m for fixed n [**Br6, Br7**]; this is what we shall do here.

RATIONALITY ASSUMPTION 2.10. *Fix $n \in 2\mathbb{Z}^+$, and make the following assumption on (n, A): there is a universal function $u_A(m)$ depending only on the formal polynomial expression for the leading symbol of A, such that the coefficients in the formal polynomial expression for $u_A(m)U_n[A]$ are rational in m for $m \in 2\mathbb{Z}^+$, $m \notin \mathcal{M}_A$.*

For example, if A is second order and has scalar leading symbol $\sigma_2(A)(x,\xi) = |\xi|^2 \operatorname{Id}_{\mathbb{V}_x}$ (where $\mathbb{V}$ is the tensor-spinor bundle in which A acts), we may take

$u_A(m) = (4\pi)^{m/2}(\dim \mathbb{V})^{-1}$ as long this is meromorphic in m. The restriction to even m above is meant to handle this problem in the case of the spinor bundle, which has fiber dimension $2^{[m/2]}$.

Now the idea is to take the quantity which is formally

$$2\ell u_A(m)^{-1}\left\{u_A(m)(m-n)^{-1}(a_n[A_\omega]-a_n[A_0])\right\}\big|_{m=n}. \tag{2.4}$$

The quantity $u_A(m)$ is inserted to allow rational continuation (meromorphic continuation on the Riemann sphere, with limit point ∞). By (2.1), this quantity will solve the same variational problem as the quantity in (2.3). Both quantities vanish at $\omega = 0$. But this variational problem solution is unique, since it agrees with the initial value ODE problem solutions gotten by taking conformal curves $g_{\varepsilon\omega}$ for fixed ω. This proves:

THEOREM 2.11 [**Br6, Br7**]. *Under the assumptions of Theorem 2.9, plus the rationality assumption 2.10, the quantities in (2.3) and (2.4) agree.*

We shall evaluate and extremize determinant quotients on the spheres S^m. To put these formulas and the corresponding estimates in the proper context, we need some representation theory.

3. The complementary series

DEFINITION 3.1. *A conformal transformation on a Riemannian manifold (M,g) is a diffeomorphism $h: M \to M$ with $h \cdot g = \Omega_h^2 g$ for some $0 < \Omega_h \in C^\infty(M)$. A conformal vector field is a vector field X with $\mathcal{L}_X g = 2\omega_X g$ for some $\omega_X \in C^\infty(M)$.*

Here $h\cdot$ is the natural action of a diffeomorphism on tensors, or of a conformal transformation on tensor-spinors. (On covariant tensors like g, $h\cdot = (h^{-1})^*$.) $\mathcal{L}_X$ is the Lie derivative. The conformal transformations form a group $\mathbf{C}(M,g)$, and the conformal vector fields form a Lie algebra $\mathbf{c}(M,g)$; these statements are equivalent to the *cocycle conditions*

$$\Omega_{h_1 \circ h_2} = \Omega_{h_1}(h_1 \cdot \Omega_{h_2}), \qquad \omega_{[X_1,X_2]} = X_1\omega_{X_2} - X_2\omega_{X_1}.$$

By the cocycle conditions, the maps

$$\begin{array}{cc} u_a^{\mathbb{V}(\lambda)}: \mathbf{C}(M,g) \to \operatorname{Aut} C^\infty(M, \mathbb{V}^0(\lambda)) & U_a^{\mathbb{V}(\lambda)}: \mathbf{c}(M,g) \to \operatorname{End} C^\infty(M, \mathbb{V}^0(\lambda)) \\ h \mapsto \Omega_h^a h\cdot & X \mapsto \mathcal{L}_X + a\omega_X \end{array}$$

are, respectively, group and Lie algebra homomorphisms for all $a \in \mathbb{C}$. By setting the internal conformal weight to 0 (using $\mathbb{V}^0(\lambda)$), we have implicitly used the actions of $h\cdot$ and $\mathcal{L}_X$ on densities. Alternatively, we could have described $u_a(h)$ and $U_a(X)$ as $h\cdot$ and $\mathcal{L}_X$ on $\mathbb{V}^a(\lambda)$ be reweighting as in Remark 1.2.

The behavior of conformal covariants under conformal transformations and vector fields can be understood by viewing a conformal transformation as a composition

$$(M,g) \xrightarrow{h} (M, \Omega_h^2 g) \xrightarrow{\mathrm{Id}} (M,g)$$

of an isometry and a conformal change of metric. Since conformal covariants are isometry invariants, a covariant $D: \mathbb{V}^0(\lambda) \to \mathbb{V}^0(\sigma)$ of reduced bidegree (a,b) has

$$D(\Omega_h^a h \cdot \varphi) = \Omega_h^b h \cdot (D\varphi), \qquad D(\mathcal{L}_X + a\omega_X)\varphi = (\mathcal{L}_X + b\omega_X)D\varphi \tag{3.1}$$

for all $f \in C^\infty(M, \mathbb{V}^0(\lambda))$.

On the sphere S^m with the standard round metric g_0, all conformal vector fields integrate globally, and $\mathbf{C}_0(S^m, g_0) \cong \mathrm{SO}_0(m+1,1) =: G$ (where the subscript 0 indicates identity component) as follows. Let $y = (y_0 \dots, y_m)$ be homogeneous coordinates on S^m; if $T \in G$, we take the linear action of T on the $(m+2)$-tuple $(y, 1)$, then divide by $(T(y,1))_{m+1} > 0$ to get a new element $(T \cdot y, 1)$ of $S^m \times \{1\}$. The maximal compact subgroup K of G is the isometry group of S^m, namely the $\mathrm{SO}(m+1)$ that acts in the $(y_0, \dots, y_m)$ variables.

REMARK 3.2. Let $\mathbb{V} = \mathbb{V}(\lambda)$. The restriction of $u_a^{\mathbb{V}}$ to K is independent of a; we use $u^{\mathbb{V}}$ as a common notation for these restrictions. By standard semisimple theory, or more elementary arguments adapted to the present situation, all information on the representations $u_a^{\mathbb{V}}$ of G is encoded in the $(\mathfrak{g}, K)$ module representation $(U_a^{\mathbb{V}}, u^{\mathbb{V}})$. $\mathfrak{g} \cong \mathbf{c}(S^m, g_0)$ has the Cartan decomposition $\mathfrak{g} = \mathfrak{k} + \mathfrak{s}$, where $\mathfrak{s}$ is generated by the homogeneous coordinate derivatives $Y_i := \partial_i$, $i = 0, \dots, m$. The adjoint representation of K on $\mathfrak{s}$ is a copy of the defining representation of $\mathrm{SO}(m)$. The Y_i generate the *proper* conformal vector fields, i.e. those $X \in \mathbf{c}(S^m, g_0)$ with $\omega_X \neq 0$. In fact,

$$\omega_{Y_i} = y_i\,, \qquad dy_i = -(Y_i)_\flat\,, \tag{3.2}$$

where $\flat$ is the metric identification $TS^m \to T^*S^m$.

We shall be interested in *intertwining operators* between representations $u_a^{\mathbb{V}(\lambda)}$ and $u_b^{\mathbb{V}(\sigma)}$, especially in the case $\lambda = \sigma$. By (3.1), conformal covariants give rise to differential intertwinors. By Remark 3.2, an intertwinor is a K-invariant operator A with $AU_a^{\mathbb{V}(\lambda)}(Y_i) = U_b^{\mathbb{V}(\sigma)}(Y_i)A$. Before computing intertwinors, we need some information on the K-spectrum of the sections of $\mathbb{V}(\lambda)$.

Suppose λ is irreducible and recall the parameterization (1.2). Note that if the components of λ are half-integral, we need to work with $\bar{K} = \mathrm{Spin}(m+1)$; that is, we must look at S^m as $\mathrm{Spin}(m+1)/\mathrm{Spin}(m) = \bar{K}/\bar{M}$ instead of as $\mathrm{SO}(m+1)/\mathrm{SO}(m) = K/M$. With this tacitly in mind, note that the irreducible representations of K are parameterized as in (1.2), just changing m to $m+1$ and ℓ to $L := [(m+1)/2]$. By Frobenius reciprocity, if α is the highest weight of an irreducible representation R_α of $\bar{K}$, and $\mathcal{E}_\lambda$ is the space of $\bar{K}$-finite sections of $\mathbb{V}_\lambda$, there is a natural identification

$$\mathrm{Hom}_{\bar{K}}(R_\alpha, \mathcal{E}_\lambda) \cong \mathrm{Hom}_{\bar{M}}(R_\alpha, V_\lambda).$$

Thus we know the $\bar{K}$-spectrum of $\mathbb{V}_\lambda$ if we know the *branching rule* showing how each R_α restricts to $\bar{M}$. This branching rule is as follows [**Bo**]. $m(\alpha, \lambda) := \dim \mathrm{Hom}_{\bar{M}}(R_\alpha, V_\lambda)$ takes only the values 0 and 1, and it is 1 iff $\alpha_1 - \lambda_1 \in \mathbb{Z}$ and

$$\begin{aligned} &\alpha_1 \geq \lambda_1 \geq \alpha_2 \geq \cdots \lambda_\ell \geq |\alpha_{\ell+1}|, && n \text{ odd},\\ &\alpha_1 \geq \lambda_1 \geq \alpha_2 \geq \cdots \lambda_{\ell-1} \geq \alpha_\ell \geq |\lambda_\ell|, && n \text{ even}. \end{aligned} \tag{3.3}$$

We use $\alpha \downarrow \lambda$ or $\lambda \uparrow \alpha$ as an abbreviation for (3.3); note that

$$\mathcal{E}_\lambda \cong_{\bar{K}} \bigoplus_{\alpha \downarrow \lambda} R_\alpha\,. \tag{3.4}$$

DEFINITION 3.3. *On a Riemannian spin manifold* (M, g), *the operator* $\nabla^*\nabla = (\nabla^*\nabla)_{\mathbb{V}}$ *on a tensor-spinor bundle* $\mathbb{V}$ *is called the Bochner Laplacian.*

In more detail, the covariant derivative ∇ carries $\mathbb{V}$ to $TM \otimes \mathbb{V}$; ∇^* is the formal adjoint of ∇; the leading symbol of $\nabla^*\nabla$ is $|\xi|^2$. Since $\nabla^*\nabla$ is an H-operator, it can be reweighted to act between tensor-spinor density bundles; thus $\nabla^*\nabla : \mathbb{V}^r(\lambda) \to \mathbb{V}^r(\lambda)$ makes sense, and its spectrum is independent of r.

THEOREM 3.4 [**Br5**]. *On* (S^m, g_0), $(\nabla^*\nabla)_{\mathbb{V}_\lambda}$ *has the eigenvalue*

$$\kappa(\alpha, \lambda) := \alpha(\mathrm{Cas}_{\mathfrak{k}}) - \lambda(\mathrm{Cas}_{\mathfrak{m}}) = \langle \alpha + 2\rho_{\mathfrak{k}}, \alpha \rangle_{\mathbb{R}^L} - \langle \lambda + 2\rho_{\mathfrak{m}}, \lambda \rangle_{\mathbb{R}^\ell}$$

on the R_α *summand in (3.4). Here "Cas" denotes the Casimir element of the universal enveloping algebra,* $\ell = [m/2]$, $L = [(m+1)/2]$, $2\rho_{\mathfrak{m}} = (m-2, m-4, \ldots, m-2\ell)$, $2\rho_{\mathfrak{k}} = (m-1, m-3, \ldots, m+1-2L)$, *and the inner products are standard.*

COROLLARY 3.5. $\Delta y_i = m y_i$.

REMARK 3.6. Let $\mathbb{V} = \mathbb{V}^r(\lambda)$, and consider the effect of ∇_{Y_i} and $\mathcal{L}_{Y_i}$ on smooth sections φ of $\mathbb{V}$. For general (M, g) and X, the symmetry of ∇ implies that $(\mathcal{L}_X - \nabla_X)|_{\mathbb{V}}$ depends only on ∇X. On the other hand, $\mathcal{L}_X - \nabla_X$ has order 0 as a differential operator; thus $(\mathcal{L}_X - \nabla_X)|_{\mathbb{V}} = Z_{\mathbb{V}}(\nabla X)$ for some smooth section $Z_{\mathbb{V}}$ of $\mathbb{V} \otimes \mathbb{V}^* \otimes T^*M \otimes TM$. But according to paragraph 1.6, ∇X breaks up into $dX_\flat$, $\delta X_\flat$, and $SX_\flat$ contributions, where S is the gradient with target bundle $\mathbb{V}(2)$. The condition $SX_\flat = 0$ is equivalent to the assertion that X is conformal, and in our situation above, the $d(Y_i)_\flat = 0$. Thus $Z_{\mathbb{V}}(\nabla Y_i)$ is multiplication by a constant multiple of $\delta(Y_i)_\flat = -\delta d y_i = -\Delta y_i = -m y_i$ (where the last equality is Corollary 3.5); say $Z_{\mathbb{V}}(\nabla Y_i) = \alpha_{\mathbb{V}} \mu(y_i)$. Since the bundle $\mathbb{V}$ has internal conformal weight r, the natural metric $g_{\mathbb{V}}$ induced from g has $\mathcal{L}_X g_{\mathbb{V}} = -2r\omega_X g_{\mathbb{V}}$ for X a conformal vector field. Thus for a smooth section φ of $\mathbb{V}$,

$$\begin{aligned} 0 &= (\mathcal{L}_{Y_i} - \nabla_{Y_i})(g_{\mathbb{V}}(\varphi, \varphi)) \\ &= 2g_{\mathbb{V}}(\varphi, (\mathcal{L}_{Y_i} - \nabla_{Y_i})\varphi) - 2r\omega_{Y_i} g_{\mathbb{V}}(\varphi, \varphi) \\ &= (2\alpha_{\mathbb{V}} - 2r) y_i g_{\mathbb{V}}(\varphi, \varphi). \end{aligned}$$

This shows $\alpha_{\mathbb{V}} = r$, whence

$$(\mathcal{L}_{Y_i} - \nabla_{Y_i})|_{\mathbb{V}^r(\lambda)} = \mu(r y_i). \tag{3.5}$$

Fixing λ, we can express all the representations $U_a^{\mathbb{V}(\lambda)}$ in terms of the cocycle ω_X .

THEOREM 3.7. *If* $Y \in \mathfrak{s}$, *then* $[(\nabla^*\nabla)_{\mathbb{V}(\lambda)}, \mu(\omega_Y)] = 2U_{m/2}^{\mathbb{V}(\lambda)}(Y)$.

PROOF. The commutator is in the sense of differential operators; recall that $\mu(\omega_Y)$ is multiplication by the function ω_Y . We need only check the relation at $Y = Y_i$. If φ is a smooth section of $\mathbb{V}$,

$$[(\nabla^*\nabla)_{\mathbb{V}(\lambda)}, \omega_{Y_i}]\varphi = (\Delta\omega_{Y_i})\varphi - 2\iota(d\omega_{Y_i})\nabla\varphi,$$

where ι is interior multiplication: if $\eta \otimes \psi$ is a simple element of $T_x^*M \otimes \mathbb{V}_x$ and $\tau \in T_x^*M$ is dual to $T \in T_xM$ under the metric identification, then

$$\iota(\tau)(\eta \otimes \psi) = g^\sharp(\tau, \eta)\psi = \langle T, \eta \rangle \psi = \iota(T)(\eta \otimes \psi).$$

By (3.2), $-d\omega_{Y_i}$ and Y_i are dual under the metric identification. This and Corollary 3.5 show that

$$[(\nabla^*\nabla)_{\mathbb{V}(\lambda)}, \omega_{Y_i}] = m\omega_{Y_i} + 2\iota(Y_i)\nabla = m\omega_{Y_i} + 2\nabla_{Y_i}\,.$$

By Remark 3.6, the right-hand side of this is $2U^{\mathbb{V}(\lambda)}_{m/2}(Y_i)$.

COROLLARY 3.8. *If $\alpha, \beta \downarrow \lambda$ and $Y \in \mathfrak{s}$, then*

$$\operatorname{Proj}_\beta U_{m/2}(Y)|_\alpha = \tfrac{1}{2}(\kappa(\beta,\lambda) - \kappa(\alpha,\lambda))\operatorname{Proj}_\beta \mu(\omega_Y)|_\alpha\,,$$

where the restriction and projection refer, respectively, to the R_α and R_β summands in (3.4). As a consequence,

$$\operatorname{Proj}_\beta U_a(Y)|_\alpha = \tfrac{1}{2}(\kappa(\beta,\lambda) - \kappa(\alpha,\lambda) + 2a - m)\operatorname{Proj}_\beta \mu(\omega_Y)|_\alpha\,. \tag{3.6}$$

REMARK 3.9. Of course, not many β can be reached from a given α with one application of $\omega(\mathfrak{s}) = \{\mu(\omega_Y) : Y \in \mathfrak{s}\}$. Since $\mathfrak{s}$ carries the defining representation of K and

$$Y \otimes \varphi \mapsto \omega_Y\varphi, \qquad \mathfrak{s} \otimes \mathcal{E}_\lambda \to \mathcal{E}_\lambda$$

is a $\bar{K}$-map, the only β that can really be reached are the summands of $(1) \otimes \alpha$. These are given by (1.4) if we just shift the dimension: m (resp. ℓ, λ, σ) becomes $m+1$ (resp. L, α, β). We say $\alpha \leftrightarrow \beta$ if and only if α and β are dominant $\bar{K}$-weights and

$$\begin{array}{ll} \beta = \alpha \pm e_a\,, & \text{some } a \in \{1, \dots, L\}, \\ \underline{\text{or}} & m+1 \text{ is odd}, \ \alpha_L \neq 0, \ \beta = \alpha. \end{array} \tag{3.7}$$

Note that the relation $\leftrightarrow$ is symmetric. (3.7) is called a *selection rule*. We have:

$$\operatorname{Proj}_\beta U_a(Y)|_\alpha = \operatorname{Proj}_\beta \mu(\omega_Y)|_\alpha = 0 \qquad \text{unless} \qquad \alpha \leftrightarrow \beta.$$

REMARK 3.10. Now consider the problem of getting intertwinors on $\mathbb{V}(\lambda)$, i.e. $\bar{K}$-operators A with

$$AU^{\mathbb{V}(\lambda)}_a(Y_i) = U^{\mathbb{V}(\lambda)}_b(Y_i)A$$

for $a, b \in \mathbb{C}$. Because the $\bar{K}$-types in (3.4) occur with multiplicity 1, such an A is described by a list of eigenvalues:

$$A : \varphi = \sum_{\alpha\downarrow\lambda} \varphi_\alpha \mapsto \sum_{\alpha\downarrow\lambda} T(\alpha,\lambda)\varphi_\alpha\,, \tag{3.8}$$

where $\varphi \in \mathcal{E}_\lambda$ (so the sums are finite), and φ_α is in the R_α summand in (3.4). If $\operatorname{Proj}_\beta \mu(\omega_Y)|_\alpha \neq 0$, (3.6) together with (3.8) constrains the list $\{T_\alpha\}$:

$$(\kappa(\beta,\lambda) - \kappa(\alpha,\lambda) + 2a - m)T(\beta,\lambda) = (\kappa(\beta,\lambda) - \kappa(\alpha,\lambda) + 2b - m)T(\alpha,\lambda).$$

Switching the roles of α and β, one quickly sees that apart from trivialities, we must have $a = m - b$.

REMARK 3.11. Fixing $r = b - m/2 = m/2 - a$, we have

$$(\kappa(\beta,\lambda) - \kappa(\alpha,\lambda) - 2r)T(\beta,\lambda) = (\kappa(\beta,\lambda) - \kappa(\alpha,\lambda) + 2r)T(\alpha,\lambda). \tag{3.9}$$

By the selection rule (3.7) and the branching rule (3.3), the $\bar{K}$-decomposition (3.4) of each $\mathcal{E}_\lambda$ is connected under $\leftrightarrow$. Thus if we choose a particular $\alpha^0 \downarrow \lambda$ and step to the other K-types of $\mathcal{E}_\lambda$, using (3.9) to convert differences of Bochner Laplacian eigenvalues into quotients of intertwinor eigenvalues. We run into a consistency problem only if it is possible to have $\alpha \leftrightarrow \alpha$.

THEOREM 3.12. *Suppose it is not the case that m is even and $\lambda_{(m-2)/2} \neq 0$. Suppose*

$$r \notin \mathcal{K}_\lambda := \{\tfrac{1}{2}(\kappa(\beta,\lambda) - \kappa(\alpha,\lambda)) \,:\, \alpha,\beta \downarrow \lambda,\ \alpha \leftrightarrow \beta\}.$$

Choose $\alpha^0 \downarrow \lambda$. Then the operator $A =: A_{2r,\lambda}$ described by (3.8) with

$$T(\alpha,\lambda) =: T_{2r}(\alpha,\lambda) = \prod_{a=1}^{[(m+1)/2]} \frac{\Gamma(\frac{m}{2}+1-a+\alpha_a+r)\Gamma(\frac{m}{2}+1-a+\alpha_a^0-r)}{\Gamma(\frac{m}{2}+1-a+\alpha_a-r)\Gamma(\frac{m}{2}+1-a+\alpha_a^0+r)}$$

intertwines $u^{\mathbb{V}(\lambda)}_{m/2-r}$ and $u^{\mathbb{V}(\lambda)}_{m/2+r}$.

REMARK 3.13. We avoid the set $\mathcal{K}_\lambda$ so that $T_{2r}(\alpha,\lambda)$ will avoid the values 0 and ∞. From the point of view of representation theory, $\mathcal{K}_\lambda$ is an extremely interesting set; here the $u^{\mathbb{V}(\lambda)}_{m/2\pm r}$ have nontrivial composition structure, indicated by the poles and zeros of $T_{2r}(\alpha,\lambda)$, viewed as a rational function of r.

REMARK 3.14. The first assumption in Theorem 3.12 avoids the situation where $\alpha \leftrightarrow \alpha$ is possible, but it can be weakened somewhat. We actually need only avoid the case where m is even and $\lambda_{m/2} \neq 0$. For if m is even and $\lambda_{(m-2)/2} \neq 0 = \lambda_{m/2}$, we can "upgrade" to a representation of $G_1 = \mathrm{O}(m+1,1)$, or G_1 the corresponding extension of $\mathrm{Spin}(n+1,1)$. If $\alpha_{m/2} \neq 0$, there are two representations $\alpha^\pm$ of K_1 which restrict to α (labelled $\pm$ according to the effect of $-I \in \mathrm{O}(m+1)$). Since I acts as -1 in the adjoint representation of K_1 on $\mathfrak{s}$, the representations $\alpha^\pm$ are interchanged upon tensoring with $\mathfrak{s}$. For a given α, multiplicity one implies that only one of $\alpha^\pm$ can occur in the K-decomposition of $\mathcal{E}_\lambda$; but the selection rule $\alpha \leftrightarrow \alpha$ can only implement $\alpha^+ \leftrightarrow \alpha^-$. Thus it must be the case that $\mathrm{Proj}_\alpha\, \mu(\omega(\mathfrak{s}))|_\alpha = 0$; it is not necessary to satisfy (3.9) for $\beta = \alpha$, and the consistency problem disappears. If m is even and $\lambda_{m/2} \neq 0$, the consistency problem is in the nature of things: the general theory of intertwinors, which involves more Lie theory than we care to introduce here, predicts intertwinors between $u^{\mathbb{V}(\lambda')}_{m/2+r'}$ and $u^{\mathbb{V}(\lambda)}_{m/2+r}$ when $(\lambda, r) = w(\lambda', r')$ for some element w of the $(\mathfrak{g},\mathfrak{a})$ *Weyl group*. As it happens, we can find a w to reflect r without moving λ unless m is even and $\lambda_{m/2} \neq 0$; here the least we can disturb λ is to reflect $\lambda_{m/2}$, thereby intertwining $u^{\mathbb{V}(\tilde{\lambda})}_{m/2-r}$ and $u^{\mathbb{V}(\lambda)}_{m/2+r}$ for $\tilde{\lambda} = (\lambda_1, \ldots, \lambda_{(m-2)/2}, -\lambda_{m/2})$. Examples of this phenomenon have already appeared in paragraphs 1.6 and 1.7: in even dimensions, the Dirac operator interchanges positive and negative spinors, and the second-order operator $D_{2,m/2}$ interchanges the two eigenbundles of the Hodge $\star$.

For a given λ, if we can show that the compression $\mathrm{Proj}_\beta\, \mu(\omega(\mathfrak{s}))|_\alpha$ is nonzero in all instances where $\beta \leftrightarrow \alpha \downarrow \lambda$ and $\beta \neq \alpha$, then $\mathcal{E}_\lambda$ will be irreducible under $u^{\mathbb{V}(\lambda)}_{m/2-r}$ for all $r \notin \mathcal{K}_\lambda$. This will show that (3.9) is necessary, as well as sufficient, for the construction of an intertwinor, and thus will show that the intertwinor of Theorem 3.12 is unique. To get information on the compressions, consider the quantities

$$t(\alpha, \beta; \lambda) := \sum_i \operatorname{Proj}_\alpha \mu(\omega_{Y_i})|_\beta \operatorname{Proj}_\beta \mu(\omega_{Y_i})|_\alpha .$$

As a $\bar{K}$-map on the R_α summand of (3.4), $t(\alpha, \beta; \lambda)$ is a constant times the identity; we abuse the notation just introduced by denoting the constant $t(\alpha, \beta; \lambda)$. Since $t(\alpha, \beta; \lambda)$ is the trace of a sum of operators of the form L^*L (in any normalization of the inner products on the K-types), we have

$$t(\alpha, \beta; \lambda) \geq 0 \qquad \text{with equality iff } \operatorname{Proj}_\beta \mu(\omega(\mathfrak{s}))|_\alpha = 0.$$

Comparing traces of the just-mentioned sums of the L^*L and the corresponding LL^*, we get

$$t(\alpha, \beta; \lambda) \dim \alpha = t(\beta, \alpha; \lambda) \dim \beta. \tag{3.10}$$

Since $\sum_{i=0}^m y_i^2 = 1$,

$$\sum_{\beta \leftrightarrow \alpha} t(\alpha, \beta; \lambda) = 1 \qquad \text{for fixed } \alpha. \tag{3.11}$$

There is another, slightly more subtle, relation among the $t(\alpha, \beta; \lambda)$ along the same lines. Evaluate both sides of (3.6) at $a = 0$, switch the roles of α and β, compose the role-switched formula with the original one, and sum over $Y = Y_i$ to get

$$-(\kappa(\beta, \lambda) - \kappa(\alpha, \lambda))^2 t(\alpha, \beta; \lambda) = 4 \sum_i \operatorname{Proj}_\alpha U_{m/2}^{\mathbb{V}(\lambda)}(Y_i)|_\beta \operatorname{Proj}_\beta U_{m/2}^{\mathbb{V}(\lambda)}(Y_i)|_\alpha .$$

Summing over $\beta \leftrightarrow \alpha$ for fixed α, we get

$$\begin{aligned}\sum_{\beta \leftrightarrow \alpha} (\kappa(\beta, \lambda) - \kappa(\alpha, \lambda))^2 t(\alpha, \beta; \lambda) &= -4 \operatorname{Proj}_\alpha U_{m/2}^{\mathbb{V}(\lambda)}(\mathrm{Cas}_\mathfrak{g} - \mathrm{Cas}_\mathfrak{k})|_\alpha \\ &= -4 \operatorname{Proj}_\alpha U_{m/2}^{\mathbb{V}(\lambda)}(\mathrm{Cas}_\mathfrak{g})|_\alpha + 4\alpha(\mathrm{Cas}_\mathfrak{k}).\end{aligned}$$

By a classical formula for the Casimir operator in an induced representation,

$$U_{m/2}^{\mathbb{V}(\lambda)}(\mathrm{Cas}_\mathfrak{g}) = -(m/2)^2 + \lambda(\mathrm{Cas}_\mathfrak{m}).$$

(See [**Br6, Br7, BÓØ**] for the normalizations.) Thus

$$\sum_{\beta \leftrightarrow \alpha} (\kappa(\beta, \lambda) - \kappa(\alpha, \lambda))^2 t(\alpha, \beta; \lambda) = m^2 - 4\lambda(\mathrm{Cas}_\mathfrak{m}) + 4\alpha(\mathrm{Cas}_\mathfrak{k}) \qquad (\text{fixed } \alpha),$$

so by Theorem 3.4,

$$\sum_{\beta \leftrightarrow \alpha} (\kappa(\beta, \lambda) - \kappa(\alpha, \lambda))^2 t(\alpha, \beta; \lambda) = m^2 + 4\kappa(\alpha, \lambda) \qquad (\text{fixed } \alpha). \tag{3.12}$$

If the $\bar{K}$-spectrum is not too complicated, we can use (3.10), (3.11), and (3.12) to actually compute the $t(\alpha, \beta; \lambda)$, and thus get total information on irreducibility questions. For example, consider the bundles $\mathbb{V}(p)$ of trace-free symmetric p-tensors.

THEOREM 3.15. *Let $m \geq 4$ and $p \geq 0$. Then $\mathcal{E}_{(p)}$ is irreducible under $u^{\mathbb{V}(p)}_{m/2-r}$ provided $\pm r \notin \frac{m}{2} - 1 + \mathbb{N}$. Thus $A_{2r,(p)}$ is, up to constant multiple, the unique intertwinor between $u^{\mathbb{V}(p)}_{m/2-r}$ and $u^{\mathbb{V}(p)}_{m/2+r}$.*

PROOF. By (3.4), the K-type decomposition is

$$\mathcal{E}_{(p)} \cong_K \bigoplus_{\substack{j \in \mathbb{N} \\ 0 \leq q \leq p}} R_{(p+j,q)} \,.$$

Note that Theorem 3.12 applies directly only if $m \geq 5$, but we can include $m = 4$ by Remark 3.14. The relevant Casimir data are

$$\begin{aligned} &\kappa((p+j,q),(p)) = (p+j)(p+j+m-1) + q(q+m-3), \\ &\kappa((p+j+1,q),(p)) - \kappa((p+j,q),(p)) = m + 2(p+j), \\ &\kappa((p+j,q+1),(p)) - \kappa((p+j,q),(p)) = m - 2 + 2q. \end{aligned} \tag{3.13}$$

By Weyl's dimension formula, the relevant dimension quotients are

$$\frac{\dim(p+j+1,q)}{\dim(p+j,q)} = \frac{(p+j+q+m-1)(p+j-q+2)(p+j+m-2)(p+j+\frac{m+1}{2})}{(p+j+q+m-2)(p+j-q+1)(p+j+2)(p+j+\frac{m-1}{2})},$$

$$\frac{\dim(p+j,q+1)}{\dim(p+j,q)} = \frac{(p+j+q+m-1)(p+j-q)(q+m-3)(q+\frac{m-1}{2})}{(p+j+q+m-2)(p+j-q+1)(q+1)(q+\frac{m-3}{2})}.$$

This gives, via recursive solution of (3.10), (3.11), and (3.12),

$$t((p+j,q),(p+j+1,q);(p)) = \frac{(j+1)(j+m+2p-1)(j+p+m-2)}{(p+j-q+1)(2p+2j+m-1)(p+j+m+q-2)},$$

$$t((p+j,q),(p+j,q+1);(p)) = \frac{(m+q-3)(p-q)(m+p+q-2)}{(m+2q-3)(p+j-q+1)(p+j+m+q-2)}.$$

These quantities $t(\alpha,\beta;\lambda)$ are nonzero, as are the quantities $t(\beta,\alpha;\lambda)$ related by (3.10). In view of (3.13), the condition $\pm r \notin \frac{m}{2} - 1 + \mathbb{N}$ is exactly that needed to avoid $r \in \mathcal{K}_{(p)}$.

Looking at the above proof, we can improve the result in the case $p = 0$:

COROLLARY 3.16. *If $m \geq 3$ and $\pm r \notin \frac{m}{2} + \mathbb{N}$, then $\mathcal{E}_{(0)}$ is irreducible under $u^{\mathbb{V}(0)}_{m/2-r}$. Thus $A_{2r} := A_{2r,(0)}$ is, up to constant multiple, the unique intertwinor between $u^{\mathbb{V}(0)}_{m/2-r}$ and $u^{\mathbb{V}(0)}_{m/2+r}$.*

From our point of view, the main use of $A_{2r,\lambda}$ will be as an invariant pre-Hilbert space inner product.

THEOREM 3.17. *If $0 < \min\{|r| \,:\, r \in \mathcal{K}_\lambda\} =: K_\lambda$, then for r in the real interval $(-K_\lambda, K_\lambda)$,*

$$\langle \varphi, \psi \rangle_{m/2-r} = (\varphi, A_{2r,\lambda}\psi)_{L^2}$$

is a $u^{\mathbb{V}(\lambda)}_{m/2-r}$ invariant, positive definite inner product on $\mathcal{E}_\lambda$.

PROOF. For invariance, it is sufficient to check $\tilde{K}$ invariance, and $\mathfrak{g}$-skewness. The first of these is immediate. For the second, note that the Bochner Laplacian and $\mu(\mathfrak{s})$ are formally self-adjoint operators; thus by Theorem 3.7, the commutator $U_{m/2}^{\mathbb{V}(\lambda)}(Y)$ is formally skew-adjoint for $Y \in \mathfrak{s}$. This and the intertwining relation satisfied by A_{2r} justify the following calculation:

$$\begin{aligned}
\langle\varphi, U_{m/2-r}^{\mathbb{V}(\lambda)}(Y)\psi\rangle_{m/2-r} &= (\varphi, A_{2r,\lambda}U_{m/2-r}^{\mathbb{V}(\lambda)}(Y)\psi)_{L^2} \\
&= (\varphi, U_{m/2+r}^{\mathbb{V}(\lambda)}(Y)A_{2r,\lambda}\psi)_{L^2} \\
&= (\varphi, (U_{m/2}^{\mathbb{V}(\lambda)}(Y) + r\omega_Y)A_{2r,\lambda}\psi)_{L^2} \\
&= ((-U_{m/2}^{\mathbb{V}(\lambda)}(Y) + r\omega_Y)\varphi, A_{2r,\lambda}\psi)_{L^2} \\
&= -(U_{m/2-r}^{\mathbb{V}(\lambda)}(Y)\varphi, A_{2r,\lambda}\psi)_{L^2} \\
&= -\langle U_{m/2-r}^{\mathbb{V}(\lambda)}(Y)\varphi, \psi\rangle_{m/2-r},
\end{aligned}$$

where "L^2" abbreviates $L^2(S^m, \mathbb{V}(\lambda))$. For positive definiteness, note that the factors

$$\frac{T(\beta,\lambda)}{T(\alpha,\lambda)} = \frac{\kappa(\beta,\lambda) - \kappa(\alpha,\lambda) + 2r}{\kappa(\beta,\lambda) - \kappa(\alpha,\lambda) - 2r} \tag{3.14}$$

from (3.9) are positive for $r \in (-K_\lambda, K_\lambda)$; the expression for $T_{2r}(\alpha,\lambda)$ in Theorem 3.12 is a product of such factors. Note that Theorem 3.12 applies, since the assumption $K_\lambda > 0$ rules out the possibility of $\alpha \leftrightarrow \alpha$.

In fact, this inner product gives a particular L^2 Sobolev topology.

THEOREM 3.18. *Under the assumptions of Theorem 3.17, the norm* $\|\varphi\|_{m/2-r}^2 = \langle\varphi,\varphi\rangle_{m/2-r}$ *is equivalent to the* $L_r^2(S^m, \mathbb{V}(\lambda))$ *norm.*

PROOF. The only component of α that can take on infinitely many values is α_1; thus by Theorem 3.12 and the positivity of the factors (3.14), there are positive constants c and C depending on m, λ, and r such that

$$c\,\frac{\Gamma(\frac{m}{2}+\alpha_1+r)}{\Gamma(\frac{m}{2}+\alpha_1-r)} \le T_{2r}(\alpha,\lambda) \le C\,\frac{\Gamma(\frac{m}{2}+\alpha_1+r)}{\Gamma(\frac{m}{2}+\alpha_1-r)}.$$

Thus by Stirling's formula, $T_{2r}(\alpha,\lambda)$ grows like α_1^{2r} as $\alpha_1 \uparrow \infty$. Comparing this to the growth rate of Bochner Laplacian eigenvalues from Theorem 3.4, we find that A_{2r} and $(\nabla^*\nabla)^r$ define the same equivalence class of norms, that being L_r^2.

COROLLARY 3.19. *The norms* $\|\varphi\|_{m/2-r}^2$ *give the* L_r^2 *Sobolev topology for* $m \ge 4$, $\lambda = (p)$, *and* $r \in (-(m-2)/2, (m-2)/2)$; *and for* $m \ge 3$, $\lambda = (0)$, *and* $r \in (-m/2, m/2)$.

For the moment, we shall be especially interested in the case of scalar functions; i.e. the case $\lambda = (0)$. Here, by Theorem 3.4, the Laplacian Δ takes the value $j(m-j+1)$ on the space E_j of jth-order spherical harmonics; this identifies the "quantum number" j as the operator

$$B := \sqrt{\Delta + ((m-1)/2)^2}.$$

This allows us to write:

THEOREM 3.20 ([**Br3**], Remark 2.23). *On scalar functions for $r \in (-m/2, m/2)$, $m \geq 3$,*

$$A_{2r} = \frac{\Gamma(B + \frac{1}{2} + r)\Gamma(\frac{m}{2} - r)}{\Gamma(B + \frac{1}{2} - r)\Gamma(\frac{m}{2} + r)}.$$

A_{2r} is a differential operator if and only if $r \in \mathbb{N}$.

PROOF. We have proved all but the assertion about differential operators. The order of A_{2r} is a natural number if and only if $r \in \frac{1}{2}\mathbb{N}$. If $r \in \mathbb{N}$, then A_{2r} is a constant times a product of operators of the form $(B + a)(B - a)$, and thus is a polynomial in Δ. If $r \in \frac{1}{2} + \mathbb{N}$, then A_{2r} is a constant times B times a product of $(B + a)(B - a)$, thus has leading symbol a constant multiple of $|\xi|^{\text{odd}}$, and thus is not differential.

4. Sharp inequalities and extremals of the functional determinant

In a fundamental work, Lieb [**L**] used symmetrization arguments to find the best constants and extremals for the Sobolev embeddings $L^{2m/(m+2r)}(\mathbb{R}^m) \hookrightarrow L^2_{-r}(\mathbb{R}^m)$, $r \in (0, m/2)$. This is a norm computation for the convolution operator $r^{2r-m}*$ (r being the radial parameter in $\mathbb{R}^m$), the problem being

$$\text{maximize } |(r^{2r-m} * f, f)_{L^2}| \text{ subject to } \|f\|_{L^{2m/(m+2r)}} = 1.$$

Integral operators like these are the basis of the Knapp-Stein theory of intertwining operators for semisimple Lie groups [**KS**], and in fact, it is *exactly* the operators $r^\lambda *$ that appear if the group is $\mathrm{SO}_0(m+1, 1)$. Thus it is natural that these intertwinors in the "compact picture", described in the last section, should come into play when Lieb's inequalities are moved to S^m via the stereographic projection. The stereographic projection itself is a special case of a Lie-theoretic construct, the map $\bar{N} \to G/MAN$ arising from a Langlands decomposition $G = \bar{N}MAN$.

Beckner [**Be**] transferred Lieb's results to the S^m setting:

THEOREM 4.1 [**Be**]. *If $r \in (0, m/2)$,*

$$\max_{\varphi \in C^\infty(S^m)} \frac{|(A_{-2r}\varphi, \varphi)_{L^2(S^m, d\xi)}|}{\|\varphi\|^2_{L^{2m/(m+2r)}(S^m, d\xi)}}$$

is attained exactly when φ is a nonzero constant multiple of some $\Omega_h^{m/2+r}$, $h \in \mathbf{C}(S^m, g_0)$. Here $d\xi$ is normalized Lebesgue measure (i.e., the normalized Riemannian measure for the round metric).

In particular, one of the extremals is the constant function 1, so the maximum value is $A_{-2r}1 = 1$. By Theorem 3.18, this describes the embedding $L^{2m/(m+2r)}(S^m) \hookrightarrow L^2_{-r}(S^m)$. The dual embedding is $L^2_r(S^m) \hookrightarrow L^{2m/(m-2r)}(S^m)$. In the diagram

$$\begin{array}{ccc} L^2_r(S^m) & \xrightarrow{\iota} & L^{2m/(m-2r)}(S^m) \\ \uparrow{\scriptstyle A_{-2r}} & & \uparrow{\scriptstyle A_{-2r}} \\ L^2_{-r}(S^m) & \xleftarrow[\iota]{} & L^{2m/(m+2r)}(S^m) \end{array}$$

the horizontal maps are Sobolev embeddings, the vertical map on the left is an isometry, and the vertical map on the right is bounded as a result of the boundedness of the other maps.

THEOREM 4.2 [**Be**]. *If* $r \in (0, m/2)$,

$$\min_{\varphi \in C^\infty(S^m)} \frac{|(A_{2r}\varphi, \varphi)_{L^2(S^m, d\xi)}|}{\|\varphi\|^2_{L^{2m/(m-2r)}(S^m, d\xi)}}$$

is attained exactly when φ *is a nonzero constant multiple of some* $\Omega_h^{m/2-r}$, $h \in \mathbf{C}(S^m, g_0)$.

THEOREM 4.3 [**Br7**]. *Let* $m \geq 3$, *let* $r < m/2$ *be a positive integer, and let* P_{2r} *be the order* $2r$ *operator of Graham, Jenne, Mason, and Sparling described in paragraph 1.8. Then if* $\varphi \in C^\infty(S^m)$,

$$\begin{aligned}\|\varphi\|^2_{L^{2m/(m-2r)}(S^m, d\xi)} &\leq ((P_{2r})_0 1)^{-1}((P_{2r})_0\varphi, \varphi)_{L^2(S^m, d\xi)} \\ &= \frac{\Gamma(\frac{m-2r}{2})}{\Gamma(\frac{m+2r}{2})}((P_{2r})_0\varphi, \varphi)_{L^2(S^m, d\xi)},\end{aligned}$$

with equality iff φ *is a constant multiple of some* $\Omega_h^{m/2-r}$, $h \in \mathbf{C}(S^m, g_0)$. *In addition, the quantity*

$$\|(P_{2r})_\omega 1\|_{L^{m/2r}(S^m, (d\xi)_\omega)}$$

is minimized exactly when e^ω *is a constant multiple of some* Ω_h. *Here* $(P_{2r})_\omega$ *is the evaluation of* P_{2r} *in the metric* $g_\omega = e^{2\omega} g_0$, *and* $(d\xi)_\omega = (dv)_\omega / v_m$, *where* $(dv)_\omega$ *is the Riemannian measure of* g_ω *and* v_m *is the volume of the standard metric on* S^m.

PROOF. For the sharp inequality, we just need to verify that $((P_{2r})_0 1)^{-1}(P_{2r})_0 = A_{2r}$. By the uniqueness of the intertwinor (Corollary 3.16), this is true up to a constant multiple; but both sides fix the constant function 1, so the normalization is correct. For the value of $((P_{2r})_0 1)^{-1}$ in terms of Γ, we use Theorem 3.20. For the extremal problem, let $c = (P_{2r})_0 1$, $f = e^{(m-2r)\omega/2}$, and $T = (P_{2r})_0 - (P_{2r})_0 1$. By Hölder's inequality,

$$c\|f\|^2_{2m/(m-2r)} \leq \int_{S^m} f(T+c) f \, d\xi \leq \|f^2\|_{m/(m-2r)} \left\| \frac{Tf}{f} + c \right\|_{m/2r}.$$

Here $\|\cdot\|_p$ denotes the $L^p(S^m, d\xi)$ norm. Since $\|f\|^2_{2m/(m-2r)} = \|f^2\|_{m/(m-2r)}$, this is the desired result.

COROLLARY 4.4. *If* $m \geq 3$,

$$\left\| \frac{(m-2)\tau_\omega}{4(m-1)} \right\|_{L^{m/2}(S^m, (d\xi)_\omega)} \geq \frac{m(m-2)}{4},$$

with equality iff e^ω *is a constant multiple of some* Ω_h. *Here, as in Sec. 1,* τ *is the scalar curvature.*

PROOF. This is the case $r = 1$; the estimate is on the constant term in the conformal Laplacian P_2.

We can use the formula for the Paneitz operator P_4, described in paragraph 1.8, to get another corollary. In the general setting of Riemannian manifolds (M, g), let

$$\begin{aligned}J &:= \tau/2(m-1), \qquad m \neq 1, \\ V &:= (\rho - Jg)/(m-2), \qquad m \neq 1, 2,\end{aligned}$$

where ρ is the Ricci tensor. Then [**Pa, Br2**]

$$P_4 = \Delta^2 + \delta T d + (m-4)Q/2, \qquad m \neq 1, 2,$$

where

$$T = (m-2)J - 4V\cdot\,, \qquad Q = \frac{m}{2}J^2 - 2|V|^2 + \Delta J. \tag{4.1}$$

Here $V\cdot$ is the natural action of a two-tensor on one-forms, $(V\cdot\varphi)_j = V_j^i\varphi_i$.

COROLLARY 4.5. *If $m \geq 5$,*

$$\left\| \frac{m-4}{2}\left(\frac{m}{2}J^2 - 2|V|^2 + \Delta J\right)_\omega \right\|_{L^{m/4}(S^m,(d\xi)_\omega)} \geq \frac{(m+2)m(m-2)(m-4)}{16},$$

with equality iff e^ω is a constant multiple of some Ω_h.

Note that for $m = 8$, this is an L^2 estimate; we shall meet this estimate again in Sec. 6.

Using an endpoint differentiation argument, Beckner [**Be**] obtained an inequality, generalizing the Moser-Trudinger inequality on S^2, which describes the embedding of $L^2_{m/2}$ into the Orlicz class e^L. This result was also obtained independently by Carlen and Loss [**CL**].

THEOREM 4.6 ([**Be**]; see also [**CL**]). *Let $\bar\omega = \int_{S^m} \omega d\xi$ be the mean value of $\omega \in C^\infty(S^m)$. Then*

$$\log \int_{S^m} e^{m(\omega-\bar\omega)} d\xi \leq \frac{m}{2\Gamma(m)}(P_m\omega, \omega)_{L^2(S^m,d\xi)},$$

with equality iff e^ω is a constant multiple of some Ω_h.

Remarkably, in dimensions 2 and 4, the conformally invariant inequalities above provide a complete solution to the extremal problems for the functional determinants of the conformal Laplacian L and the square $\nabla\!\!\!/^2$ of the Dirac operator, in the standard conformal class on the sphere. The inequalities do not just provide the estimates needed; more than this, the determinant formulas are expressions in exactly the quantities asserted positive by Theorems 4.6 and 4.3. To get these determinant formulas in a systematic way, it is useful to look more deeply into the algebraic structure of the operators P_n.

THEOREM 4.7 [**Br7**]. *In the conformally flat category, P_n has the form $P_n^0 + \frac{m-n}{2}Q_n$, where P_n^0 annihilates constants, and the coefficients of P_n^0 and Q_n are rational in m. P_n and P_n^0 are formally self-adjoint.*

The conformal covariance relation for P_n, applied to the function 1, gives

$$\frac{m-n}{2}(Q_n)_\omega e^{\frac{m+n}{2}\omega} = \left(P_n^0 + \frac{m-n}{2}Q_n\right)_0 e^{\frac{m-n}{2}\omega}.$$

Since P_n^0 annihilates constants, this says (setting $\beta = (m-n)/2$)

$$\beta(Q_n)_\omega e^{m-\beta}\omega = (P_n^0)_0(e^{\beta\omega} - 1) + \beta(Q_n)_0 e^{\beta\omega}. \tag{4.2}$$

Dividing by β and evaluating at $\beta = 0$, we get

$$(P_m)_0\omega + (Q_m)_0 = (Q_m)_\omega e^{m\omega}, \qquad m \text{ even}.$$

(For $m = 2$, this is the Gauss curvature prescription equation.) Since $(dv)_\omega = e^{m\omega}(dv)_0$,

$$((P_m)_0\omega + (Q_m)_0)(dv)_0 = (Q_m dv)_\omega\,, \qquad m \text{ even.}$$

Since P_m^0 is natural and annihilates constants, it has the form Sd; since it is self-adjoint, this equals δS^*. Thus by Stokes' Theorem,

$$\int (Q_m dv)_\omega = \int (Q_m dv)_0\,. \tag{4.3}$$

That is, $\int Q_m$ is a conformal invariant.

We get more information by computing to higher order in the β parameter. By (4.2),

$$\left\{(P_n^0)_0\left(\frac{e^{\beta\omega}-1}{\beta}\right) + (Q_n)_0 e^{\beta\omega}\right\}(dv)_0 = (Q_n)_\omega e^{-\beta\omega}(dv)_\omega\,.$$

Thus

$$\begin{aligned}(Q_n dv)_\omega - (Q_n dv)_0 = \Bigg\{&(P_n^0)_0\left(\frac{e^{\beta\omega}-1}{\beta}\right) + (e^{\beta\omega}-1)(P_n^0)_0\left(\frac{e^{\beta\omega}-1}{\beta}\right)\\ &+ (e^{2\beta\omega}-1)(Q_n)_0\Bigg\}(dv)_0\,.\end{aligned}$$

Integrating, we lose the first term on the right by Stokes' Theorem. Dividing by 2β and evaluating at $\beta = 0$, we get

$$\int \left.\frac{(Q_n dv)_\omega - (Q_n dv)_0}{m-n}\right|_{m=n} = \tfrac{1}{2}\int \omega (P_m)_0\omega(dv)_0 + \int \omega(Q_n)_0(dv)_0\,. \tag{4.4}$$

By Theorem 2.11, the behavior of the functional determinant in dimension 2 is governed by the formula for the heat invariant a_2. By [**G2**], an operator $\nabla^*\nabla - \mathcal{E}$ on a bundle $\mathbb{V}$, $\mathcal{E}$ being a bundle endomorphism, will produce the heat invariant

$$6(4\pi)^{m/2}a_2[\nabla^*\nabla - \mathcal{E}] = \int \operatorname{tr}_{\mathbb{V}}\{\tau + 6\mathcal{E}\}dv,$$

where $\operatorname{tr}_{\mathbb{V}}$ is the fiber trace. Thus

$$6(4\pi)^{m/2}a_2[L] = (4-m)\int J dv,$$

and by the Lichnerowicz formula,

$$6(4\pi)^{m/2}2^{-m/2}a_2[\nabla\!\!\!/^{\,2}] = (1-m)\int J dv, \qquad m \text{ even.}$$

Since $J = Q_2$, $L = P_2$, and L is just Δ in dimension 2,

$$\int \left.\frac{(J dv)_\omega - (J dv)_0}{m-2}\right|_{m=2} = \tfrac{1}{2}\int \omega\Delta_0\omega(dv)_0 + \int \omega J_0(dv)_0\,.$$

This and Theorem 2.11 allow us to write a formula for the determinant.

THEOREM 4.8. *On S^2, for the scale-invariant determinant functional $\mathcal{D}$ of Sec. 2,*

$$-\log \frac{\mathcal{D}(\Delta, g_\omega)}{\mathcal{D}(\Delta, g_0)} = \log \frac{\mathcal{D}(\nabla\!\!\!/^{\,2}, g_\omega)}{\mathcal{D}(\nabla\!\!\!/^{\,2}, g_0)} = \frac{1}{3}\left(-\log \int_{S^2} e^{2(\omega-\bar{\omega})} d\xi + \int_{S^2} \omega(\Delta_0 \omega) d\xi\right).$$

As a result, $\mathcal{D}(\Delta, g_\omega)$ is maximized, and $\mathcal{D}(\nabla\!\!\!/^{\,2}, g_\omega)$ is minimized, exactly when e^ω is a constant multiple of some Ω_h.

This was originally proved by Onofri [**O**] in the case of the ordinary Laplacian Δ. (In computing the exponential term in this case, one must keep in mind that Δ has a one-dimensional null space, the constant functions.) In fact, since any Riemannian metric on S^2 is diffeomorphic to one in the standard conformal class (the space of metrics modulo the action of the group $\mathrm{Diffeo}(S^2) \ltimes \mathbf{C}(S^2, g_0)$ is just a point), this solves the problem of extremizing over *all* metrics.

Thus in dimension 2, the functional determinant "is" exactly the quantity asserted by the Moser-Trudinger inequality to have a definite sign. From another point of view, spectral geometry, through the spectral invariant $\mathcal{D}$, encodes the embedding $L^2_1 \hookrightarrow e^L$. It was shown by Branson, Chang, Ørsted, and Yang [**BØ3, BCY**] that a similar phenomenon is at work in dimension 4: $\mathcal{D}$ is a combination of the quantities which describe the embeddings $L^2_2 \hookrightarrow e^L$ and $L^2_1 \hookrightarrow L^4$. First note that the heat invariant that will enter the calculation via Theorem 2.11 is a_4. Any operator A satisfying the analytic and naturality assumptions 2.1 and 2.3 will have $a_4[A]$ a linear combination of the integrals of J^2, $|V|^2$, and $|C|^2$, where C is the Weyl conformal curvature tensor. In the conformally flat category, C vanishes identically. Under the conformal assumptions 2.6, there will be one linear relation in dimension 4 between the J^2 and $|V|^2$ coefficients as a consequence of the conformal index property, Theorem 2.8. By (4.3), this relation is the same as that satisfied by $Q = Q_4$. By (4.1), this says

$$a_2[A] = \text{const} \int (J^2 - |V|^2), \qquad m = 4,$$

so that under the rationality assumption 2.10,

$$\begin{aligned} a_2[A] &= u_A(m) \int \{k_1(m)(J^2 - |V|^2) + (m-4)c_1(m)B(m)\} dv \\ &= u_A(m) \int \{k(m)Q + (m-4)c(m)B(m)\} dv, \end{aligned}$$

where B is any fixed rational linear combination of J^2 and $|V|^2$, and the coefficients $k_1(m), c_1(m), k(m), c(m)$ are rational. It is convenient to make the choice $B(m) = J^2$. By [**Br7**],

$$\begin{aligned} 360(4\pi)^{m/2} a_4[L] &= (m-2)(m-6) \int Q dv + 2(m-4)(m-6) \int J^2 dv, \\ 360(4\pi)^{m/2} \cdot 2^{-m/2} &= \tfrac{1}{2}(m-2)(2m+3) \int Q dv + \tfrac{1}{2}(m-4)(6m-17) \int J^2 dv. \end{aligned}$$

By (4.4) and Theorem 2.11, we have:

THEOREM 4.9 [**BØ3**]. *On* S^4,

$$360 \log \frac{\mathcal{D}(L, g_\omega)}{\mathcal{D}(L, g_0)} = 6 \left(- \log \int_{S^4} e^{4(\omega - \bar{\omega})} d\xi + \tfrac{1}{3} \int_{S^4} \omega(\Delta_0(\Delta_0 + 2)\omega) d\xi \right)$$
$$+ 4 \int_{S^4} \{(J^2 dv)_\omega - (J^2 dv)_0\},$$
$$360 \cdot 2^{-m/2} \log \frac{\mathcal{D}(\nabla\!\!\!\!/^{\,2}, g_\omega)}{\mathcal{D}(\nabla\!\!\!\!/^{\,2}, g_0)} =$$
$$- \tfrac{33}{2} \left(- \log \int_{S^4} e^{4(\omega - \bar{\omega})} d\xi + \tfrac{1}{3} \int_{S^4} \omega(\Delta_0(\Delta_0 + 2)\omega) d\xi \right)$$
$$- \tfrac{7}{2} \int_{S^4} \{(J^2 dv)_\omega - (J^2 dv)_0\}.$$

Both quantities are linear combinations of the quantity from Theorem 4.6, whose nonnegativity describes the embedding $L^2_2 \hookrightarrow e^L$, and the quantity from Corollary 4.4, whose nonnegativity originates in the embedding $L^2_1 \hookrightarrow L^4$ given by Theorem 4.3. The two inequalities have the same extremals, and in each case (L and $\nabla\!\!\!\!/^{\,2}$), the signs of the coefficients of the two nonnegative quantities are the same. Thus the extremal problem is solved:

THEOREM 4.10 [**BCY**]. *$\mathcal{D}(L, g_\omega)$ is minimized, and $\mathcal{D}(\nabla\!\!\!\!/^{\,2}, \omega)$ maximized, exactly when e^ω is a constant multiple of some Ω_h.*

In [**Br7**], an invariant theoretic argument is made to explain why log-determinant functionals of the type studied here must be linear combinations of

$$\mathcal{F}_2(\omega) := - \log \int_{S^4} e^{4(\omega - \bar{\omega})} d\xi + \tfrac{1}{3} \int_{S^4} \omega(\Delta_0(\Delta_0 + 2)\omega) d\xi,$$
$$\mathcal{F}_1(\omega) := \int_{S^4} \{(J^2 dv)_\omega - (J^2 dv)_0\}.$$

What is harder to explain is the sign agreement of c_2 and c_1 in the expressions $c_2\mathcal{F}_2 + c_1\mathcal{F}_1$. As it stands, this information is simply buried in the heat invariants of the operators L and $\nabla\!\!\!\!/^{\,2}$. When we look at dimension 6, the sign miracles persist, even more improbably (we get sign agreement of four coefficients for each operator). It is a challenge for future work to explain why these sign miracles occur, allowing solution of the extremal problem. The underlying reason is more delicate than just some *a priori* convexity of the determinant: we get maxima and minima according to the "checkerboard" pattern of Table 4.1.

TABLE 4.1

	$\det L_\omega$	$\det \nabla\!\!\!\!/^{\,2}_\omega$
S^2	max	min
S^4	min	max
S^6	max	min

In the 6-dimensional case (Sec. 6), one of the inequalities used describes an intrinsically tensor-valued $L_1^2 \hookrightarrow L^3$ embedding; this comes into play even though the operator whose determinant is being analyzed (e.g. L) may only be scalar valued. The inequality in question is a conformally invariant lower bound for the bottom eigenvalue of a tensor-valued conformally covariant operator, in fact, one of the Wünsch operators mentioned in paragraph 1.9.

5. Second-order conformal covariants

In order to derive the tensor-valued Sobolev inequality mentioned in the last paragraph, put it in its proper context, and derive other inequalities of its ilk, we present here a classification of second-order conformal covariants on tensor-spinor bundles, on Riemannian manifolds of dimension $m \geq 3$. By the unitary trick applied to the structure group $\mathrm{SO}(m)$ or $\mathrm{Spin}(m)$, this will also be a classification for pseudo-Riemannian manifolds. These results were first distributed in lecture note form in [**Br4**].

First note that it is sufficient to work with irreducible bundles $\mathbb{V}(\lambda)$. We begin by giving a formula for a conformal covariant on each $\mathbb{V}(\lambda)$. Ultimately, we will find that this formula often gives zero, or an order zero operator. However, it will also turn out that all second-order conformal covariants, modulo actions of the Weyl tensor C, are given by the formula. We stress the following are general results on Riemannian and pseudo-Riemannian manifolds, not just on spheres.

Recall the gradients of paragraph 1.6; we shall use the notation $\lambda \leftrightarrow \sigma$ for the relation (1.4). Given λ, let $\{\sigma_u\}_{u=1}^{N_\lambda}$ be the list of σ with $\lambda \leftrightarrow \sigma$, and consider operators of the form

$$\tilde{D}_\lambda := \sum_{u=1}^{N_\lambda} a_u G_u^* G_u \,,$$

where $G_u := G_{\lambda\sigma_u}$, and a_u are constants. We normalize our formal adjoints by taking the usual product Riemannian structure on the bundle $T^*M \otimes \mathbb{V}(\lambda)$ (of which the $\mathbb{V}(\sigma_u)$ are subbundles). Note that G_u^* is (a realization of) the gradient $G_{\sigma_u\lambda}$. By (1.3), $G_u^* G_u = \nabla^* P_u \nabla$, where P_u in the bundle projection to $\mathbb{V}(\sigma_u)$, so

$$\tilde{D}_\lambda = \nabla^* \left(\sum_{u=1}^{N_\lambda} a_u P_u \right) \nabla.$$

We seek a conformal covariant of reduced bidegree $((m-2)/2, (m+2)/2)$, of the form $D_\lambda = \tilde{D}_\lambda + Z$, where Z is an order zero curvature correction. Everything is local, so we may assume the structure group to be $H = \mathrm{Spin}(m)$. It is convenient to work with particular CH-bundles, i.e. to set certain internal conformal weights. We do this by viewing G_u and G_u^* as follows:

$$\mathbb{V}^{(m-2)/2}(\lambda) \xrightarrow{G_u} \mathbb{V}^{m/2}(\sigma_u) \xrightarrow{G_u^*} \mathbb{V}^{(m+2)/2}(\lambda). \tag{5.1}$$

By the result of [**F**] described in paragraph 1.6, together with Remark 1.2, G_u (resp. G_u^*) has conformal bidegree $(c_u\,, c_u)$ (resp. $(-c_u\,, -c_u)$) in (5.1), where

$$c_u = \tfrac{1}{2}(1 + \langle 2\rho + \lambda + \sigma_u, \lambda - \sigma_u \rangle). \tag{5.2}$$

Thus (5.1) is not necessarily a diagram of invariant operators. However, taking a conformal curve of metrics $g_{\varepsilon\omega} = e^{2\varepsilon\omega} g_0$, and denoting $(d/d\varepsilon)|_{\varepsilon=0}$ by a dot, we have

(5.3)
$$\begin{aligned}\dot G_u(\omega) &= c_u[G_u\,,\mu(\omega)] = c_u\zeta_u(d\omega),\\ (G_u^*)^{\cdot}(\omega) &= -c_u[G_u^*\,,\mu(\omega)] = c_u\zeta_u(d\omega)^*,\\ (G_u^*G_u)^{\cdot}(\omega) &= c_u(G_u^*\zeta_u(d\omega)+\zeta_u(d\omega)^*G_u) = c_u(\nabla^*P_u\tau(d\omega)+\tau(d\omega)^*P_u\nabla),\end{aligned}$$

where, for η a one-form and φ a section of $\mathbb{V}(\lambda)$,

$$\tau(\eta)\varphi = \eta\otimes\varphi, \qquad \zeta_u(\eta) = \mathrm{Proj}_{\sigma_u}\tau(\eta).$$

Note that $\zeta_u(d\omega) = [G_u\,,\mu(\omega)]$, and that $\sqrt{-1}\zeta_u$ is the leading symbol of G_u.

If some c_u vanishes, (5.3) and Remark 1.3 imply that the corresponding $G_u^*G_u$ is conformally invariant; in fact, this is the case in which Fegan's result shows that (5.1) *is* a diagram of invariant operators. By (5.2), this happens if and only if

Case II. m is even, $\lambda_\ell = 0 \neq \lambda_{\ell-1}$, $\sigma_u = \lambda\pm e_\ell$.

We continue the discussion in the complementary case.

Case I. We are not in Case II; thus all $c_u \neq 0$.

Here
$$\left(\sum_{u=1}^{N_\lambda} c_u^{-1}G_u^*G_u\right)^{\cdot}(\omega) = \nabla^*\tau(d\omega)+\tau(d\omega)^*\nabla = \mu(\Delta\omega).$$

We can thus compensate the conformal variation of this choice ($a_u = c_u^{-1}$) of $\tilde D_\lambda$ with the scalar curvature: by the conformal covariance relation for the conformal Laplacian, the normalization J used in the last section satisfies

(5.4)
$$\dot J(\omega) = \Delta\omega.$$

Thus:

THEOREM 5.1.
$$D_\lambda = \begin{cases} J - \sum_{u=1}^{N_\lambda} c_u^{-1}G_u^*G_u & \text{in Case I,}\\ G_{\lambda,\lambda\pm e_\ell}^*G_{\lambda,\lambda\pm e_\ell} & \text{in Case II}\end{cases}$$

is conformally covariant of reduced bidegree $((m-2)/2,(m+2)/2)$.

THEOREM 5.2. *In Case II, if*
$$D'_\lambda := J - \sum_{\sigma_u\neq\lambda\pm e_\ell} c_u^{-1}G_u^*G_u\,,$$

then $(\varphi, D'_\lambda\psi)_{L^2}$ *is an invariant form on the null space* $\mathcal{N}(G_{\lambda,\lambda+e_\ell})\cap\mathcal{N}(G_{\lambda,\lambda-e_\ell})$ *in* $\mathbb{V}^{((m-2)/2}(\lambda)$.

PROOF. Infinitesimal invariance is given by the calculation

$$\begin{aligned}
(\varphi, D'_\lambda\psi)^\cdot &= \left(\varphi, \left(\Delta\omega - \sum_{\sigma_u \neq \lambda \pm e_h} (G_u^*\zeta_u(d\omega) + \zeta_u(d\omega)^* G_u)\right)\psi\right) \\
&= (\varphi, (\Delta\omega)\psi) - \sum_{\sigma_u \neq \lambda \pm e_h} \{(G_u\varphi, \zeta_u(d\omega)\psi) + (\varphi, \zeta_u(d\omega)^* G_u\psi)\} \\
&= (\varphi, (\Delta\omega)\psi) - \sum_{u=1}^{N_\lambda} \{(G_u\varphi, \zeta_u(d\omega)\psi) + (\varphi, \zeta_u(d\omega)^* G_u\psi)\} \\
&= \left(\varphi, \left(\Delta\omega - \sum_{u=1}^{N_\lambda} (G_u^*\zeta_u(d\omega) + \zeta_u(d\omega)^* G_u)\right)\psi\right) \\
&= 0.
\end{aligned}$$

(The bundle weights $(m-2)/2 + (m+2)/2 = m$ balance the variation of the Riemannian measure.) Finite invariance now follows from an argument like Remark 1.3.

The importance of Theorems 5.1 and 5.2 for estimation problems is the following.

COROLLARY 5.3. *In Case I, if the form* $(\varphi, D_\lambda\psi)_{L^2}$ *is (positive or negative) definite or semidefinite in some metric* g_0, *it is similarly definite in any conformal metric* $e^{2\omega}g_0$. *The same is true in Case II for the form* $(\varphi, D'_\lambda\psi)_{L^2}$ *on* $\mathcal{N}(G_{\lambda,\lambda+e_\ell}) \cap \mathcal{N}(G_{\lambda,\lambda-e_\ell})$.

It is important to note that in Case I, D_λ can well be zero or have order zero. In fact, much of the work in classifying second-order conformal covariants will be the determination of when this happens.

EXAMPLE 5.4. The spinor bundles are $\Sigma = \mathbb{V}((\frac{1}{2})_\ell)$ if m is odd, and $\Sigma_\pm = \mathbb{V}((\frac{1}{2})_{\ell-1}, \pm\frac{1}{2})$ if m is even. The gradient targets are

$$\begin{cases} \Sigma \longrightarrow \mathbb{T} := \mathbb{V}(\frac{3}{2}, (\frac{1}{2})_{\ell-1}) \text{ and } \Sigma, & m \text{ odd}, \\ \Sigma_\pm \longrightarrow \mathbb{T}_\pm := \mathbb{V}(\frac{3}{2}, (\frac{1}{2})_{\ell-2}, \pm\frac{1}{2}) \text{ and } \Sigma_\mp, & m \text{ even}. \end{cases}$$

The gradients with targets $\mathbb{T}$ or $\mathbb{T}_\pm$ are sometimes called *twistor operators*. Labelling the targets 1 and 2 in the order given, we have

$$c_1 = -(m-1)/2, \qquad c_2 = 1/2.$$

The operator

$$D = J + \frac{2}{m-1} G_1^* G_1 - 2G_2^* G_2,$$

being conformally covariant, has vanishing *second* conformal variation:

$$\begin{aligned}
0 &= \ddot{D}(\omega, \eta) \\
&= \dot{\Delta}(\eta)\omega - \sum_{u=1}^{2}\{(G_u^*)^{\cdot}(\eta)\zeta_u(d\omega) + \zeta_u(d\omega)^*\dot{G}_u(\eta)\} \\
&= \dot{\Delta}(\eta)\omega + \frac{m-1}{2}\{\zeta_1(d\eta)^*\zeta_1(d\omega) + \zeta_1(d\omega)^*\zeta_1(d\eta)\} \\
&\quad - \tfrac{1}{2}\{\zeta_2(d\eta)^*\zeta_2(d\omega) + \zeta_2(d\omega)^*\zeta_2(d\eta)\}.
\end{aligned}$$

The conformal covariance relation for the conformal Laplacian L computes $\dot{\Delta}(\omega)$, and we get

(5.5)

$$0 = \tfrac{1}{2}\{\zeta_1(d\eta)^*\zeta_1(d\omega) + \zeta_1(d\omega)^*\zeta_1(d\eta)\} - \frac{m-1}{2}\{\zeta_2(d\eta)^*\zeta_2(d\omega) + \zeta_2(d\omega)^*\zeta_2(d\eta)\}.$$

But this asserts the vanishing of the leading symbol of D: when $\eta = \omega$, the right side of (5.5) is

$$\frac{m-1}{2}\sigma_2(D)(d\omega).$$

By homogeneity considerations, D cannot have order 1; thus it has order zero. By naturality, it is an action of the Riemann tensor; by conformal invariance, it is an action of the Weyl tensor. But the Weyl tensor, which is a section of $\mathbb{V}(2,2)$ for $m \geq 4$, or of $\mathbb{V}(2,2) \oplus \mathbb{V}(2,-2)$ for $m = 4$, cannot act on the bundle Σ, since $\mathbb{V}(2,2) \otimes \Sigma$ is a direct sum of bundles of highest weights $\mu \geq$ with $m_1 \geq \frac{3}{2}$; similarly for $\mathbb{V}(2,-2)$ when $m = 4$. Thus $D = 0$.

REMARK 5.5. One can obtain two important identities directly from the fact that $D = 0$ and $G_1^*G_1 + G_2^*G_2 = \nabla^*\nabla$ in the spinor case above. The first is the Lichnerowicz formula

$$mG_2^*G_2 = \nabla^*\nabla + \frac{\tau}{4}.$$

Since the Clifford relations show $\sigma_2(\nabla\!\!\!/^2) = \sigma_2(\nabla^*\nabla)$, we also obtain the normalization $mG_2^*G_2 = \nabla\!\!\!/^2$. The second identity is

$$\frac{m}{m-1}G_1^*G_1 = \nabla^*\nabla - \frac{\tau}{4(m-1)}\,.$$

It is important in the study of harmonic spinors that the operator on the right is semidefinite; this is true by virtue of the representation on the left.

EXAMPLE 5.6. The conformal Laplacian L is just (a normalization of) $D_{(0)}$. For the only gradient target is $\mathbb{V}(1)$, so

$$D_{(0)} = J + \frac{2}{m-2}\nabla^*\nabla = \frac{2}{m-2}L.$$

EXAMPLE 5.7. Consider the bundle Λ^k of k-forms, and suppose for now that $0 < k < m/2$, $k \neq (m-2)/2$; then $\lambda = (1_k)$ and we are in Case I. The gradient targets are

$$\mathbb{F}^k = \mathbb{V}(2, 1_{k-1}),\ \Lambda^{k+1},\ \Lambda^{k-1}.$$

Labelling these 1,2,3 in the order given, a normalization chase shows that

$$G_2 = d_k/(k+1), \qquad G_3 = \delta_k/(m-k+1), \qquad G_2^* = \delta_{k+1}\,, \qquad G_3^* = d_{k-1}\,.$$

(The fact that the standard form inner product $(k!)^{-1}\varphi^{i_1\dots i_k}\psi_{i_1\dots i_k}$ is in conflict with our "default" inner product, given by the product Riemannian structure on $T^*M\otimes\Lambda^k$, makes this chase somewhat nontrivial.) In the case $k=1$, the operator G_1 is important in elasticity and quasiconformal theory. Setting $s=(m-2k)/2$, we get

$$c_1=-m/2, \qquad c_2 = 1-s, \qquad c_3 = 1+s.$$

Let W be the *Weitzenböck operator* $\delta d + d\delta - \nabla^*\nabla$; then

$$\begin{aligned} D_{(1_k)} &= J + \frac{2}{m}\left(\nabla^*\nabla - \frac{\delta d}{k+1} - \frac{d\delta}{m-k+1}\right) + \frac{\delta d}{(s-1)(k+1)} \\ &\quad - \frac{d\delta}{(s+1)(m-k+1)} \\ &= J + \frac{2}{m}\left\{-W + \frac{s}{s-1}\delta d + \frac{s}{s+1}d\delta\right\}. \end{aligned}$$

In a local frame and dual coframe,

$$W = -R^i{}_j{}^p{}_q\varepsilon^q\iota_p\varepsilon^j\iota_i\,,$$

where ε and ι are exterior and interior multiplication. A local calculation now shows that $m(s+1)(s-1)D_{(1_k)}/2s$ is the operator D_k introduced in [**Br1**], up to an order zero action of the Weyl tensor C (which is separately conformally covariant):

$$(5.6)\quad D_k = \frac{m-2k+2}{2}\delta d + \frac{m-2k-2}{2}d\delta + \frac{(m-2k+2)(m-2k-2)}{4}(J-2V\cdot),$$

where locally $V\cdot = V^i{}_j\varepsilon^j\iota_i$. Note that the Ricci tensor, which is not explicit in Theorem 5.1, makes its appearance when we write the "highest" $G_u^*G_u$ (the one corresponding to the gradient target of the maximal highest weight, here $\mathbb{F}^k$) in terms of "lower" $G_u^*G_u$.

EXAMPLES 5.8. For even m, $\Lambda^{(m-2)/2}$ fits under Case II and yields the Maxwell operator δd on "vector potentials". The cases $k>m/2$ are, of course, dual to the cases above under the (conformally covariant) Hodge $\star$ operator. This leaves the case in which m is even and $k=m/2$. Running our argument with $\mathrm{O}(m)$ gradients, or using (5.6), we get a conformal covariant which interchanges the subbundles $\Lambda_+^{m/2}=\mathbb{V}(1_\ell)$ and $\Lambda_-^{m/2}=\mathbb{V}(1_{\ell-1}\,,-1)$.

EXAMPLE 5.9. Assume $m\geq 5$, and consider the bundle $\mathrm{TFS}^p=\mathbb{V}(p)$ of trace-free symmetric p-tensors. The gradient targets are

$$\mathbb{V}(p+1),\ \mathbb{V}(p,1),\ \mathbb{V}(p-1);$$

we label them 1,2,3 in the order given, and compute

$$-c_1 = c_3 = (m+2p-2)/2, \qquad c_2 = -(m-4)/2.$$

We have

$$D_{(p)} = J - \nabla^*\left(-\frac{2}{m-4}I + \frac{4(p+1)}{(m+2p-2)(m-4)}P_1 + \frac{4(m+p-3)}{(m+2p-2)(m-4)}P_3\right)\nabla,$$

using $\sum P_u = 1$. To get an explicit tensorial formula, we just need such formulas for P_1 and P_3. If φ is a section of $\mathrm{TFS}^p M$, the $\mathrm{TFS}^{p+1} M$ part of φ has components

$$\frac{1}{p+1}\sum_{s=0}^{p}\nabla_{i_s}\varphi_{i_0\ldots\hat{i}_s\ldots i_p} - a\sum_{s<t} g_{i_s i_t}\nabla^l\varphi_{l i_0\ldots\hat{i}_s\ldots\hat{i}_t\ldots i_p},$$

where a is a constant determined by the fact that the $i_0 i_1$ trace (and thus all traces) vanish:

$$a = 2/(p+1)(m+2p-2).$$

The $\mathrm{TFS}^{p-1} M$ part has components of the form

$$b_1\sum_{0<s<t} g_{i_s i_t}\nabla^l\varphi_{l i_0\ldots\hat{i}_s\ldots\hat{i}_t\ldots i_p} + b_2\sum_{0<s} g_{i_0 i_s}\nabla^l\varphi_{l i_1\ldots\hat{i}_s\ldots i_p},$$

where b_1 and b_2 are constants. The condition that the $i_1 i_2$ trace (and thus all $i_s i_t$ traces with $0<s<t$) vanish implies that

$$b_2 = -(m+2p-4)b_1/2.$$

b_1 can then be determined by the projection condition $P_2^2 = P_2$:

$$b_1 = -2/(m+2p-2)(m+p-3).$$

Now a local calculation gives

$$\begin{aligned}\frac{(m+2p-2)(m-4)}{2(m+2p-4)}(D_{(p)}\varphi)_{i_1\ldots i_p} &= -\nabla^l\nabla_l\varphi_{i_1\ldots i_p}\\ &+\frac{4p}{m+2p-2}\operatorname{Symm}\nabla_{i_1}\nabla^l\varphi_{l i_2\ldots i_p}\\ &-\frac{4p(p-1)}{(m+2p-2)(m+2p-4)}\operatorname{Symm} g_{i_1 i_2}\nabla^l\nabla^q\varphi_{l q i_3\ldots i_p}\\ &+\frac{2p}{m+2p-4}\operatorname{Symm}\rho^q{}_{i_1}\varphi_{q i_2\ldots i_p} - \frac{2p(p-1)}{m+2p-4}\operatorname{Symm} C^l{}_{i_1}{}^q{}_{i_2}\varphi_{l q i_3\ldots i_p}\\ &+\frac{(m+2p-2)(m-4)}{4(m+2p-4)(m-1)}\tau\varphi_{i_1\ldots i_p},\end{aligned}\tag{5.7}$$

where "Symm" is symmetrization in $i_1\ldots i_p$. Subtracting the C action, which is separately covariant, we get -1 times the operator introduced by Wünsch [**W**].

EXAMPLES 5.10. When $m=4$, the c_u corresponding to the targets $\mathbb{V}(p,\pm 1)$ vanish; we are in Case II, and we get the covariant $G^*_{(p),(p+1)}G_{(p),(p+1)}$. This operator is $\frac{1}{2}\nabla^*(I-P_1-P_3)\nabla$, and so is $p/2(p+1)$ times the operator on the right in (5.7). When $m=3$, the targets $\mathbb{V}(p+1)$, $\mathbb{V}(p)$, $\mathbb{V}(p-1)$ have $-c_1=c_3=(2p+1)/2$, $c_3=\frac{1}{2}$. The upshot is that (5.7) is still a formula for $D_{(p)}$, though $C=0$, when $m=3$.

Consider now the problem of classifying *all* second-order conformal covariants. "Second-order" should mean "universally second-order", excluding, for example, $\tau\Delta$, which vanishes identically for some metrics. We shall classify modulo order zero actions of the Weyl tensor C. By Weyl's invariant theory, a second-order conformal covariant thus has parallel leading symbol, and a difference T of two conformal covariants with the same leading symbol is (by homogeneity considerations) zeroth order. As such, it is an action of the Riemann tensor. But by (5.4) and

$$\dot{V}(\omega) = -\operatorname{Hess}\omega, \tag{5.8}$$

T is an action of the Weyl tensor (or its self- or anti-self-dual parts if $m = 4$). We claim further that second-order conformal covariants which begin and end in the same Spin(m)-bundle $\mathbb{V}(\lambda)$ must have leading terms of the form $\sum a_u G_u^* G_u$ studied above, and must have reduced bidegree $((m-2)/2, (m+2)/2)$.

For this, let $H = \operatorname{Spin}(m)$, and note that the leading symbol of a Spin(m)-operator on $\mathbb{V}(\lambda)$ is an element of

$$\begin{aligned}\operatorname{Hom}_H(\operatorname{Symm}(\mathbb{V}(1)\otimes\mathbb{V}(1)), \operatorname{End}\mathbb{V}(\lambda)) &\subset \operatorname{Hom}_H(\mathbb{V}(1)\otimes\mathbb{V}(1), \operatorname{End}\mathbb{V}(\lambda)) \\ &\cong \operatorname{End}_H(\mathbb{V}(1)\otimes\mathbb{V}(\lambda)).\end{aligned}$$

The latter space is described by the rule $\sigma \leftrightarrow \lambda$; it is a copy of $\mathbb{C}^{N_\lambda}$, and has a natural basis $\{E_u\}$, where E_u is the identity on the $\mathbb{V}(\sigma_u)$ summand, and 0 on the $\mathbb{V}(\sigma_v)$ summands for $v \neq u$. The map

$$\operatorname{Op} : \operatorname{End}_H(\mathbb{V}(1)\otimes\mathbb{V}(\lambda)) \to \operatorname{DO}(\mathbb{V}(\lambda)), \qquad \operatorname{Op}\left(\sum a_u E_u\right) = \sum a_u G_u^* G_u$$

produces an operator with leading symbol $\mathcal{S}(\sum a_u E_u)$, where $\mathcal{S}$ is the projection

$$\mathcal{S} : \mathbb{V}(1)\otimes\mathbb{V}(1)\otimes\mathbb{V}(\lambda) \to \operatorname{Symm}(\mathbb{V}(1)\otimes\mathbb{V}(1))\otimes\mathbb{V}(\lambda).$$

Let j be the identification

$$\operatorname{Hom}_H(\mathbb{V}(1)\otimes\mathbb{V}(1), \operatorname{End}_H\mathbb{V}(\lambda)) \xrightarrow{j} \operatorname{End}_H(\mathbb{V}(1)\otimes\mathbb{V}(\lambda))$$

used above; then

$$\mathcal{N}(\operatorname{Op}) = j(\operatorname{Alt}(\mathbb{V}(1)\otimes\mathbb{V}(1))\otimes\mathbb{V}(\lambda)) = j(\Lambda^2\otimes\mathbb{V}(\lambda)).$$

This proves:

LEMMA 5.11. *A natural second-order differential operator on* $\mathbb{V}(\lambda)$ *has the form* $\sum a_u G_u^* G_u + Z$, *where* Z *is an order zero action of the Riemann tensor.*

Further, we may assume that a second-order conformal covariant is formally self-adjoint, by the following two lemmas.

LEMMA 5.12. *Suppose* $\tilde{D}$ *is a natural, formally self-adjoint differential operator on* $\mathbb{V}(\lambda)$ *of level* r *and order* $s > 0$. *If* $D = \tilde{D} + Z$ *is conformally covariant of reduced bidegree* (p,q), *where* Z *is natural and of order* $\leq s-2$, *then necessarily* $p = (m-r)/2$, $q = (m+r)/2$, *and* $\frac{1}{2}(D + D^*)$ *is conformally covariant of reduced bidegree* $((m-r)/2, (m+r)/2)$.

PROOF. Covariance under uniform dilations shows that Z has level r, so $q = p + r$. The pre-L^2 inner product on $C^\infty(\mathbb{V}^{m/2}(\lambda))$ is conformally invariant, so for operators on this space, the conformal variation commutes with the taking of the formally self-adjoint part. Viewing D as an operator on this space,

$$\dot{D}(\omega) = -r\omega D + (p - m/2)[D, \mu(\omega)]. \tag{5.9}$$

Since D is formally self-adjoint, the skew part of this equation is

$$(Z - Z^*)^{\cdot}(\omega) = (2p - m + r)[\tilde{D}, \mu(\omega)] + (-r - p + m/2)(\omega Z - Z^*\mu(\omega)) \\ + (p - m/2)(Z\mu(\omega) - \omega Z^*).$$

Thus, unless $p = (m - r)/2$, the order of $[D, \mu(\omega)]$ is at most $s - 2$. But

$$\sqrt{-1}\sigma_{s-1}([\tilde{D}, \mu(\omega)])(d\omega) = s\sigma_s(\tilde{D})(d\omega)$$

cannot vanish identically, since ω is arbitrary. Thus p, and consequently q, have the indicated values. Given this, the self-adjoint part of (5.9) gives

$$(D + D^*)(\omega) = -r\omega(D + D^*) - \frac{r}{2}[D + D^*, \mu(\omega)],$$

as desired for conformal covariance.

LEMMA 5.13. *A second-order conformal covariant D has the form $D' + Z$, where D' is a formally self-adjoint second-order conformal covariant, and Z is an action of the Weyl tensor.*

PROOF. By the last two lemmas, we have the result with Z a conformally covariant action of the Riemann tensor. By (5.4) and (5.8), this can only involve the Weyl tensor.

Thus a second-order conformal covariant must be of the form

$$\sum a_u G_u^* G_u + bJ + Z_V ,$$

where Z_V is an order zero action of the normalized Ricci tensor V.

In Case I, we would now like to study the map

$$\tilde{\mathcal{C}} : \mathrm{Hom}_H(\mathbb{V}(1) \otimes \mathbb{V}(1), \mathrm{End}\, \mathbb{V}(\lambda)) \to \mathrm{Hom}_H(\mathrm{Alt}(\mathbb{V}(1) \otimes \mathbb{V}(1)), \mathrm{End}\, \mathbb{V}(\lambda)) \\ = \mathrm{Hom}_H(\Lambda^2, \mathrm{End}\, \mathbb{V}(\lambda))$$

induced by the conformal variation via the composition

$$(5.10) \qquad j^{-1}E_u \overset{j}{\longmapsto} E_u \overset{\mathrm{Op}}{\longmapsto} G_u^* G_u \overset{\cdot}{\longmapsto} c_u(G_u^*\zeta_u(d\omega) + \zeta_u(d\omega)^* G_u) \\ \overset{\sigma_1}{\longmapsto} -\sqrt{-1}c_u(\zeta_u(\xi)^*\zeta_u(d\omega) - \zeta_u(d\omega)^*\zeta_u(\xi)).$$

The last expression is antisymmetric in ξ and $d\omega$, so $\tilde{\mathcal{C}} = \mathrm{Alt} \circ j^{-1} \circ \mathcal{C} \circ j$, where $\mathcal{C}$ is the invertible map

$$\sum_u a_u E_u \mapsto \sum_u c_u a_u E_u ,$$

and $\tilde{\mathcal{C}}$ vanishes on $\mathrm{Hom}_H(\Lambda^2, \mathrm{End}\, \mathbb{V}(\lambda))$. We introduce the notation

$$\mathcal{A}_\lambda := \mathrm{Hom}_H(\Lambda^2, \mathrm{End}\, \mathbb{V}(\lambda)), \qquad \mathcal{S}_\lambda := \mathrm{Hom}_H(\mathrm{Symm}(\mathbb{V}(1) \otimes \mathbb{V}(1)), \mathrm{End}\, \mathbb{V}(\lambda)).$$

Since $j^{-1} \circ \mathcal{C} \circ j$ carries $\mathcal{A}_\lambda$ to $\mathcal{S}_\lambda$, we have:

LEMMA 5.14. *In Case I,*

$$\left\{\sigma_2\left(\sum a_u G_u^* G_u\right) : \left(\sum a_u G_u^* G_u\right)^{\bullet}(\omega) \text{ has order } 0 \text{ for all } \omega\right\}$$
$$\cong (j^{-1}\circ\mathcal{C}\circ j)^{-1}(\mathcal{S}_\lambda)/\mathcal{A}_\lambda \cong (j^{-1}\circ\mathcal{C}\circ j)^{-1}(\mathcal{S}_\lambda)\cap\mathcal{S}_\lambda\,.$$

In particular, this space has dimension $\dim\mathcal{S}_\lambda - \dim\mathcal{A}_\lambda$. *In particular,* $\dim\mathcal{S}_\lambda \geq \dim\mathcal{A}_\lambda$.

When $(\sum a_u G_u^* G_u)^{\bullet}(\omega)$ has order zero, Weyl's invariant theory shows it is an action of $\operatorname{Hess}\omega$. By (5.8), this can be compensated by the conformal variation of a unique action of the normalized Ricci tensor V. Thus:

LEMMA 5.15. *In Case I, the dimension* d_λ *of the space of second-order conformal covariants, modulo actions of the Weyl tensor, is* $\dim\mathcal{S}_\lambda - \dim\mathcal{A}_\lambda$.

If λ is an irreducible $\mathrm{Spin}(m)$-module and μ is a finite-dimensional $\mathrm{Spin}(m)$ module, let $m_\lambda\mu$ be the multiplicity of λ as a summand in μ. We may then extend m_λ to act on the representation ring of $\mathcal{R}$ of $\mathrm{Spin}(m)$. If $S^2 = \mathrm{Symm}(\mathbb{V}(1)\otimes\mathbb{V}(1))$, then

$$a_\lambda := \dim\mathcal{A}_\lambda = m_\lambda\Lambda^2\otimes\lambda, \qquad s_\lambda := \dim\mathcal{S}_\lambda = m_\lambda S^2\otimes\lambda.$$

We have

$$N_\lambda = s_\lambda + a_\lambda\,, \qquad d_\lambda = s_\lambda - a_\lambda = m_\lambda((S^2\ominus\Lambda^2)\otimes\lambda).$$

We can now take Cases I and II together to get:

THEOREM 5.16. *The dimension of the space of second-order conformal covariants on* $\mathbb{V}(\lambda)$, *modulo actions of the Weyl tensor, is*

$$d_\lambda = \begin{cases} 1, & m \text{ odd}, \ \lambda_\ell \neq \frac{1}{2}\,, \\ 1, & m \text{ even}, \ \lambda_\ell = 0, \\ 0 & \text{otherwise}. \end{cases}$$

PROOF. Kostant's formula [**Kost**] allows us to calculate $m_\lambda(\sigma\otimes\tau)$ for any irreducible λ,σ,τ if all the weights of σ are known. Let $\mathcal{W}$ be the Weyl group of $\mathfrak{so}(m)$, which acts on weights by permutation and sign change of entries (arbitrarily many sign changes if m is odd, an even number if m is even). The *sign* $\operatorname{sgn} w$ of $w\in\mathcal{W}$ is the sign of the permutation involved, times -1 to the number of sign changes. If μ and ν are weights, let

$$F_\nu(\mu) = F_\mu(\nu) = \sum_{w\in\mathcal{W}} (\operatorname{sgn} w)\delta^\mu_{w\nu}\,.$$

If $q_\sigma(\mu)$ is the multiplicity of μ as a weight in σ, then

$$\sum_\mu q_\sigma(\mu)F_{\rho+\tau-\mu} = \sum_\lambda m_\lambda(\sigma\otimes\tau))F_{\rho+\lambda}\,.$$

Applying this formula to $\tilde\lambda = \rho+\lambda$ for a given dominant weight λ, we get

$$m_\lambda(\sigma\otimes\tau) = \sum_\mu q_\sigma(\mu)F_{\rho+\tau-\mu}(\tilde\lambda) = \sum_\mu q_\sigma(\mu)F_{\tilde\lambda}(\rho+\tau-\mu), \tag{5.11}$$

since the fact that $\tilde\lambda$ is *strictly* dominant, $\tilde\lambda_1 > \ldots > \tilde\lambda_{\ell-1} > |\lambda_\ell|$, implies that only the identity in $\mathcal{W}$ can fix $\tilde\lambda$. By the additivity of both sides of (5.11) in the σ

argument, we can dispense with one of our irreducibility assumptions, and take σ to be in the representation ring $\mathcal{R}$.

We are interested in the situation $\tau = \lambda$, $\sigma = S^2 \ominus \Lambda^2$. The weights of S^2, respectively Λ^2, are easily computed:

$0,$	with multiplicity	$m - \ell,$	resp.	$\ell,$
$2e_a\,,$	with multiplicity	$1,$	resp.	$0,$
$-\,2e_a\,,$	with multiplicity	$1,$	resp.	$0,$
$e_a - e_b\,,\ a \neq b,$	with multiplicity	$1,$	resp.	$1,$
$e_a + e_b\,,\ a \neq b,$	with multiplicity	$1,$	resp.	$1,$
$-\,e_a - e_b\,,\ a \neq b,$	with multiplicity	$1,$	resp.	$1,$
$e_a\,,$	with multiplicity	$m - 2\ell,$	resp.	$m - 2\ell,$
$-\,e_a\,,$	with multiplicity	$m - 2\ell,$	resp.	$m - 2\ell.$

Thus the weights of $S^2 \ominus \Lambda^2$ are

$0,$	with multiplicity	$m - 2\ell,$
$2e_a\,,$	with multiplicity	$1,$
$-\,2e_a\,,$	with multiplicity	$1.$

Thus

$$m_\lambda((S^2 \ominus \Lambda^2) \otimes \lambda) = m - 2\ell + \sum_{a=1}^{\ell} \{F_{\tilde{\lambda}}(\tilde{\lambda} + 2e_a) + F_{\tilde{\lambda}}(\tilde{\lambda} - 2e_a)\}. \tag{5.12}$$

For a term involving $\pm 2e_a$ to make a contribution to (5.12), we need

$$\{|\tilde{\lambda}_1|, \ldots, |\tilde{\lambda}_\ell|\} = \{|\tilde{\lambda}_1|, \ldots, |\lambda_a \pm 2|, \ldots, |\tilde{\lambda}_\ell|\}.$$

By the strict dominance of $\tilde{\lambda}$, this can only happen if $|\tilde{\lambda}_a \pm 2| = |\tilde{\lambda}_a|$. This, in turn, can only happen in three situations: (*i*) m is odd, $a = \ell$, $\lambda_\ell = \frac{1}{2}$; (*ii*) m is even, $a = \ell$, $\lambda_\ell = \mp 1$; or (*iii*) m is even, $a = \ell - 1$, $\lambda_{\ell-1} = 0$. In situation (*ii*), there is no contribution anyway, since all entries of $\tilde{\lambda}$, $\tilde{\lambda} \pm 2e_\ell$ are nonzero, but the parities of the numbers of negative entries do not agree. In situation (*i*), the w involved switches the sign of the last entry, so -1 is contributed. In situation (*iii*), the w involved switches the signs of the last two entries, so $+1$ is contributed. This shows that

$$m_\lambda((S^2 \ominus \Lambda^2) \otimes \lambda) = \begin{cases} 1, & m \text{ odd}, \lambda_\ell \neq \frac{1}{2}\,, \\ 1, & m \text{ even}, \lambda_{\ell-1} = 0, \\ 0 & \text{otherwise}. \end{cases}$$

By Lemma 5.15, this completes the proof in Case I.

In Case II, we claim first that

$$G_+^* G_+ := G_{\lambda,\lambda+e_\ell}^* G_{\lambda,\lambda+e_\ell} \qquad \text{and} \qquad G_-^* G_- := G_{\lambda,\lambda-e_\ell}^* G_{\lambda,\lambda-e_\ell}$$

are the same operator, up to an action of the Weyl tensor. A careful study of the example $\lambda^0 := (1_{\ell-1}\,, 0) = \Lambda^{\ell-1}$ is useful in working out the general case. Here $G_\pm = P_\pm \circ d$, where d is the exterior derivative $\Lambda^{\ell-1} \to \Lambda^\ell$, and $P_\pm$ are the projections onto the two eigenbundles of the Hodge $\star$ for middle-forms. This gives, for some complex units u, u',

$$G_+^* G_+ - G_-^* G_- = u \star d \star \star d = u' \star dd = 0.$$

For a general λ in Case II,

$$\lambda = \lambda' \boxtimes \lambda^0, \qquad \lambda' = (\lambda_1 - 1, \dots, \lambda_{\ell-1} - 1, 0),$$

where $\boxtimes$ is the *Cartan product*, which selects the irreducible constituent of the tensor product containing the higest weight vector. We view $\mathbb{V}(\lambda)$ as a bundle of $(\ell-1)$-forms with values in the coefficient bundle $\mathbb{V}(\lambda')$. The operators in question are $(G_\pm^\nabla)^* G_\pm^\nabla$, where the connection ∇ on the coefficient bundle is just the Riemannian connection on $\mathbb{V}(\lambda')$. The difference of these works out, as above, to $u' \star d^\nabla d^\nabla$, which is an action of the coefficient bundle curvature, thus an action of the Riemann tensor R, thus (by conformal covariance and (5.8)) an action of the Weyl tensor C. Now an analysis as in Case I takes over to show that there are no other linearly independent second-order conformal covariants, modulo actions of C. Specifically, the map $\mathcal{C}$ constructed in Case I is invertible on the quotient of $\mathrm{End}_H(\mathbb{V}(1) \otimes \mathbb{V}(\lambda))$ by $\mathbb{C}E_+ \oplus \mathbb{C}E_-$, where $E_\pm$ is the projection corresponding to the gradient $G_\pm$. This shows that $d_\lambda = 1$ in Case II, and completes the proof of the theorem.

This leaves one loose end – when there is a conformal covariant, is it given by the formula for D_λ in Theorem 5.1? Since the Ricci curvature corrections constructed above are unique, D_λ fails to generate the space of second-order conformal covariants modulo Weyl tensor actions if and only if $d_\lambda = 1$ and $\sigma_2(D_\lambda) = 0$. This does not happen in Case II. In Case I, a case-by-case analysis shows:

LEMMA 5.17. *For λ in Case I with $d_\lambda = 1$,*

$$\{u : \sigma_u > \lambda\} = \{u : c_u < 0\}, \qquad \{u : \sigma_u \le \lambda\} = \{u : c_u > 0\},$$

where the weight inequalities are with respect to the lexicographic ordering.

The relevance of this to our question about $\sigma_2(D_\lambda)$ is as follows. Let φ be a highest weight vector for λ, and consider the real cotangent vector $\xi = \alpha + \bar{\alpha}$, where $\bar{\alpha}$ has the weight e_ℓ in the defining representation for $\mathfrak{so}(m)$. By weight considerations, $\sigma_1(G_u)(\xi)\varphi$ can have a component in σ_u only if $\sigma_u > \lambda$, $\sigma_u = \lambda$, or $\sigma_u = \lambda - e_\ell$. Thus:

LEMMA 5.18. *For m even, λ in Case I, $d_\lambda = 1$, and φ, ξ as above,*

$$\sigma_2(D_\lambda)(\xi)\varphi = - \sum_{c_u < 0} c_u^{-1} \zeta_u(\xi)^* \zeta_u(\xi)\varphi,$$

and $\zeta_u(\xi)^*\zeta_u(\xi)\varphi = 0$ whenever $c_u \ge 0$. Thus

$$(\sigma_2(D_\lambda)(\xi)\varphi, \varphi) \ge \left(\min_{c_u<0} (-c_u^{-1}) \right) |\xi|^2 |\varphi|^2 = \frac{2|\xi|^2|\varphi|^2}{m + 2\lambda_1 - 2} > 0.$$

In particular, $\sigma_2(D_\lambda) \ne 0$.

This completes the job in even dimensions: we know for which λ there is a second-order conformal covariant; we know that it is unique; and we have a formula for it. In odd dimensions, the question of whether D_λ gives a formula for the covariant, when it exists, cannot be completely treated without a deeper analysis. The case m odd, $\lambda_\ell = 0$ can be treated as above, using the vector of weight 0 in the defining representation of $\mathfrak{so}(m)$ in place of ξ above. Summing up, we get:

THEOREM 5.19. *If $d_\lambda = 1$ and*

$$(m \text{ is even}) \text{ or } (m \text{ is odd and } \lambda_\ell = 0),$$

the D_λ of Theorem 5.1 generates the space of second-order conformal covariants modulo actions of the Weyl tensor.

6. Geometric inequalities based on tensor-valued conformal covariants

We shall now use second-order conformal covariants to generate sharp forms of tensor-valued inequalities of Sobolev embedding type. One of these, together with the inequalities of Sec. 4, is enough to solve the extremal problems for $\det L$ and $\det \nabla^2$ on S^6.

The Riemann curvature tensor R is a section of $\mathbb{V}(0)\oplus\mathbb{V}(2)\oplus\mathbb{V}(2,2)$ for $m \geq 5$, the summands holding the scalar curvature, trace-free Ricci, and Weyl tensor parts respectively. When $m = 4$, the last summand is replaced by $\mathbb{V}(2,2)\oplus\mathbb{V}(2,-2)$; this reflects the splitting of the Weyl tensor into self- and anti-self-dual parts. When $m = 3$, R is a section of $\mathbb{V}(0) \oplus \mathbb{V}(2)$; when $m = 2$, of $\mathbb{V}(0)$.

By (1.4), ∇R is *a priori* a section of $2\mathbb{V}(1)\oplus 2\mathbb{V}(2,1)\oplus V(2,2,1)\oplus V(3)\oplus V(3,2)$ for $m \geq 7$, with similar formulas for $m < 7$. But the content of the second Bianchi identity is that the components of ∇R in some of these summands are missing [**S**]: ∇R is a section of a bundle which is $\mathrm{SO}(m)$-isomorphic to

$$\begin{array}{ll}
\mathbb{V}(1) \oplus \mathbb{V}(3) \oplus \mathbb{V}(2,1) \oplus \mathbb{V}(3,2), & m \geq 5, \\
\mathbb{V}(1) \oplus \mathbb{V}(3) \oplus \mathbb{V}(2,1) \oplus \mathbb{V}(2,-1) \oplus \mathbb{V}(3,2) \oplus \mathbb{V}(3,-2), & m = 4, \\
\mathbb{V}(1) \oplus \mathbb{V}(2) \oplus \mathbb{V}(3), & m = 3, \\
\mathbb{V}(1), & m = 2.
\end{array}$$

In particular, the summands in this bundle occur with multiplicity one. For example, one could build a tensor living in a $\mathbb{V}(1)$ bundle by forming $\nabla_i \tau$ or $\nabla^j \rho_{ij}$; by the Bianchi identity, these are linearly dependent.

One inequality of the type we are interested in here arises from the conformal Laplacian.

THEOREM 6.1. *On S^6, for $g_\omega = e^{2\omega} g_0$,*

$$\int_{S^6} \left(\{|dJ|^2 + 2J^3\}\, d\xi\right)_\omega \geq 6 \left(\int_{S^6} (|J|^3 d\xi)_\omega\right)^{2/3} \geq 54, \tag{6.1}$$

with equality in both $\geq$ iff e^ω is a constant multiple of some Ω_h.

PROOF. For $m \geq 3$, the Yamabe functional in the metric g_ω, at a function $u \in C^\infty(S^m)$, is

$$\mathcal{Y}(\omega, u) = \frac{\left(\left(\Delta + \frac{m-2}{2} J\right)_\omega u, u\right)_{L^2((d\xi)_\omega)}}{\|u\|^2_{L^{2m/(m-2)}((d\xi)_\omega)}},$$

where u is not identically zero. The conformal covariance relation for L implies that

$$\mathcal{Y}(\omega + \eta, u) = \mathcal{Y}(\omega, e^{2\eta} u);$$

by the solution of the Yamabe problem on the sphere we have

$$\mathcal{Y}(\omega,u)\geq\mathcal{Y}(0,1)=(m-2)J_0/2=m(m-2)/4.$$

Applying this to $u=J_\omega$ in the case $m=6$, we get the first $\geq$ in (6.1). Corollary 4.4 gives the second $\geq$, and shows that equality holds there exactly for constant multiples of conformal factors. But any g_ω with $e^{2\omega}=\alpha^2\Omega_h^2$ has $J\equiv 3\alpha^{-2}$ and $\int_{S^6}(d\xi)_\omega=\alpha^6$, showing that the integral on the left in (6.1) is 54.

Another such inequality arises when we apply the second-order conformal covariant $D_{(2)}$ on $\mathbb{V}(2)$ to the Einstein (normalized trace-free Ricci) tensor $b:=V-Jg/m$. In analogy with the above argument, which analyzes the behavior of $(LJ,J)_{L^2}$, we analyze the behavior of $(D_{(2)}b,b)_{L^2}$. If A is a 2-tensor, we set $\operatorname{tr}A^3=A^i{}_jA^j{}_kA^k{}_i$.

THEOREM 6.2. *Suppose $m\geq 5$, and let g_0 be the round metric on S^m. For $g_\omega=e^{2\omega}g_0$, $\omega\in C^\infty(S^m)$,*
(6.2)

$$\int_{S^m}\left(\left\{\frac{2(m-1)(m-2)}{m(m+2)^2}|dJ|^2-\frac{1}{m}J^3+\frac{m+4}{m+2}J|V|^2-\frac{2m}{m+2}\operatorname{tr}V^3\right\}d\xi\right)_\omega\geq 0,$$

with equality iff e^ω is a constant multiple of some Ω_h.

PROOF. Fix a metric g_ω conformal to the round one. By paragraph 5.9, the $\mathbb{V}(3)$ part of any three-tensor (φ_{ijk}) which is trace free and symmetric in the last two arguments is

$$(\operatorname{Proj}_{(3)}\varphi)_{ijk}=\frac{1}{3}(\varphi_{ijk}+\varphi_{jki}+\varphi_{kij})-\frac{2}{3(m+2)}(g_{ij}\varphi_{llk}+g_{ki}\varphi_{llj}+g_{jk}\varphi_{lli}),$$

and the $\mathbb{V}(1)$ part is

$$(\operatorname{Proj}_{(1)}\varphi)_{ijk}=-\frac{2}{(m+2)(m-1)}g_{jk}\varphi_{lli}+\frac{m}{(m+2)(m-1)}(g_{ij}\varphi_{llk}+g_{ik}\varphi_{llj}).$$

Here we take indices with respect to a local orthonormal frame, and so are entitled to write all indices in the down position. The Weyl tensor is

$$C_{ijkl}=R_{ijkl}+V_{jk}g_{il}-V_{jl}g_{ik}+V_{il}g_{jk}-V_{ik}g_{jl}\,.$$

On a conformally flat manifold,

$$C_{ijkl|i}=(m-3)(V_{jl|k}-V_{jk|l}),$$

so the Bianchi identity yields

$$b_{ij|k}-b_{ik|j}=\frac{1}{m}(J_{|j}g_{ik}-J_{|k}g_{ij}),\qquad b_{ij|j}=\frac{m-1}{m}J_{|i}\,,$$

where indices after a bar denote covariant differentiations. As a result,

$$\nabla b=G_{(2)(3)}b+G_{(2)(1)}b,\qquad \text{so } G_{(2)(2,1)}b=0.$$

By paragraph 5.9,

$$D_{(2)}b=\left(J+\frac{2}{m+2}\nabla^*\nabla-\frac{4}{m+2}G^*_{(2)(1)}G_{(2)(1)}\right)b.$$

Thus, using the Bianchi and Ricci identities and integrating by parts,

$$\begin{aligned}\|\nabla b\|^2 &= \int \left(\frac{m-1}{m} |dJ|^2 + J|V|^2 - m \operatorname{tr} V^3 \right) dv, \\ \|G_{(2)(1)} b\|^2 &= \frac{2(m-1)}{m(m+2)} \int |dJ|^2 dv.\end{aligned}$$

This shows that $(b, D_{(2)} b)_{L^2}$ is the quantity on the left in (6.2).

We now bring the round metric g_0 into play. By Theorem 5.19, the second-order conformal covariant on $\mathbb{V}(2)$ is unique up to a constant factor (and up to actions of the Weyl tensor, here zero). By Theorem 3.15, which asserts uniqueness of the principal series intertwining operator, $A_{2,(2)}$ and $D_{(2)}$ agree up to a constant factor at g_0 :

$$(D_{(2)})_0 = k A_{2,(2)} \,.$$

Consider the K-type $\mathcal{E}_{(2)}(2)$ in $\mathcal{E}_{(2)}$ with highest weight (2). By the branching rule (3.3), there is no constituent in the K-decomposition of $\mathcal{E}_{(3)}$ or $\mathcal{E}_{(2,1)}$ having highest weight (2). Thus

$$(D_{(2)})_0 \varphi = \left(J - \frac{2}{m+2} \nabla^* \nabla \right)_0 \varphi = \frac{(m+4)(m-2)}{2(m+2)} \varphi \,, \qquad \varphi \in \mathcal{E}_{(2)}(2),$$

since $J_0 = m/2$ and, by Theorem 3.4, $\nabla^* \nabla \varphi = 2\varphi$. By Theorem 3.12, the spectrum of $A_{2,(2)}$ is given by

$$T_2((2+j,q),(2)) = T_2((2),(2)) \frac{(m+4+2j)(m+2+2j)(m+2q-2)(m+2q-4)}{(m+4)(m+2)(m-2)(m-4)} \,.$$

Thus the spectrum of $(D_{(2)})_0$ is

$$(D_{(2)})_0|_{(2+j,q)} = \frac{(m+4+2j)(m+2+2j)(m+2q-2)(m+2q-4)}{2(m+2)^2(m-4)} > 0.$$

The quadratic form $(\psi, D_{(2)} \psi)_{L^2}$ is positive definite in the metric g_0 , and thus by Corollary 5.3, in the metric g_ω . In particular, $((b, D_{(2)} b)_{L^2})_\omega \geq 0$. This proves inequality (6.2).

For equality to hold in (6.2), we need $b_\omega = 0$; that is, we need g_ω to be an Einstein metric. Einstein metrics have constant scalar curvature for $m > 2$; thus by Obata's Theorem, $e^{2\omega} g_0 = \alpha^2 h \cdot g_0 = \alpha^2 \Omega_h^2 g_0$ for some $0 < \alpha \in \mathbb{R}$ and some $h \in \mathbf{C}(S^m, g_0)$. This completes the proof of the theorem.

When $m = 6$, a convenient form of this inequality is obtained when we shift by a multiple of the quantity $\int Q_6$ from Theorem 4.7; by (4.3), this quantity is a conformal invariant.

THEOREM 6.3 [**Br7**]. *On conformally flat manifolds,*

$$\int Q_6 dv = \int \left\{ \frac{m-6}{2} |dJ|^2 - \frac{(m+2)(m-2)}{4} J^3 - 4mJ|V|^2 + 16 \operatorname{tr} V^3 \right\} dv.$$

COROLLARY 6.4. *If $m = 6$,*

$$\int_{S^6} \left(\left\{ |dJ|^2 + \frac{28}{5} J^3 - \frac{48}{5} J |V|^2 \right\} d\xi \right)_\omega \geq 108,$$

with equality iff e^ω is a constant multiple of some Ω_h.

PROOF. We add $\frac{9}{10} \int_{S^6} (Q_6 d\xi)_\omega$ on the left, and $\frac{9}{10} \int_{S^6} (Q_6 d\xi)_0$ on the right, in (6.2).

To estimate the determinant on S^6, we need one more inequality:

THEOREM 6.5. *If $m \geq 3$, $\int_{S^m} (|dJ|^2 d\xi)_\omega \geq 0$, with equality iff e^ω is a constant multiple of some Ω_h.*

PROOF. The inequality is obvious; equality holds iff J is constant, iff (by Obata's Theorem) $e^{2\omega}$ has the form $\alpha^2 \Omega_h^2$.

The determinant on S^6 is computed by specializing Gilkey's formula [**G2**] for the heat invariant a_6 to L and $\nabla\!\!\!/^2$, and using Theorem 2.11. For ease of notation, we express it here in terms of quantities already estimated above by sharp inequalities. Let $\mathcal{B}_6(\omega)$ be the quantity estimated by the Beckner-Moser-Trudinger inequality, Theorem 4.6:

$$\mathcal{B}_6(\omega) := -\log \int_{S^6} e^{6(\omega - \bar{\omega})} (d\xi)_0 + \tfrac{1}{40} \int_{S^6} \omega(\Delta_0(\Delta_0 + 4)(\Delta_0 + 6)\omega)(d\xi)_0 .$$

Let $\mathcal{L}_1(\omega)$, $\mathcal{L}_2(\omega)$, $\mathcal{L}_3(\omega)$ be the quantities estimated by Theorem 6.5, Theorem 6.1, and Corollary 6.4 respectively. Let $\mathcal{C}_i(\omega) = \mathcal{L}_i(\omega) - \mathcal{L}_i(0)$. In analogy with Theorems 4.8 and 4.9, we have:

THEOREM 6.6 [**Br7**]. *On S^6,*

$$-3 \cdot 7! \cdot \tfrac{1}{2} \log \frac{\mathcal{D}(L, g_\omega)}{\mathcal{D}(L, g_0)} = \frac{10}{3} \mathcal{B}_6(\omega) + \frac{1}{60} \left(2\mathcal{C}_1(\omega) + \frac{23}{2} \mathcal{C}_2(\omega) + \frac{10}{3} \mathcal{C}_3(\omega) \right),$$

$$-72 \cdot 7! \cdot \tfrac{1}{2} \cdot 2^{-3} \log \frac{\mathcal{D}(\nabla\!\!\!/^2, g_\omega)}{\mathcal{D}(\nabla\!\!\!/^2, g_0)} = -382 \mathcal{B}_6(\omega) + \frac{1}{60} \left(-\frac{93}{2} \mathcal{C}_1(\omega) - 278 \mathcal{C}_2(\omega) - \frac{365}{2} \mathcal{C}_3(\omega) \right).$$

As a result, $\mathcal{D}(L, g_\omega)$ is maximized, and $\mathcal{D}(\nabla\!\!\!/^2, g_\omega)$ is minimized, exactly when e^ω is a constant multiple of some Ω_h.

Thus our "luck" has held out; all coefficients, in each case, have the same sign, and the checkerboard pattern of Table 4.1 persists. The coefficients involved are purely algebraic objects, implicit in the universal formulas for heat invariants; there would seem to be no *a priori* reason for all these signs to fall into place. What has happened in six dimensions is analogous to what happened in four dimensions, but is more improbable: in four dimensions, the signs of two quantities estimated by sharp embeddings $L_2^2 \hookrightarrow e^L$ and $L_1^2 \hookrightarrow L^4$ agreed; in six dimensions, four quantities estimated by the sharp $L_3^2 \hookrightarrow e^L$ embedding and various $L_1^2 \hookrightarrow L^3$ embeddings (including one tensor-valued embedding) have signs that agree. Table 4.1 would seem to rule out any simple approach in general even dimension; the signs depend on the operator whose determinant is being taken, and on m modulo 4. Thus there is no *a priori* principle asserting, for example, that the determinant functional is automatically convex.

However, the approach to sharp inequalities sketched in this section offers promise for such estimation problems in special dimensions. For example, by Corollary 4.5,

$$\int_{S^8}((4J^2 - 2|V|^2 + \Delta J)^2 d\xi)_\omega \geq 3600, \tag{6.3}$$

with equality iff e^ω is a constant multiple of some Ω_h. We can complement this with various inequalities asserting the nonnegativity of the components of $\nabla\nabla\rho$ in various O(8) bundles, plus inequalities obtained by applying second-order conformal covariants to the components of ∇R. As we found out above, there are only two of these in the conformally flat case, namely the section dJ of $\mathbb{V}(1)$, and $\mathrm{Proj}_{(3)}\nabla b$, the components of which are

$$\begin{aligned}\psi_{ijk} := (\mathrm{Proj}_{(3)}\nabla b)_{ijk} &= \nabla_i b_{jk} + \frac{2}{m(m+2)}J_{|i}g_{jk} - \frac{1}{m+2}(J_{|j}g_{ik} + J_{|k}g_{ij}) \\ &= \nabla_i V_{jk} - \frac{1}{m+2}(J_{|i}g_{jk} + J_{|j}g_{ik} + J_{|k}g_{ij}).\end{aligned} \tag{6.4}$$

Applying the operator D_1 of (5.6) to dJ, we get

$$(dJ, D_1 dJ)_{L^2} = \frac{m-4}{2}\int(\Delta J)^2 dv + \frac{m(m-4)}{4}\int(J|dJ|^2 - 2V_{ij}J_{|i}J_{|j})dv. \tag{6.5}$$

By uniqueness of the second-order conformal covariant, uniqueness of the principal series intertwinor, Theorem 3.12, and Corollary 5.3, the quadratic form $(\eta, D_1\eta)$ is positive definite for $m \geq 5$, at any g_ω on S^m. If $(dJ, D_1 dJ) = 0$, then J is constant, so e^ω is a constant multiple of some Ω_h by Obata's Theorem. If $m = 4$, the Case II statement in Corollary 5.3 applies, and we get positive definiteness of the form $(df, D_1' df)$ on functions, where D_1' is the appropriate normalization of the $D_{(1)}'$ of Theorem 5.2. The upshot is that we get a dimension-regularized version of the inequality expressing definiteness of (6.5):

THEOREM 6.7. *If* $m \geq 4$,

$$\int_{S^m}((\Delta J)^2 d\xi)_\omega + \frac{m}{2}\int_{S^m}((J|dJ|^2 - 2V_{ij}J_{|i}J_{|j})d\xi)_\omega \geq 0,$$

with equality iff e^ω *is a constant multiple of some* Ω_h.

The invariants in (6.3) and Theorem 6.7 can be related by various identities obtained by integration by parts and the Bianchi identity; for example,

$$\int(\Delta J)J^2 dv = 2\int J|dJ|^2, \qquad \int(\Delta J)|V|^2 dv = -2\int(J_{|ik}V_{ji}V_{jk}dv + J_{|i}J_{|j}V_{ij})dv.$$

Another inequality can be derived by using (5.7) to apply $D_{(3)}$ to the tensor ψ of (6.4): (3.13) and the construction leading to Theorem 3.12 show that $D_{(3)}$ is positive definite as long as

$$2 < \min_{\alpha\leftrightarrow\beta\downarrow(3)}|\kappa(\beta,(3)) - \kappa(\alpha,(3))| = m-2;$$

i.e. $m > 4$.

An approach to the extremal problem in general even dimension that is likely to work in the long run would involve estimates for $\zeta(s)$ for small real s, followed by an endpoint differentiation argument (using the fact that $\zeta(0)$ is a conformal invariant) to derive the requisite estimate on $\zeta'(0)$. That is, instead of appealing to the Beckner-Moser-Trudinger inequality to handle the leading term in $\zeta'(0)$, then scrambling to pick up the trailing terms with borderline Sobolev embedding estimates, one could try to recapitulate the endpoint differentiation argument that leads to the Beckner-Moser-Trudinger inequality. The estimates on $\zeta(s)$ are not yet in place; one difficulty is that the values in question are in the analytic continuation of the zeta function. To get around this, Ryszard Nest suggests the use of modern operator algebraic techniques involving generalized traces, to redo the zeta function theory in such a way that there are no analytic continuations. Alternatively, one could take inspiration in a different way from the process leading to the Beckner-Moser-Trudinger inequality. Here one first derives the sharp form of the dual embedding $L \log L \hookrightarrow L^2_{-m/2}$. The idea, suggested by Carlo Morpurgo [**M2**], is to do the analogue of this by looking at values of the heat kernel trace. Two advantages of this are that the heat kernel requires no analytic continuation, and that work estimating the relevant values already exists [**M1**].

References

[Be] W. Beckner, *Sharp Sobolev inequalities on the sphere and the Moser-Trudinger inequality*, Annals of Math. **138** (1993), 213–242.

[Bo] H. Boerner, *Darstellungen von Gruppen*, Springer-Verlag, Berlin, 1955.

[Br1] T. Branson, *Conformally covariant equations on differential forms*, Comm. Partial Differential Equations **7** (1982), 392–431.

[Br2] T. Branson, *Differential operators canonically associated to a conformal structure*, Math. Scand. **57** (1985), 293–345.

[Br3] T. Branson, *Group representations arising from Lorentz conformal geometry*, J. Funct. Anal. **74** (1987), 199–291.

[Br4] T. Branson, *Second-order conformal covariants*, University of Copenhagen preprint series, nos. 2,3 (1989).

[Br5] T. Branson, *Harmonic analysis in vector bundles associated to the rotation and spin groups*, J. Funct. Anal. **106** (1992), 314–328.

[Br6] T. Branson, *The Functional Determinant*, Lecture Note Series, number 4, Global Analysis Research Center, Seoul National University, 1993.

[Br7] T. Branson, *Sharp inequalities, the functional determinant, and the complementary series*, Trans. Amer. Math. Soc. **347** (1995), 3671–3742.

[BCY] T. Branson, S.-Y. A. Chang, and P. Yang, *Estimates and extremals for zeta function determinants on four-manifolds*, Commun. Math. Phys. **149** (1992), 241–262.

[BGP] T. Branson, P. Gilkey, and J. Pohjanpelto, *Invariants of locally conformally flat manifolds*, Trans. Amer. Math. Soc. **347** (1995), 939–954.

[BÓØ] T. Branson, G. Ólafsson, and B. Ørsted, *Spectrum generating operators, and intertwining operators for representations induced from a maximal parabolic subgroup*, to appear, J. Funct. Anal.

[BØ1] T. Branson and B. Ørsted, *Conformal indices of Riemannian manifolds*, Compositio Math. **60** (1986), 261–293.

[BØ2] T. Branson and B. Ørsted, *Conformal geometry and global invariants*, Diff. Geom. Appl. **1** (1991), 279–308.

[BØ3] T. Branson and B. Ørsted, *Explicit functional determinants in four dimensions*, Proc. Amer. Math. Soc. **113** (1991), 669–682.

[CL] E. Carlen and M. Loss, *Competing symmetries, the logarithmic HLS inequality and Onofri's inequality on S^n*, Geometric and Functional Analysis **2** (1992), 90–104.

[F] H. D. Fegan, *Conformally invariant first order differential operators*, Quart. J. Math. Oxford **27** (1976), 371–378.

[G1] P. Gilkey, *Smooth invariants of a Riemannian manifold*, Adv. in Math. **28** (1978), 1–10.

[G2] P. Gilkey, *Recursion relations and the asymptotic behavior of the eigenvalues of the Laplacian*, Compositio Math. **38** (1979), 201–240.

[G3] P. Gilkey, *Invariance Theory, the Heat Equation, and the Atiyah-Singer Index Theorem*, Publish or Perish, Wilmington, Delaware, 1984.

[GJMS] C. R. Graham, R. Jenne, L. Mason, and G. Sparling, *Conformally invariant powers of the Laplacian, I: existence*, J. London. Math. Soc. **46** (1992), 557–565.

[J] R. Jenne, *A construction of conformally invariant differential operators*, Ph.D. dissertation, University of Washington, 1988.

[Kosm] Y. Kosmann, *Dérivées de Lie des spineurs*, Ann. Mat. Pura Appl. (4) **XCI** (1972), 317–395.

[Kost] B. Kostant, *A formula for the multiplicity of a weight*, Trans. Amer. Math. Soc. **93** (1959), 53–73.

[KS] A. Knapp and E. Stein, *Analytic continuation of intertwining operators*, Inventiones Math. **60** (1980).

[L] E. Lieb, *Sharp constants in the Hardy-Littlewood-Sobolev and related inequalities*, Annals of Math. **118** (1983), 349– 374.

[M1] C. Morpurgo, *Local extrema of traces of heat kernels on S^2*, to appear, J. Funct. Anal.

[M2] C. Morpurgo, *The logarithmic Hardy-Littlewood-Sobolev inequality and extremals of zeta functions on S^n*, to appear, Geom. and Funct. Anal.

[O] E. Onofri, *On the positivity of the effective action in a theory of random surfaces*, Commun. Math. Phys. **86** (1982), 321–326.

[OPS1] B. Osgood, R. Phillips, and P. Sarnak, *Extremals of determinants of Laplacians*, J. Funct. Anal. **80** (1988), 148–211.

[OPS2] B. Osgood, R. Phillips, and P. Sarnak,. *Compact isospectral sets of surfaces*, J. Funct. Anal. **80** (1988), 212–234.

[Pa] S. Paneitz, *A quartic conformally covariant differential operator for arbitrary pseudo-Riemannian manifolds*, preprint, 1983.

[Po1] A. Polyakov, *Quantum geometry of Bosonic strings*, Phys. Lett. B **103** (1981), 207–210.

[Po2] A. Polyakov, *Quantum geometry of Fermionic strings*, Phys. Lett. B **103** (1981), 211–213.

[SW] E. Stein and G. Weiss, *Generalization of the Cauchy-Riemann equations and representations of the rotation group*, Amer J. Math. **90** (1968), 163–196.

[S] R. Strichartz, *Linear algebra of curvature tensors and their covariant derivatives*, Canad. J. Math. **40** (1988), 1105–1143.

[W] V. Wünsch, *On conformally invariant differential operators*, Math. Nachr. **129** (1986), 269–281.

[Y] H. Yamabe, *On a deformation of Riemannian structures on compact manifolds*, Osaka J. Math. **12** (1960), 21–37.

DEPARTMENT OF MATHEMATICS, THE UNIVERSITY OF IOWA, IOWA CITY IA 52242 USA
E-mail address: `branson@math.uiowa.edu`

Proceedings of Symposia in Pure Mathematics
Volume 59, 1996

EXISTENCE THEOREM FOR SOLUTIONS OF EINSTEIN'S EQUATIONS WITH 1 PARAMETER SPACELIKE ISOMETRY GROUPS

YVONNE CHOQUET-BRUHAT AND VINCENT MONCRIEF

INTRODUCTION

It has been shown by one of us (V.Moncrief), by reduction of the Hamiltonian, that the Einstein equations for a four dimensional space-time admitting a one parameter space-like isometry group can be written (up eventually to the inclusion of a harmonic 1-form) as the equation for a harmonic map u into hyperbolic 2-space, from a 3-manifold $\Sigma \times \mathbf{R}$ with a pseudo Riemannian metric $\tilde{g}$ satisfying Einstein's 3 dimensional equations with a source given by the stress energy tensor of the harmonic map. Using the fact that the data of a general metric on a 3 dimensional space time $\Sigma \times \mathbf{R}$ is equivalent to the data on each $\Sigma_t = \Sigma \times \{t\}, t \in \mathbf{R}$ of a 2 dimensional metric $\bar{g}_t$ together with a function N, the lapse, and a vector field ν, the shift, it has been shown that the determination of $\tilde{g}$ can be obtained by solution of an elliptic system whose coefficients depend on u and its conjugate variable $\dot{u}$. The solution of this elliptic, non-linear system has been studied.

We shall first show in this paper how one can obtain by direct methods, without making use of a Hamiltonian formalism, the splitting of part of the vacuum Einstein equations with 1 parameter isometry group into a system elliptic on each spacelike slice on the one hand, and on the other hand a hyperbolic system which is essentially a harmonic map equation from a pseudo-Riemannian manifold, whose metric depends on the solution of the elliptic system, into the 2-dimensional symmetric Riemannian space $\mathcal{H}$.

We then prove global in space, local in time existence and uniqueness of solutions of the Cauchy problem for the hyperbolic-elliptic system so obtained, when $\Sigma = S^2$, in appropriate functional spaces. We show that in that case, the obtained solution

1991 *Mathematics Subject Classification.* Primary 83C20; Secondary 83C10, 53C80. V. Moncrief's research supported in part by NSF grants INT 9015153 and PHY 9201196.

This paper is the symposium contribution of Y. Choquet-Bruhat.

satisfies the original Einstein equations. We finally discuss the general solution of the system, in the case where Σ is not simply connected. We show that when the genus of Σ is greater than 1, we can determine $\bar{g}_t$ so that it is conformal to a metric σ_t which remains in some chosen cross section of a bundle over the Teichmüller space of Σ (cf. an analogous method in Moncrief [10] for the 3-dimensional Einstein vacuum equations). Here the evolution of σ_t is determined by requiring the mixed elliptic-hyperbolic system obtained previously to be a solution of the original, 4-dimensional, Einstein equations.

1. Definitions

We consider a space time $(V, \underline{g})$, with V a smooth 4 dimensional manifold of the type $M \times \mathbf{R}$, $\underline{g}$ a pseudo-Reimannian metric of hyperbolic signature such that the submanifolds $M_t \equiv M \times \{t\}$ are space-like and the lines $\{x\} \times \mathbf{R}$ are time-like.

We suppose that M is a principal fiber bundle with 2 dimensional base Σ and group G_1, a one dimensional Lie group. V is therefore a principal fiber bundle with base $\Sigma \times \mathbf{R}$ and group G_1. We suppose that $\underline{g}$ is invariant under the right action of G_1 on V. Note that the corresponding Killing vectors are tangent to the submanifolds M_t, hence space-like. It results from the hypotheses that the metric $\underline{g}$ reads (Kaluza Klein ansatz)

$$\underline{g} \equiv \pi^* g + (\pi^* e^{2\gamma})\underline{\theta}^2 \tag{1.1}$$

where g is a pseudo-Riemannian meric and γ a function on $\Sigma \times \mathbf{R}$, both pulled back by the bundle projection $\pi : V \equiv M \times \mathbf{R} \to \Sigma \times \mathbf{R}$ while $\underline{\theta}$ is a G_1 connection 1-form on $M \times \mathbf{R}$. The hypothesis on the signature of $\underline{g}$ and on time and space are satisfied if g is of Lorentzian signature on $\Sigma \times \mathbf{R}$, with Σ_t space-like and $\{\xi\} \times \mathbf{R}$ time-like.

We shall take local coordinates on V adapted to its product and bundle structure: we denote by $x^i, i = 1, 2$, coordinates in a domain of a chart U_Σ of Σ, we set $(x^\alpha) = (x^1, x^2, t), t \in \mathbf{R}$, and denote by x^3 a local coordinate in G_1 corresponding to a trivialization of M over U_Σ. In such coordinates, $\pi^* g$ and g have the same expression

$$\pi^* g = g = g_{\alpha\beta} dx^\alpha dx^\beta,$$

while

$$\underline{\theta} \equiv \theta^3 \equiv dx^3 + A_\alpha dx^\alpha,$$

with $g_{\alpha\beta}, \gamma, A_\alpha$ independent of x^3. Note that $A_\alpha dx^\alpha$ is a 1-form on $U_\Sigma \times \mathbf{R}$, representing the connection but does not extend to a 1-form on $\Sigma \times \mathbf{R}$ if the bundle $M \to \Sigma$ is not trivial.

2. Kaluza Klein Formulas

The Ricci tensor of $\underline{g}$ can be expressed in terms of the Ricci tensor of g, the scalar function γ and the curvature F of the connection A (cf. the formulas for an arbitrary Lie isometry group in CB BM II, Ref. [5]). They read as follows in the adapted frame $dx^\alpha, \theta^3 \equiv \underline{\theta}$

$$\underline{R}_{\alpha\beta} \equiv R_{\alpha\beta} - \partial_\alpha\gamma\partial_\beta\gamma - \nabla_\alpha\partial_\beta\gamma - \frac{1}{2}e^{2\gamma}F_{\alpha\lambda}F_\beta{}^\lambda$$

$$\underline{R}_{\alpha 3} \equiv -\frac{1}{2}e^{-\gamma}\nabla_\lambda(e^{3\gamma}F_\alpha{}^\lambda)$$

$$\underline{R}_{33} \equiv -e^{2\gamma}(g^{\alpha\beta}\nabla_\alpha\partial_\beta\gamma + g^{\alpha\beta}\partial_\alpha\gamma\partial_\beta\gamma - \frac{1}{4}e^{2\gamma}F_{\alpha\beta}F^{\alpha\beta}).$$

∇ is the covariant derivative in the metric g (note $\partial_\alpha\gamma = \partial\gamma/\partial x^\alpha$ because $\partial\gamma/\partial x^3 = 0$), F is a 2 form on $\Sigma \times \mathbf{R}$, representing the curvature $\underline{F}$ of the G_1 connection $\underline{\theta}$:

$$\underline{F} = d\underline{\theta} = \pi^* F.$$

The expression of F in the frame dx^α is

$$F_{\alpha\beta} \equiv \nabla_\alpha A_\beta - \nabla_\beta A_\alpha.$$

3. Twist Potential

The equations $\underline{R}_{\alpha 3} = 0$ express the vanishing of a 1-form on $\Sigma \times \mathbf{R}$, the codifferential in the metric g of the 2-form $e^{3\gamma}F$. Equivalently, it requires the vanishing of the differential of the adjoint E of this form, i.e.

$$dE = 0,$$

with $E \equiv e^{3\gamma}\, {}^*F$, $({}^*F)_\alpha \equiv \frac{1}{2}\eta_{\alpha\beta\lambda}F^{\beta\lambda}$, (hence $F_{\alpha\beta} \equiv e^{-3\gamma}\eta_{\alpha\beta\lambda}E^\lambda$), where η is the volume 3 form of $\Sigma \times \mathbf{R}$ in the metric g. The general solution of $dE = 0$ is

$$e = d\omega + H,$$

with ω a scalar function on $\Sigma \times \mathbf{R}$ called the *twist potential,* and H a representative of a 1-cohomology class of $\Sigma \times \mathbf{R}$. There are b_1, (the first Betti number of Σ) linearly independent such representatives. If Σ is compact, we can take for H the pull back by the trivial projection $\Sigma \times \mathbf{R} \to \Sigma$ of a harmonic 1-form on (Σ, m), m any given Riemannian metric on Σ. If Σ is not compact, an analogous result holds under appropriate hypothesis "at infinity". If the 2 dimensional manifold Σ is compact and orientable, then $b_1 = 2G$, where G is the genus of Σ.

When g, γ and ω are known on $\Sigma \times \mathbf{I}$, $\mathbf{I}$ an interval of $\mathbf{R}$, we also know the 2-form F on $\Sigma \times \mathbf{I}$ and its pull back $\underline{F} \equiv \pi^* F$ on $M \times \mathbf{I}$. We can deduce from $\underline{F}$ a connection 1-form $\underline{\theta}$ on $M \times \mathbf{I}$, with local representative A on $U_\Sigma \times \mathbf{I}$ satisfying there $dA = F$ if and only if

1) Local condition

$$dF = 0 \text{ equivalently } \nabla \cdot ({}^*F) = 0.$$

That is the twist potential ω must satisfy the equation

$$\nabla_\alpha(e^{-3\gamma}g^{\alpha\beta}\partial_\beta\omega) + \nabla_\alpha(e^{-3\gamma}H^\alpha) = 0. \tag{3.1}$$

2) Global condition. In the case Σ compact, $G_1 \equiv U(1)$ this condition reads (the integral does not depend on t when F is closed)

$$n = \frac{1}{2\pi}\int_{\Sigma_t} F \tag{i.1}$$

The integer n is the Chern number of the $U(1)$ bundle $M \times I$ (cf. Cameron [4]), for a trivial bundle $n = 0$.

4. Conformal Formulas. Wave Map

When g and H are given, the equations (3.1) for the twist potential together with $\underline{R}_{33} = 0$ are a system of non linear equations for γ and ω which are hyperbolic if g is of Lorentzian signature. These equations take the form of a wave map if one introduces the conformal metric $\tilde{g}$

$$g = e^{-2\gamma}\tilde{g}$$

and denotes by a twidle geometric elements relative to $\tilde{g}$. Standard computations (cf. for instance CB-DM I, Ref. [5]), with F replaced by its value in terms of ω, show that $e^{-4\gamma}\underline{R}_{33} = 0$ reads

$$\tilde{\nabla}^\alpha\partial_\alpha\gamma + \frac{1}{2}e^{-4\gamma}\tilde{g}^{\alpha\beta}\partial_\alpha\omega\partial_\beta\omega = -e^{-4\gamma}\tilde{g}^{\alpha\beta}(\partial_\alpha\omega H_\beta + \frac{1}{2}H_\alpha H_\beta), \tag{Ia}$$

while the equation (3.1) for the twist potential ω gives

$$\tilde{\nabla}^\alpha\partial_\alpha\omega - 4\tilde{g}^{\alpha\beta}\partial_\alpha\gamma\partial_\beta\omega = -e^{4\gamma}\tilde{\nabla}^\alpha(e^{-4\gamma}H_\alpha). \tag{Ib}$$

Inspection of (Ia), (Ib) show that their left hand side is the *wave map operator from* $(\Sigma \times I, g)$ *into* $(\mathbf{R}^2, 2d\gamma\, d\gamma + \frac{1}{2}e^{-4\gamma}d\omega d\omega)$, i.e. *the 2-space $\mathcal{H}$ of constant negative curvature.*In particular, we have:

Theorem 1. *The equations $\underline{R}_{\alpha 3} = 0, \underline{R}_{33} = 0$ are satisfied on $(\Sigma \times \mathbf{I})$ if (γ, ω) defines a harmonic map from $(\Sigma \times \mathbf{I}, \tilde{g})$ into $\mathcal{H}$. The condition is necessary if the first Betti number of Σ is zero.*

5. Conformal Formulas. 3 Dimensional Einstein Equations with Sources

We use the general formulas (cf. CB-DM I, Ref. [5]) linking the Ricci tensors of two conformal metrics to show that, when F is expressed in terms of ω and H,

$$\underline{R}_{\alpha\beta} + g_{\alpha\beta}\underline{R}_{33} \equiv \tilde{R}_{\alpha\beta} - \rho_{\alpha\beta} \tag{5.1}$$

with

$$\rho_{\alpha\beta} \equiv \frac{1}{2}\{e^{-4\gamma}\partial_\alpha\omega\partial_\beta\omega + 4\partial_\alpha\gamma\partial_\beta\gamma + e^{-4\gamma}(\partial_\alpha\omega H_\beta + \partial_\beta\omega H_\alpha + H_\alpha H_\beta)\}.$$

We deduce from (5.1) the following theorem

Theorem 2. *When (γ, ω) satisfy equations (I), the equations $\underline{R}_{\alpha\beta} = 0$ reduce to Einstein's equations on $\Sigma \times I$ with a specified source. If the 1-form H is chosen zero the source is the stress energy tensor of the wave map (γ, ω).*

6. 2 + 1 Decomposition. Constraints on Σ_t

In any dimension $n+1 \geq 3$, the Einstein equations on $\Sigma \times I$ are known to split between constraints, generally written as an elliptic system on a fixed manifold $\Sigma_0 \equiv \Sigma \times \{0\}$, and evolution equations which depend on the choice of gauge. This choice can for instance fix the shift ν to be zero and the lapse as linked to the mean curvature of the Σ_t in one manner or another (cf, C-B and Ruggeri 1983, Christodoulou and Klainerman 1991). In the case $n = 2$, the solution of the constraints on each Σ_t and the gauge fixing conditions determine essentially the 3-metric, as we shall see.

Let $\tilde{g}$ be a Lorentzian metric on $\Sigma \times \mathbf{R}$; denote by N and ν its lapse and shift relative to the foliation; $\tilde{g}$ then reads

$$\tilde{g} = -N^2 dt^2 + \bar{g}_{ab}(dx^a + \nu^a dt)(dx^b + \nu^b dt)$$

where $\bar{g}_t \equiv \bar{g}_{ab} dx^a dx^b$ is a properly Riemannian metric on $\Sigma_t \sim \Sigma$, $x^a, a = 1, 2$ local coordinates on Σ.

We denote by k_t the extrinsic curvatrure of Σ_t in $(\Sigma \times \mathbf{R}, \tilde{g})$, i.e. with $\partial_t \equiv \frac{\partial}{\partial t}$

$$k_t = \frac{1}{2N}(-\partial_t \bar{g}_t + \mathcal{L}_\nu \bar{g}), i.e., k_{ab} = \frac{1}{2N}(-\partial_t \bar{g}_{ab} + \bar{\nabla}_a \nu_b + \bar{\nabla}_b \nu_a).$$

We denote by τ the mean curvature of Σ_t as submanifold of $(\Sigma \times \mathbf{R}, \tilde{g})$, that is

$$\tau = tr_{\bar{g}} k_t = k_a^a.$$

We use the coframe $\theta^a = dx^a + \nu^a dt, \theta^0 = dt$, hence $\partial_0 f = \partial_t f + \nu^a \partial f / \partial x^a$, $\partial_a f = \partial f / \partial x^a$ (recall that $\partial_t = \partial / \partial t$) and obtain the identities

$$\tilde{R}_{ab} \equiv \bar{R}_{ab} + N^{-1}(-\partial_t k_{ab} + \mathcal{L}_\nu k_{ab}) - 2k_{ac}k_b^c + \tau k_{ab} - N^{-1}\bar{\nabla}_a \partial_b N, \tag{6.1}$$

$$\tilde{R}_{0a} \equiv N(-\bar{\nabla}_b k_a^b + \partial_a \tau) \tag{6.2}$$

$$\tilde{R}_{00} \equiv N\bar{\nabla}^a \partial_a N + N\partial_t \tau - N^2 k_b^a k_a^b \tag{6.3}$$

The equations (5.1) imply on each Σ_t the following equations called constraints, because they do not depend on second derivatives transversal to Σ_t

$$N^{-1}\tilde{R}_{0a} \equiv -\bar{\nabla}_b k_a^b + \partial_a \tau = J_a \equiv N^{-1}\{\frac{1}{2}(\partial_a \omega + H_a)\partial_0 \omega + 2\partial_a \gamma \partial_0 \gamma\} \tag{6.4}$$

$$2N^{-2}\tilde{S}_{00} \equiv 2N^{-2}(\tilde{R}_{00} - \frac{1}{2}\tilde{g}_{00}\tilde{R}) \equiv \bar{R} - k_b^a k_a^b + \tau^2 = Q, \tag{6.5}$$

$$Q \equiv \frac{3}{2}N^{-2}\{e^{-4\gamma}(\partial_0 \omega)^2 + 4(\partial_0 \gamma)^2\} + \bar{g}^{ab}\{e^{-4\gamma}(\partial_a \omega + H_a)(\partial_b \omega + H_b) + 4\partial_a \gamma \partial_b \gamma\}$$

We express the constraints as an elliptic system on each Σ_t by using the conformal method (cf. CB-Y [6]): that is, we choose some metric σ_t on Σ_t and we set

$$\bar{g}_t = e^{2\lambda_t}\sigma_t \quad i.e. \quad \bar{g}_{ab} = e^{2\lambda}\sigma_{ab},$$
$$k^b_a = e^{-2\lambda}h^b_a + \frac{1}{2}\delta^t_a\tau, \quad h \text{ is then traceless.}$$

The equations (6.5), (6.6) become, with D the covariant derivative, Δ and ρ the Laplacian and the Riemann scalar curvature of the metric σ_t,

(IIa) $$D_c h^c_a = \frac{1}{2}e^{2\lambda}\partial_a\tau - e^{2\lambda}J_a \equiv L_a, h^a_a = 0.$$

(IIb) $$2\Delta\lambda = \rho - e^{-2\lambda}(h^a_b h^b_a + A) + \frac{1}{2}e^{2\lambda}\tau^2 - B$$

where

$$A \equiv N^{-2}e^{4\lambda}\{\frac{1}{2}e^{-4\gamma}(\partial_t\omega + H_0)^2 + 2(\partial_t\gamma)^2\},$$
$$B \equiv \sigma^{ab}\{\frac{1}{2}e^{-4\gamma}(\partial_a\omega + H_a)(\partial_b\omega + H_b) + 2\partial_a\gamma\partial_b\gamma\}$$

The equation (IIb) is a non-linear elliptic equation for λ.

The left hand side of (IIa) is a linear operator on traceless tensors h; its adjoint, acting on vector fields, is the conformal Killing operator

$$(\mathcal{L}_{\sigma_t}v) \equiv D_a v_b + D_b v_a - \sigma_{ab}D_c v^c.$$

When Σ is compact (IIa) will admit solutions only if

(i.2) $$\int_{\Sigma_t} L_a v^a d\mu(\sigma) = 0, \text{ for all } v \text{ a conformal Killing field of } (\Sigma, \sigma_t).$$

7. Equations for lapse and shift

The $2+1$ decomposition shows that

$$\underline{R}_{00}N^{-1} - N^2 e^{-2\gamma}\underline{R}_{33} = 0$$

is equivalent to

$$\bar{\Delta}N - \bar{q}N = -\partial_t\tau,$$
$$\bar{q} \equiv |k|^2 + M, |k|^2 \equiv k^a_b k^b_a, M \equiv 2(\partial_t\gamma)^2 + \frac{1}{2}e^{-4\gamma}(\partial_t\omega + H_0)^2$$

i.e. in terms of Σ and h on each Σ_t the linear elliptic equation for N

(IIIa) $$\Delta N - qN = -e^{2\lambda}\partial_t\tau, \quad q \equiv e^{-2\lambda}|h|^2 + \frac{1}{2}e^{2\lambda}\tau^2 + e^{2\lambda}M.$$

Shift fixing. The definition of k_t gives in covariant components, where

$$h_{ab} = \bar{k}_{ab} - \frac{1}{2}\bar{g}_{ab}\tau, \bar{\nu}_a = e^{2\lambda}\nu_a, \mathcal{L}_{\bar{g}}\nu = e^{2\lambda}\mathcal{L}_\sigma\nu,$$

(IIIb) $$(\mathcal{L}_{\sigma_t}\nu)_{ab} = f_{ab} \equiv 2Ne^{-2\lambda}h_{ab} + \partial_t\sigma_{ab} - \frac{1}{2}\sigma_{ab}\sigma^{cd}\partial_t\sigma_{cd}.$$

The left hand side of (IIIb) is a linear operator on ν with injective symbol. The kernel of its dual is the space of transverse traceless symmetric 2-tensors on (Σ_t, σ_t). The equation (IIIb) will have solutions when Σ is compact if and only if the right hand side is orthogonal to this kernel, i.e.

(i.3) $$\int_{\Sigma_t} f_{ab}T^{ab}d\mu(\sigma_t) = 0, \text{ for all } T \text{ such that } T^a_a = 0, D_aT^{ab} = 0.$$

8. Mixed elliptic-hyperbolic system. Cauchy Problem

We consider as given on each Σ_t the scalar τ and the metric σ_t. If Σ is compact, we choose a harmonic 1-form H.

The unknowns on $\Sigma \times I$ are $\omega, \gamma, \lambda, N$ and ν. They determine a space-time metric $\underline{g}$ satisfying the vacuum Einstein equations only if they satisfy the system (I), (II), (III) and, if Σ is compact, the integral condition (i.1). The principal operator in (I) is hyperbolic, in (II) and (III) it is elliptic, but all equations depend on all the unknowns and are mostly non linear. It appears convenient, before developing an iteration scheme for solving the system to introduce the "conjugate variables" in the sense of Hamilton (cf. Moncrief [1]) of ω and γ. Namely, denoting $u \equiv (\omega, \gamma)$, we set

$$\dot{u} \equiv e^{2\lambda}\partial_0 u, \quad \partial_0 \equiv N^{-1}(\partial_t - \nu^a\partial_a).$$

The equations (II) read then, with a dot the scalar product in $\mathcal{H}$ and $|\ |$ the norm in σ_t

(IIa) $$D_b h^b_a = L_a, L_a \equiv -\frac{1}{2}D_a u \cdot \dot{u} + \frac{1}{2}e^{2\lambda}\partial_a\tau, \quad h^a_a = 0.$$

(IIb) $$\Delta\lambda = F(x, \lambda) \equiv p_1 e^{2\lambda} - p_2 e^{-2\lambda} + p_3$$

$$p_1 = \frac{1}{4}\tau^2 \geq 0, p_2 = \frac{1}{2}|h|^2 + \frac{1}{8}\dot{u} \cdot \dot{u} \geq 0, p_3 = \frac{1}{2}\rho - Du \cdot Du$$

while the coefficient q in (IIIa) reads

$$q \equiv e^{-2\lambda}|h|^2 + \frac{1}{2}e^{2\lambda}\tau^2 + \frac{1}{8}e^{-2\lambda}|\dot{u}|^2 \geq 0.$$

Considering τ and σ_t as given on $\Sigma \times I$ we have the following theorem

Theorem 3. *When $u \equiv (\omega, \gamma)$ is known on $\Sigma \times I$ together with its conjugate variable, the system (II) is for each t an elliptic system for λ and h.*

2) When $\dot{u}, \lambda$ and h are known (IIIa) is, for each t, a linear elliptic equation for N. When N is also known (IIIb) is a linear elliptic system for ν.

3) When $\tilde{g}$, determined by λ, N, ν, is known on $\Sigma \times I$ the equations (I) are a harmonic map equation $(\Sigma \times I, \tilde{g}) \to \mathcal{H}$.

Corollary. *When τ is spatially constant the system (II) splits into a linear system for h, independent of λ, and a non linear elliptic equation for λ, dependent on h.*

The theorem 3 leads naturally to an iteration scheme which will in principle determine all unknowns from the Cauchy data for u.

The Cauchy data on Σ_0 are four scalar functions $(\omega_0, \gamma_0) = u_0$ and $(\omega_0, \gamma_0) = \dot{u}_0$ which will be the values induced on Σ_0 by $u = (\omega, \gamma)$ and its conjugate variable. These data are supposed to satisfy the integral equalities met as necessary conditions in previous paragraphs

1) If n is the Chern number of the bundle $M \to \Sigma$

$$\frac{1}{2\pi} \int_{\Sigma_0} \bar{F} = n \tag{i.1}$$

$\bar{F}$, the 2-form induced on $\Sigma \times \{0\}$ by F, depends indeed only on the initial data: Let $x^a, a = 1,2$ be local coordinates in Σ and (x^a, t) coordinates in $\Sigma \times I$. Then

$$\bar{F} \equiv \frac{1}{2} \bar{F}_{ab} dx^a \wedge dx^b,$$

$$\bar{F}_{ab} \equiv \{e^{-4\gamma} \tilde{\eta}_{ab0} \tilde{g}^{0\mu} (\partial_\mu \omega + H_\mu)\}_{t=0} \equiv e^{-4\gamma_0} (\det \sigma_0)^{\frac{1}{2}} \dot{\omega}_0$$

by the definitions of $\tilde{g}$ and H.

2) For each conformal Killing vector field Z of (Σ, σ_0)

$$\int_{\Sigma_0} L_a Z^a d\mu(\sigma_0) = 0. \tag{i.2}$$

3) For each transverse traceless symmetric 2-tensor T

$$\int_{\Sigma_0} f_{ab} T^{ab} d\mu(\sigma_0) = 0. \tag{i.3}$$

9. Solution of the original equations

Supposing we have constructed a metric $\underline{g}$ which satisfies the mixed elliptic hyperbolic system (I), (II), (III) we will have to show that it satisfies the original vacuum Einstein equations on $M \times I$, i.e. that the equations

$$\underline{R}_{A3} = 0, A = \alpha, 3; \quad \underline{R}_{0\alpha} = 0, \alpha = a, 0; \quad \underline{S}_{00} = 0 \tag{9.1}$$

imply

$$\underline{R}_{ab} = 0, \quad a, b = 1, 2. \tag{9.2}$$

The equations (9.1) imply $\underline{R} = 0$ and, with the Bianchi identities

$$\underline{\nabla}_A \underline{R}^{AB} \equiv 0.$$

A straightforward computation shows that together with the previous equations, they imply on each Σ_t

$$\bar{g}_{ab}\underline{R}^{ab} = 0 \qquad \bar{\nabla}_a(\underline{R}^{ab}\mathbf{e}^{-2\gamma}N) = 0; \tag{9.3}$$

i.e., that $\underline{R}^{ab}e^{-2\gamma}N$ is transverse and traceless in the metric $\bar{g}_t$, hence $\underline{R}^{ab}$ is zero if there are no such tensor fields on Σ, and therefore that Σ is diffeomorphic to the sphere. In the general case, the equation (9.2) is a consequence of (9.1) if and only if $Ne^{-2\gamma}\underline{R}^{ab}$ is L^2 orthogonal to transverse traceless symmetric 2-tensors on $(\Sigma, \bar{g}_t)$, that is if $Ne^{-2\gamma}e^{-4\lambda}\underline{R}^{ab}$ is L^2 orthogonal to traceless transverse 2-tensors T on (Σ, σ_t):

$$\int_{\Sigma_t} Ne^{-2\gamma}e^{-4\lambda}N\underline{R}^{ab}T_{ab}d\mu(\sigma) = 0 \tag{i.4}$$

10. Case Σ Diffeomorphic to S^2. Iteration Scheme

A manifold $\Sigma \sim$ (diffeomorphic to) S^2 (of genus $G = 0$) admits no transverse traceless 2-tensors, except zero. A metric $\underline{g}$ satisfying (I), (II), (III) thus satisfies the original Einstein equations. Also the integral condition (i3) is empty.

If $\Sigma \sim S^2$ the Betti number b_1 is zero and there is no harmonic 1-form, except zero.

On $\Sigma \sim S^2$ any Riemannian metric $\bar{g}$ is conformally equivalent to the canonical metric s: it is no restriction to set

$$\bar{g}_t = e^{2\lambda_t}\varphi_t^* s,$$

with φ_t^* an arbitrary diffeomorphism $\Sigma_t \to \Sigma$. The choice of the family of diffeomorphisms φ_t is essentially equivalent to the choice of shift ν. Following Moncrief [1] we choose for φ_t the identity map. The metric $\sigma_t = s$ is then independent of t, as well as the operators D and Δ. The equation (IIIb) reduces to

$$(\mathcal{L}_s\nu)_{ab} = 2Ne^{-2\lambda}h_{ab} \tag{$IIIb'$}$$

The condition (i.2) is not empty: S^2 admits 6 linearly independent conformal Killing vectors Z.

We define as usual the Sobolev spaces H_s on the Riemannian manifold S^2.

Hypothesis on the Cauchy data $u_0, \dot{u}_0$

1) $u_0 \in H_s, \dot{u}_0 \in H_{s-1}, s \geq 4$.
2) They satisfy the integral equalities (i1), (i2).

To simplify, we suppose that τ is spatially constant and set

$$\tau = C_0(1+t), \qquad C_0 > 0. \tag{10.1}$$

We shall determine $\tilde{g}_0$, that is $\lambda_0, N_0, \nu_0 \in H_s$ by the elliptic system (II), (III) on S^2 and thus obtain the usual Cauchy data for u:

$$(u)_{t=0} = u_0, (\partial_t u)_{t=0} \equiv N_0 e^{-2\lambda_0}\dot{u}_0 + \nu_0^a \partial_a u_0.$$

The iteration scheme is the following: from $\tilde{g}_n$ and u_0, $\dot{u}_0$ we deduce a harmonic map equation for u_{n+1} on $\Sigma \times I$ and its Cauchy data on Σ_0. From u_{n+1} we deduce $\lambda_{n+1}, N_{n+1}, \nu_{n+1}$ on each Σ_t which gives $\tilde{g}_{n+1}$ on $\Sigma \times I$; for this determination, we have to use a device modifying (IIIb) because the integral equality (i2) is not in general satisfied on Σ_t at each step. After showing the convergence of the iteration, for I small enough, we prove that the obtained pair $(\tilde{g}, u)$ satisfies the original equations.

11. Solution of the elliptic system when u and $\dot{u}$ are known on $\Sigma \times I$

We suppose given $u \in C^0(I, H_3) \cap C^1(I, H_2)$, $\dot{u} \in C^0(I, H_2) \cap C^1(I, H_1)$, $I \equiv [0, T]$. We denote by M a bound of the indicated norms.

We replace (IIa) on each Σ_t by

$$D_b h_a^b = f_a \equiv L_a + \sum_{A=1}^{6} s_{ab} c_A(t) Z^{(A)b}, \quad h_a^a = 0 \tag{IIa'}$$

where the functions $c_A(t)$ are chosen so that f_a is orthogonal to the conformal Killing vector fields of S^2:

$$\int_{S^2} f_a Z^{(A)a} d\mu(s) = 0, \qquad A = 1, \dots, 6.$$

We have $L_a \equiv -\frac{1}{2} D_a u \cdot \dot{u} \in C^0(I, H_2) \cap C^1(I, H_1)$. Elliptic theory together with the non-existence of transverse traceless tensors on S^2 shows that there exists $h \in C^0(I, H_3) \cap C^1(I, H_2)$, unique, satisfying (IIa) and

$$\|h\|_{C^0(I,H_3)\cap C^1(I,H_2)} \le C(M) \tag{11.1}$$

with C a given continuous function on $\mathbf{R}$.

(IIb) is now a non-linear elliptic equation for λ on S^2

$$\Delta\lambda = F(x, \lambda) \equiv p_1 e^{2\lambda} - p_2 e^{-2\lambda} + p_3 \tag{IIb}$$

$$p_1 = \frac{1}{4}\tau^2 > 0,\ p_2 = \frac{1}{2}|h|^2 + \frac{1}{8}\dot{u}\cdot\dot{u} \ge 0, \quad p_3 = 1 - Du \cdot Du,$$
$$p_1, p_2, p_3 \in C^0(I, H_2) \cap C^1(I, H_1).$$

We suppose that

$$\int_{\Sigma_t} p_2 d\mu(s) \geq K > 0. \tag{11.2}$$

A sufficient condition for (11.2) to be satisfied is

$$\int_{\Sigma_t} \dot{u} \cdot \dot{u} d\mu(s) \geq K > 0. \tag{11.3}$$

One uses the monotonicity of F in λ to show that, if (11.2) is satisfied, (IIb) admits one solution $\lambda \in C^0(I, H_4) \cap C^1(I, H_3)$ such that

$$\|\lambda\|_{C^0(I,H_4)\cap C^1(I,H_3)} \leq C(M, K, C_0). \tag{11.4}$$

The equation (IIIa) reads now as a linear elliptic equation for N

$$\Delta N - qN = -e^{-2\lambda} C_0,$$
$$q \equiv e^{-2\lambda}|h|^2 + \frac{1}{2}e^{2\lambda}\tau^2 + \frac{1}{8}e^{-2\lambda}|\dot{u}|^2 > 0, q \in C^0(I, H_2) \cap C^1(I, H_1).$$

It has one solution $N \in C^0(I, H_4) \cap C^1(I, H_3)$ which is strictly positive:

$$\|N\|_{C^0(I,H_4)\cap C^1(I,H_3)} \leq C(M, K, C_0), \quad N \geq N_0 > 0. \tag{11.5}$$

The equation (IIIb) reads finally as the linear system
(IIIb)
$$\mathcal{L}_s\nu = 2Ne^{-2\lambda}h \in C^0(I, H_3) \cap C^1(I, H_2), \quad (\mathcal{L}_s\nu)_{ab} \equiv D_a\nu_b + D_b\nu_a - s_{ab}D_c\nu^c.$$

It has a solution $\nu \in C^0(I, H_4) \cap C^1(I, H_3)$ determined up to the addition of a conformal Killing field of S^2. We determine it uniquely by choosing it to be L^2 orthogonal to these Killing fields. We have

$$\|\nu\|_{C^0(I,H_4)\cap C^1(I,H_3)} \leq C(M, K, C_0), \quad N \geq N_0 > 0. \tag{11.6}$$

12. Solution of the harmonic map equation

Let $\lambda, N, \nu \in C^0(I, H_4) \cap C^1(I, H_3)$ be known satisfying the inequalities (11.4), (11.5) and (11.6). Set

$$\tilde{g} = -N^2dt^2 + e^{2\lambda}s_{ab}(dx^a + \nu^a dt)(dx^b + \nu^b dt).$$

The map $u : (\Sigma \times I, \tilde{g}) \to \mathcal{H}$ satisfies the wave map equation

$$\tilde{g}^{\alpha\beta}(\tilde{\nabla}_\alpha \partial_\beta u^A + \Gamma^A_{BC}(u)\partial_\alpha u^B \partial_\beta u^C) = 0. \tag{I}$$

Theorem 4. *Under the hypotheses made on* λ, N, ν *there exists* T *small enough such that (I) has one and only one solution* $u \in C^0(I, H_3) \cap C^1(I, H_2)$, *such that* $\dot{u} \equiv e^{2\lambda} N^{-1}(\partial_t u + \nu^a \partial_a u) \in C^0(I, H_2) \cap C^1(I, H_1)$, *and taking for* $t = 0$ *the Cauchy data* $u_0 \in H_4, \dot{u}_0 \in H_3$. *The solution* u *is in* $L^\infty(I, H_4)$, $\dot{u}$ *in* $L^\infty(I, H_3)$ *and it satisfies (11.3) if the Cauchy data are such that*

$$\int_\Sigma \dot{u}_0 \cdot \dot{u}_0 d\mu(s) = K_0 > K.$$

Proof. It follows the standard lines of energy estimates for iterated linear equations and convergence of the solutions of the iterates. These estimates can be obtained directly in terms of u and $\dot{u}$, and the bounds on λ, N and ν by writing the equation

$$\tilde{\nabla}^\alpha \partial_\alpha u = v$$

in the form

$$N s^{ab} D_a \partial_b u - (\partial_t - \nu^a \partial_a)\dot{u} + (D_a \nu^a)\dot{u} + s^{ab} \partial_a N \partial_b u = N e^{2\lambda} v. \tag{12.1}$$

One obtains the pth energy estimates by differentiating (12.1) p times in space directions, taking its scalar product in s with the ∂_0 derivative of the pth space derivative of u, and integrating on $\Sigma \times [0, t]$.

13. Solution of the coupled system and original Einstein equations

We can show that one can choose T small enough for the sequence $u_n \equiv (\omega_n, \gamma_n)$, $h_n, \lambda_n, N_n, \nu_n$ to converge to a solution u, h, λ, N, ν of (I), (IIa′) (IIb), (III) on $\Sigma \times [0, T]$. We show that it satisfies (IIa) by proving that if u is a wave map for $\tilde{g}$, hence has a stress energy tensor with vanishing divergence in $\tilde{g}$, the integrability condition (1.2) is satisfied for all t if it is satisfied for $t = 0$.

The condition (i.1) is also satisfied for all t, hence we can construct $\underline{g}$. This $\underline{g}$ satisfies the original Einstein equations because on S^2 there are no transverse traceless symmetric 2-tensors.

14. Manifolds of genus greater than 1

If the 2-dimensional oriented manifold Σ has a genus $G \geq 1$, the space of classes of conformally equivalent Riemannian metrics on Σ with diffeomorphisms φ homotopic to the identity is called the Teichmüller space $\mathcal{C}$ of Σ. The vector space of traceless transverse symmetric 2-tensors in a metric σ on Σ is isomorphic to the tangent space to $\mathcal{C}$ at the equivalence class of σ and it has a finite dimension N.

In the case $G = 1$ (Σ diffeomorphic to the torus T^2), every metric is conformal to a flat metric ($R = 0$) but there is a 2 parameter family of conformally inequivalent flat metrics: the Teichmüller space is $\mathbf{R}^2$, $N = 2$. In the case $G > 1$, the Teichmüller space $\mathcal{C}$ can be identified with $\mathcal{M}_{-1}/\mathcal{D}_0$, the quotient of the space of metrics with scalar curvature -1 by the group of diffeomorphisms homotopic to the identity (cf.

[8]), $\mathcal{M}_{-1} \to \mathcal{C}$ is a trivial principal fiber bundle (cf. [9,8]) whose base $\mathcal{C}$ can be endowed with the structure of the manifold $\mathbf{R}^N, N = 6G - 6$.

We choose some cross section $\mathcal{C} \to \mathcal{M}_{-1}, q \mapsto \sigma(q), q = (q^I, I = 1, \dots, N) \in \mathbf{R}^N$. Since $\partial\sigma/\partial q^I$ is in the tangent space to $\mathcal{M}_{-1}$ at $\sigma(q)$ we have

$$\frac{\partial \sigma}{\partial q^I} = \chi_I(q) + \mathcal{L}_{X_I(q)}\sigma(q)$$

where the $\chi_I(q), I = 1, \dots, N$ are symmetric 2-tensors, transverse and traceless with respect to $\sigma(q)$, which span the vector space of such tensors and where the $X_I(q)$ are vector fields on Σ.

We impose upon the metric σ_t, that we used to solve the constraints on Σ_t, that it lie in the chosen cross section, namely we set

$$\sigma_t \equiv (\sigma \circ q)(t).$$

The momentum constraint (III) is always solvable on a manifold of genus $G > 1$ since such a manifold does not admit conformal Killing vectors. Its solution is determined up to the addition of a transverse traceless tensor, i.e. reads

$$h_t = \underline{h}_t + \sum_I p^I(t)(\chi_I \circ q)(t)$$

where the $\underline{h}_t$ is a particular solution determined uniquely for example by being in the range of the conformal Killing operator. The integrability conditions (i.3), necessary and sufficient for the solvability of the shift equation (IIIb) are

$$\int_{\Sigma_t} f_{ab}\chi_I^{ab} d\mu_\sigma = 0, \quad i = 1, \dots, N$$

that is, using the expressions of $f, h, \partial\sigma/\partial q^I$ and

$$\frac{\partial \sigma}{\partial t} = \sum_I \frac{\partial q^I}{dt}\frac{\partial \sigma}{\partial q^I}$$

$$\int_{\Sigma_t} \{2Ne^{-2\lambda}(\underline{h}^{ab}\chi_{J,ab} + \sum_I p^I\chi_I^{ab}\chi_{J,ab}) + \sum_I \frac{dq^I}{dt}\chi_I^{ab}\chi_{J,ab}\} d\mu_\sigma = 0.$$

This equation determines the p^I on Σ_t as affine functions of the dq^I/dt, with coefficients depending on t.

We consider now the condition (i.4) of paragraph 9, necessary and sufficient (with (i.1)) for a solution of the reduced equations to satisfy the original equations. Looking at the formulas (5.1) and (6.1), using the vanishing of the trace of $\underline{R}_{ab}$ and the result we have just obtained, we see that equations (i.4) are a system of second order ordinary differential equations for the q^I linear in the second derivatives, whose principal term comes from $\partial_t h_{ab}$. These equations can be written

$$\frac{d^2}{dt^2}q^I = \Phi^I(t, q^I, \frac{dq^I}{dt}),$$

with Φ a C^1 function of t, smooth in q^I, a quadratic polynomial in dq^I/dt. This equation admits a C^1 solution, for t small, taking given Cauchy data

$$q^I(0) = q_0^I, \qquad (dq^I/dt)(0) = q_0'^I.$$

REFERENCES

1. V. Moncrief, *Reduction of Einstein's equations for vacuum spacetimes with $U(1)$ spacelike isometry groups*, Annals of Physics **167** (1986), 118-142.
2. ———, *Reduction of the Einstein-Maxwell and Einstein-Maxwell-Higgs equations for cosmological space times with spacelike $U(1)$ isometry groups*, Class. Quantum Grav. **7** (1990), 329-352.
3. J. Cameron and V. Moncrief, *The reduction of Einstein's vacuum equations on space times with $U(1)$ spacelike isometry groups*, Contemporary Mathematics **132** (1992), 143-169.
4. J. Cameron, *The reduction of vacuum Einstein equations with space like $U(1)$ isometry groups*, Doctoral Thesis, Yale University 1991.
5. Y. Choquet-Bruhat and C. DeWitt-Morette, *Analysis manifolds and Physics*, Part I: Basics, North Holland 1982, Part II: 92 Applications, North Holland 1989..
6. Y. Choquet-Bruhat and J. York, *The Cauchy Problem*, in General Relativity and Gravitation, A. Held ed. Plenum 1980.
7. Y. Choquet-Bruhat and V. Moncrief, *An existence theorem for the reduced Einstein equations*, C.R. Acad. Sci., t. **319**, Série I (1994), 153-159.
8. A. Fisher and A. Tromba, Math. Ann. **267** (1984), 311.
9. C. Earle and J. Eells, J. Diff. Geom. **3** (1969), 19.
10. V. Moncrief, *Reduction of Einstein equations in $2+1$ dimensions to a Hamiltonian system over Teichmuller space*, J. Math. Phys. **30** (1989), 2907-2914.

Gravitation et Cosmologie Relativistes, Université Paris VI, Paris, France
E-mail address: choquet@cicrp.jussieu.fr

Department of Mathematics, Yale University, New Haven, Connecticut 06520
E-mail address: moncrief@yalph2.physics.yale.edu

Proceedings of Symposia in Pure Mathematics
Volume **59**, 1996

Quantum Stochastic Calculus, Evolutions and Flows

R. L. Hudson

ABSTRACT. After a brief survey of quantum stochastic calculus from the point of view of quantum physics, its use in the description of noisy evolutions and flows, generalising the Schrödinger and Heisenberg pictures of quantum dynamics, is reviewed. The use of cohomology theories based on *-algebras to characterize the generators and perturbation theory of such flows is also described.

1. Introduction

The quantum stochastic calculus described in this paper may be regarded as a natural development of the Wiener–Segal duality transformation [28, 31] between the L^2-space of Brownian motion and a Fock space. Using the mathematical technology of second quantisation, the Ito stochastic calculus of Brownian motion is generalised to allow stochastic integration of operator-valued adapted processes against integrators which include separate creation and annihilation processes, whose sum corresponds to multiplication by Brownian motion under the duality transformation. Such 'quantum noises' were introduced into the physical literature by Senitzky [29]; a rigorous form of the corresponding stochastic integration theory including an Ito product formula was given by Hudson and Parthasarathy [16], where also the possibility was shown of incorporating into a unified theory the stochastic calculus of the Poisson process, based on use of the number operator in Fock space. Surprisingly, Fermionic noise can also be described in the same (Bosonic) Fock space set-up [17].

The uses of quantum stochastic calculus in physics include construction of dilations of quantum dynamical semigroups [16, 18], models of interaction of matter and radiation [2, 30], of the quantum Hall effect [19, 27], continual measurement theory [3, 5] and more recently quantum field theory, in which a scaling limit may be interpreted as a quantum functional central limit [1]. It should be noted that in most of these applications, the physical Hilbert space is not the Fock space, whose rôle is to carry the noise, but a separate 'initial' Hilbert space.

MR Classification 81S25.

In this paper after a brief survey of quantum stochastic calculus (§2), we shall describe its use in the construction of noisy Schrödinger and Heisenberg dynamics in §§3, 4. In §5 we describe the use of Hochschild cohomology of algebras, together with a non-linear extension due to Lue [23] in characterising the infinitesimal generators of such noisy dynamics and in describing their perturbations. We also describe briefly a connection with cyclic cohomology [8].

2. **Quantum stochastic calculus** [16, 26, 27]

Let $\mathcal{K}$ be a finite dimensional Hilbert space which we may think of as the ambient space of a multi-dimensional Brownian motion, while emphasising that, in common with all Hilbert spaces in this paper, it is complex. We introduce the Hilbert space

$$h = L^2(\mathbb{R}_+) \otimes \mathcal{K} = L^2(\mathbb{R}_+, \mathcal{K})$$

of vector-valued functions on $\mathbb{R}_+ = [0, \infty)$. Points $t \in \mathbb{R}_+$ label time. The Fock space

$$\mathcal{H} = \Gamma(h)$$

over h is most conveniently described axiomatically; it is a Hilbert space generated through linearity and closure by a family of *exponential vectors* $e(f)$, $f \in h$ having the property

$$\langle e(f), e(g) \rangle = \exp\langle f, g \rangle, \quad f, g \in h$$

where the inner product on the left is in $\Gamma(h)$ and that on the right in h. The Fock vacuum is $\Omega = e(0)$. In terms of the usual multi-particle space structure of Fock space given by

$$\mathcal{H} = \bigoplus_{n=0}^{\infty} \mathcal{H}_n, \qquad \mathcal{H}_n = \bigotimes{}^n_{\text{sym}}\, h,$$

$e(f)$ is the vector

$$e(f) = (1, f, (2!)^{-\frac{1}{2}} f \otimes f, (3!)^{-\frac{1}{2}} f \otimes f, (3!)^{-\frac{1}{2}} f \otimes f \otimes f, \ldots),$$

however we shall make no use of this structure beyond noting that the usual creation and annihilation operators can be extended to include exponential vectors in their domain, on which they act as

$$a^\dagger(f)e(g) = \left.\frac{d}{d\varepsilon} e(g + \varepsilon f)\right|_{\varepsilon=0}, \qquad a(f)e(g) = \langle f, g \rangle e(g).$$

We denote by $\mathcal{E}$ the dense subspace spanned by exponential vectors $e(f)$ for which $f \in L^2(\mathbb{R}_+)$ is locally bounded.

Finally, let there be given an *initial space*, a Hilbert space $\mathcal{H}_0$ which we think of as carrying the physics of some system of interest, and let

$$\tilde{\mathcal{H}} = \mathcal{H}_0 \otimes \mathcal{H}.$$

For each $t \in \mathbb{R}_+$ we form the natural direct sum decomposition

$$L^2(\mathbb{R}_+) = h = h_t \oplus h^t = L^2[0, t] \oplus L^2(t, \infty)$$

and the associated Hilbert space tensor product decomposition

$$\mathcal{H} = \mathcal{H}_t \otimes \mathcal{H}^t$$

in which

$$e(f) = e(f_t) \otimes e(f^t) \quad \text{for} \quad f = (f_t, f^t) \in h.$$

Then we can write

$$\tilde{\mathcal{H}} = \tilde{\mathcal{H}}_t \otimes \mathcal{H}^t \tag{1}$$

where $\tilde{\mathcal{H}}_t = \mathcal{H}_0 \otimes \mathcal{H}_t$, and correspondingly

$$\tilde{\mathcal{E}} = \tilde{\mathcal{E}}_t \hat{\otimes} \mathcal{E}^t \tag{2}$$

where

$$\tilde{\mathcal{E}} = \mathcal{H}_0 \hat{\otimes} \mathcal{E}, \qquad \tilde{\mathcal{E}}_t = \mathcal{H}_0 \hat{\otimes} \mathcal{E}_t,$$

$\mathcal{E}_t, \mathcal{E}^t$ are the corresponding domains in $\mathcal{H}_t$ and $\mathcal{H}^t$, and $\hat{\otimes}$ denotes the algebraic tensor product.

An *adapted process* is a family $E = (E(t),\ t \in \mathbb{R}_+)$ of operators in $\tilde{\mathcal{H}}$ defined, together with their adjoints, on the domain $\tilde{\mathcal{E}}$, with the property that each operator $E(t)$, together with the restriction $E^\dagger(t)$ of its adjoint to $\mathcal{E}$, factorises as

$$E(t) = E_t \otimes 1^t, \qquad E^\dagger(t) = E_t^\dagger \otimes 1^t$$

according to the decompositions (1), (2). In particular given vectors u and $v \in K$ and a bounded operator T on K *creation*, *annihilation* and *number* processes $A_u^\dagger$, A_v and Λ_T of strengths u, v and T respectively are defined by their actions

$$A_u^\dagger(t)\,\psi \otimes e(f) = \frac{d}{d\varepsilon}\psi \otimes e(f + \varepsilon \chi_{[0,t]}u)\Big|_{\varepsilon=0},$$

$$A_v(t)\,\psi \otimes e(f) = \int_0^t \langle f(s), v\rangle\, \psi \otimes e(f),$$

$$\Lambda_T(t)\,\psi \otimes e(f) = \frac{d}{d\varepsilon}\psi \otimes e((M_{\chi_{[0,t]}} \otimes \exp \varepsilon T + M_{\chi_{(t,\infty)}} \otimes 1_{\mathcal{K}})f)\Big|_{\varepsilon=0}.$$

Here M_{χ_S} denotes the indicator function of $S \subseteq \mathbb{R}^+$.

Stochastic integrals of the form

$$M(t) = \int_0^t (E(s)\,d\Lambda_T(s) + F(s)\,dA_u^\dagger(s) + G(s)\,dA_v(s) + H(s)\,ds)$$

can be defined, initially for adapted processes E, F, G, H which are step functions of t in the natural way, noting that the products of unbounded operators which arise, such as $E(t_1)(\Lambda_T(t) - \Lambda_T(t_1))$, are meaningful as algebraic tensor product operators of form $E_{t_1} \hat{\otimes} \Lambda^{t_1, t}$ on the domain $\tilde{\mathcal{E}} = \tilde{\mathcal{E}}_{t_1} \hat{\otimes} \mathcal{E}^{t_1}$. They satisfy the two *fundamental formulas*, deduced from the definitions of the basic processes and from the commutation relations between them, namely

$$(3)\qquad \begin{aligned}&\langle \psi\otimes e(f), M(t)\phi\otimes e(g)\rangle\\ &= \int_0^t \langle \psi\otimes e(f), (\langle f(s), Tg(s)\rangle E(s) + \langle f(s), u\rangle F(s)\\ &\qquad + \langle v, g(s)\rangle G(s) + H(s))\phi\otimes e(g)\rangle\, ds\end{aligned}$$

for arbitrary $t\in\mathbb{R}_+$, $\psi, \phi\in\mathcal{H}_0$ and $f, g\in h$, and, if also

$$M'(t) = \int_0^t (E'(s)d\Lambda_{T'}(s) + F'(s)dA_{u'}^\dagger(s) + G'(s)dA_{v'}(s) + H'(s)ds),$$

$$(4)\qquad \begin{aligned}&\langle M(t)\psi\otimes e(f), M'(t)\phi\otimes e(g)\rangle\\ &= \int_0^t [\langle M(s)\psi\otimes e(f), (\langle f(s), T_g'(s)\rangle E'(s) + \langle f(s), u'\rangle F'(s)\\ &\qquad + \langle v', g(s)\rangle G'(s) + H'(s))\phi\otimes e(g)\rangle\\ &\quad + \langle(\langle g(s), Tf(s)\rangle E(s) + \langle g(s), u\rangle F(s)\\ &\qquad + \langle v, f(s)\rangle G(s) + H(s))\psi\otimes e(f), \phi\otimes e(g)\rangle\\ &\quad + \langle Tf(s), T'g(s)\rangle\langle E(s)\psi\otimes e(f), E'(s)\phi\otimes e(g)\rangle\\ &\qquad + \langle Tf(s), u'\rangle\langle E(s)\psi\otimes e(f), F'(s)\phi\otimes e(g)\rangle\\ &\qquad + \langle u, T'g(s)\rangle\langle F(s)\psi\otimes e(f), E'(s)\phi\otimes e(g)\rangle\\ &\qquad + \langle u, u'\rangle\langle F(s)\psi\otimes e(f), F'(s)\phi\otimes e(g)\rangle]\ ds.\end{aligned}$$

Equating the two sides of the inner product (4) we obtain

$$\begin{aligned}&\|M(t)\psi\otimes e(f)\|^2\\ &= \int_0^t [2\,\mathrm{Re}\langle M(s)\psi\otimes e(f), (f(s), Tf(s)\rangle E(s) + \langle f(s), u\rangle F(s)\\ &\qquad + \langle v, g(s)\rangle G(s) + H(s))\psi\otimes e(f)\rangle\\ &\quad + \|Tf(s)\otimes E(s)\psi\otimes e(f) + u\otimes F(s)\psi\otimes e(f)\|^2]\ ds\end{aligned}$$

and hence, after some manipulations, the estimate

$$(5)\qquad \|M(t)\psi\otimes e(f)\|^2 \leqslant \alpha \max_K\{\int_0^t \|K(s)\psi\otimes e(f)\|^2 ds;\ K = E, F, G, H\}$$

holds locally uniformly for $t\in[0, t_0]$, $t_0 \geqslant 0$, where α depends on u, v, T and t_0, and on f only through $\sup_{t\in[0,t_0]} \|f(t)\|_{\mathcal{K}}$.

The *fundamental estimate* (5) permits extension of the stochastic integral from step functions of t to natural classes of integrands K which are measurable operator-valued processes E satisfying the local square-integrability requirements

$$\int_0^t \|E(s)\psi\otimes e(f)\|^2 ds, \qquad \int_0^t \|E(s)^\dagger\psi\otimes e(f)\|^2 ds < \infty.$$

The extended integral continues to satisfy the fundamental formulas and estimate and the stochastic integral process is itself integrable, allowing solution of stochastic differential equations by the iterative procedure.

In the case when all integrands and integrals comprise bounded operators, so that multiplication is possible without domain complications, the rather formidable second fundamental formula amounts to the *quantum Ito product formula* for the stochastic differential of a product

$$d(M_1 M_2) = (dM_1)M_2 + M_1 dM_2 + dM_1 dM_2. \tag{6}$$

This is reduced to a linear combination, with adapted process coefficients, of the basic differentials by the rule that the latter commute with the former, and the 'Ito correction' $dM_1 dM_2$ is evaluated by bilinear extension of the multiplication table

(7)

	$d\Lambda_T$	$dA_v^\dagger$	dA_v	dt
$d\Lambda_S$	$d\Lambda_{ST}$	$dA_{Sv}^\dagger$	0	0
$dA_u^\dagger$	0	0	0	0
dA_u	dA_{T^*u}	$\langle u, v\rangle\, dt$	0	0
dt	0	0	0	0

In particular in the case $\mathcal{K} = \mathbb{C}$ the one dimensional number, creation and annihilation processes $\Lambda = \Lambda_{id}$, $A^\dagger = A_1^\dagger$, $A = A_1$ satisfy

(8)

	$d\Lambda$	$dA^\dagger$	dA	dt
$d\Lambda$	$d\Lambda$	$dA^\dagger$	0	0
$dA^\dagger$	0	0	0	0
dA	dA	dt	0	0
dt	0	0	0	0

The Ito stochastic calculus of Brownian motion can then be included in the quantum theory by using the Wiener–Segal identification of the Fock space $\Gamma(L^2(\mathbb{R}_+))$ with the L^2-space of Wiener measure which simultaneously diagonalises the 'Brownian motion'

$$Q(t) = A(t) + A^\dagger(t), \quad t \in \mathbb{R}_+. \tag{9}$$

But one could equally well use the commuting family of essentially self-adjoint operators

$$P(t) = i(A^\dagger(t) - A(t)), \quad t \in \mathbb{R}_+. \tag{10}$$

The 'momentum process' P of course does not commute with the 'position process' Q. Poisson processes are similarly included in the theory by diagonalising the family of mutually commutative self-adjoint operators

$$\Lambda_z(t) = \Lambda(t) - zA^\dagger(t) - \bar{z}A(t) + |z|^2 t, \quad t \in \mathbb{R}_+, \tag{11}$$

which, for fixed $z \in \mathbb{C}$, realise the Poisson process of intensity measure $|z|^2 dt$ in the vacuum state.

More remarkably from a physics perspective, the stochastic differential formula [17]

$$dB = (-1)^\Lambda dA$$

generates a realisation of the Fock representation of the canonical anticommutation relations in the (Boson!) Fock space $\Gamma(h)$, and leads to the $\mathbb{Z}_2$-graded Fermionic version of stochastic calculus initially studied independently by Barnett, Streater and Wilde [4].

Accounts of quantum stochastic calculus in monograph form are now available [26]; for a more probabilistic perspective see [25].

3. Stochastic Schrödinger equation

We consider first the one-dimensional case $K = \mathbb{C}$. The stochastic differential equation

$$dU = U(L_1 d\Lambda + L_2 dA^\dagger + L_3 dA + L_4 dt), \quad U(0) = 1, \tag{12}$$

where L_1, L_2, L_3, L_4 are ampliations to $\tilde{\mathcal{H}}$ of bounded operators on the initial space, possesses a unique solution as an adapted process. Necessary conditions for this to be unitary are found by applying the Ito product rule (8) to the products $UU^* = U^*U = 1$. Thus from $U^*U = 1$ we obtain

$$\begin{aligned} 0 &= (dU^*)U + U^*dU + dU^*dU \\ &= L_1^* + L_1 + L_1^* L_1) d\Lambda + (L_3^* + L_2 + L_1^* L_2) dA \\ &\quad + (L_2^* + L_3 + L_2^* L_1) dA^\dagger + (L_4 + L_4^* + L_2^* L_3) dt. \end{aligned}$$

Using the independence of the stochastic differentials we conclude that

$$L_1^* L_1 + L_1 + L_1 = L_3^* + L_2 + L_1^* L_2 = L_2^* + L_3 + L_2^* L_1 = L_4 + L_4^* + L_2^* L_3 = 0. \tag{13}$$

There relations, together with similar ones derived from $UU^* = 1$ are satisfied if and only if

$$(L_1, L_2, L_3, L_4) = (W - 1, L, -L^*W, iH - \tfrac{1}{2}L^*L), \tag{14}$$

where W is unitary, H self-adjoint and L arbitrary on $\mathcal{H}_0$. It turns out [16] that (14) is sufficient, as well as necessary, for the unitarity of the solution of (12).

The vacuum conditional expectation $T(t)$ of such a unitary process, defined by

$$\langle \psi, T(t)\phi \rangle = \langle \psi \otimes \Omega, U(t)\phi \otimes \Omega \rangle, \tag{15}$$

is a uniformly continuous contraction semigroup in $\mathcal{H}_0$. Unitary processes of this type are characterised by the following three properties [15].

(a) $t \mapsto U(t)$ is a unitary-valued adapted process and, for $t < s$, $U(t)^*U(s)$ is the ampliation of an operator on $\mathcal{H}_0 \otimes \Gamma(L^2[s,t))$ to $\tilde{\mathcal{H}} = \mathcal{H}_0 \otimes \Gamma(L^2[0,s)) \otimes \Gamma(L^2[s,t)) \otimes \Gamma(L^2(t,\infty))$.

(b) $t \mapsto T(t)$ defined by (15) is uniformly continuous.

(c) (The cocycle condition.) For all $s, t \in \mathbb{R}_+$,

$$U(t) = \Gamma_s^\dagger U^\dagger(s) U(s+t) \Gamma_s,$$

where Γ_s is the second quantized shift, defined by

$$\Gamma_s(\psi \otimes e(f)) = \psi \otimes e(f_s), \qquad f_s(t) = \begin{cases} 0 & t < s, \\ f(t-s) & t \geqslant s. \end{cases}$$

Considerable effort has been expended in generalising this characterization to a 'stochastic Stone's theorem', in which the coefficients L_j in (12) would become unbounded, and correspondingly uniform would be replaced by strong continuity in (b). However the generalization is not straightforward [21].

We note that the condition (13) and the corresponding condition derived from $UU^* = 1$ can be expressed succinctly as

$$u^\dagger + u + u^\dagger u = u + u^\dagger + uu^\dagger = 0. \tag{16}$$

Here u is the element

$$u = L_1 \otimes d\Lambda + L_2 \otimes dA^\dagger + L_3 \otimes dA + L_4 \otimes dt \tag{17}$$

of the tensor product *-algebra $\mathcal{A} \otimes \mathcal{I}$, where $\mathcal{A}$ is the 'initial algebra' $B(\mathcal{H}_0)$ and $\mathcal{I}$ is the 'Ito algebra' spanned by $d\Lambda$, $dA^\dagger$, dA and dt with multiplication table (8). The same characterization (16) applies to generators of unitary processes based on other initial C^*-algebras and other Ito algebras, for example that with multiplication table (7).

4. Stochastic Heisenberg equation and quantum stochastic flows

Given an initial unital C^*-algebra $\mathcal{A}$ a *quantum stochastic flow* based on the Ito algebra $\mathcal{I}$ given by (8) is a family $J = (J_t \colon t \geqslant 0)$ of unital C^*-algebra morphisms $J_t \colon \mathcal{A} \to \mathcal{A} \otimes B(\mathcal{H})$ such that

(a) for each $x \in \mathcal{A}$, $t \mapsto J_t(x)$ is an integrable adapted process,

(b) there are maps λ, $\alpha^\dagger$, α and $\tau \colon \mathcal{A} \to \mathcal{A}$ such that, for each $\gamma \in \mathcal{A}$, $t \in \mathbb{R}_+$,

$$J_t(x) = x \otimes 1 + \int_0^t J_s(\lambda(x))\, d\Lambda + J_s(\alpha(x))\, dA^\dagger + J_s(\alpha^\dagger(x))\, dA + J_s(\tau(x))\, ds. \tag{18}$$

An example is obtained by defining

$$J_t(x) = U(t) x \otimes 1 U(t)^*$$

where $U(t)$ is the unitary process given by (12), (14); such a flow is called *inner*. A non-inner flow over $\mathcal{A} = B(\mathcal{H}_0)$ may be constructed (when $\dim \mathcal{H}_0 = \infty$) by taking a non-inner endomorphism σ of $\mathcal{H}_0$ and defining

$$J_t(x) = \sigma^{\Lambda(t)}(x). \tag{19}$$

Here the rhs is defined using the spectral decomposition

$$\Lambda(t) = \sum_{n=0}^{\infty} n E_n(t)$$

as

$$\sigma^{\Lambda(t)}(x) = \sum_{n=0}^{\infty} \sigma^n(x) \otimes 1 E_n(t).$$

The linearity, unitality, *-map property and multiplicativity of each J_t, together with independence of the stochastic differentials, imply that the *structure maps* $\lambda, \alpha^\dagger, \alpha, \tau$ of the flow J are linear, vanish on the unit element, and satisfy

$$\alpha^\dagger(x^*) = \alpha(x)^*, \qquad \lambda = \lambda^\dagger, \qquad \tau = \tau^\dagger \tag{20}$$

$$\lambda(xy) = \lambda(x)y + x\lambda(y) + \lambda(x)\lambda(y) \tag{21}$$

$$\alpha(xy) = \alpha(x)y + x\alpha(y) + \lambda(x)\alpha(y) \tag{22}$$

$$\alpha^\dagger(xy) = \alpha^\dagger(x)y + x\alpha^\dagger(y) + \alpha^\dagger(x)\lambda(y) \tag{23}$$

$$\tau(xy) = \tau(x)y + x\tau(y) + \alpha^\dagger(x)\alpha(y). \tag{24}$$

Conversely it can be shown [10] that given bounded maps satisfying these conditions, called *structure relations*, there is a unique flow J satisfying (18). There is also a cocycle characterization of flows [6].

The structure relation (24) may be regarded [9] as a characterization of the infinitesimal generator of a one-parameter semigroup of completely positive maps on the initial algebra. We obtain such a semigroup $\mathcal{T}$ by taking vacuum conditional expectations of the flow

$$\langle \psi, \mathcal{T}(t)(x)\phi \rangle = \langle \psi \otimes \Omega, J_t(x)\phi \otimes \Omega \rangle.$$

Thus quantum stochastic flows provide a method of dilating quantum dynamical semigroups [20]. When the cohomology of the initial algebra is trivial, the generator of the semigroup assumes the well-known Lindblad form [22]; then the dilating flow can be taken to be inner [18] by using a sufficiently large Ito algebra.

We may embody the structure maps and structure relations of a flow J over $\mathcal{A}$ with Ito algebra $\mathcal{I} = \langle d\Lambda, dA^\dagger, dA, dt \rangle$ in a single generating map $j\colon \mathcal{A} \to \mathcal{A} \otimes \mathcal{I}$ given by

$$j(x) = \lambda(x) \otimes \Lambda + \alpha(x) \otimes dA^\dagger + \alpha^\dagger(x) \otimes dA + \tau(x) \otimes dt$$

which is linear, vanishes on the unit element of $\mathcal{A}$, is a *-map from $\mathcal{A}$ to the *-algebra $\mathcal{A} \otimes \mathcal{I}$, and satisfies the relation

$$j(xy) = j(x)y + xj(y) + j(x)j(y). \tag{25}$$

Here we regard $\mathcal{A} \otimes \mathcal{I}$ as a two-sided $\mathcal{A}$-module, with actions determined by

$$x(y \otimes dN) = xy \otimes dN, \qquad (y \otimes dN)x = yx \otimes dN. \tag{26}$$

When J is inner, given by conjugation by the unitary process U with generator $u \in \mathcal{A} \otimes \mathcal{I}$, the generator j of J is given by

$$j(x) = ux + xu^\dagger + uxu^\dagger. \tag{27}$$

5. Cohomology of generators of flows

The structure relations of a quantum stochastic flow based on the Ito algebra $\mathcal{I} = \langle d\Lambda, dA^\dagger, dA, dt \rangle$ can be analysed in the language of Hochschild cohomology [12] as follows. Adding xy to both sides of (21) we see that $\sigma = \lambda + id$ is an endomorphism of $\mathcal{A}$, in terms of which (22) becomes

$$\alpha(xy) = \alpha(x)y + \sigma(x)\alpha(y). \tag{28}$$

If we regard $\mathcal{A}$ as an $\mathcal{A}$-bimodule, with right action by multiplication but left action $x \cdot y = \sigma(x)y$, α becomes a Hochschild in 1-cocycle; $\alpha \in Z_1^\sigma$. $\alpha^\dagger$ is determined from (20); (24) can be expressed as

$$b\tau = \eta_\alpha, \tag{29}$$

where b is the coboundary map for the Hochschild cohomology where again $\mathscr{A}$ is itself the bimodule but both actions are now by multiplication, and η_α is the 2-chain

$$\eta_\alpha(x, y) = -\alpha^\dagger(x)\alpha(y). \tag{30}$$

Thus this must be a 2-coboundary; $\eta_\alpha \in B_2$. It is not difficult to see that, with η_α defined by (30), $\alpha \in Z_1^\sigma$ implies $\eta_\alpha \in Z_2$ and $\alpha \in B_1^\sigma$ implies $\eta_\alpha \in B_2$. However when both the first order cohomology space $H_2^\sigma = Z_1^\sigma / B_1^\sigma$ and the second order space $H_2 = Z_2/B_2$ are non-trivial, there may exist no τ corresponding to given α and σ; an example occurs in the irrational rotation C^*-algebra [19].

A more incisive and general classification of generators of flows is obtained by using the 'non-Abelian' generalisation of Hochschild cohomology [23] in which the bi-module is itself equipped with an associative multiplication. In our case the bimodule is taken to be the *-algebra $\mathscr{A} \otimes \mathscr{I}$, with actions given by (26). Modifying Lue's notions to incorporate the involution on the bimodule, we define first the space C_0 of 0-chains to be the set of elements u of $\mathscr{A} \otimes \mathscr{I}$ satisfying (16), in other words to be the generator of unitary processes. These do not form a linear space, except in the case when the Ito algebra is trivial in which case everything reduces to Hochschild cohomology. However they do form a group, under the composition

$$u * v = u + v + uv. \tag{31}$$

Next, we introduce the space of 1-cocycles Z^1; these are linear *-maps $j: \mathscr{A} \to \mathscr{A} \otimes \mathscr{I}$ satisfying (25). Thus generators of flows are 1-cocycles. (27) tells us that generators of inner flows are coboundaries, the right hand side being precisely $bu(x)$ where b is the coboundary map $C_0 \to Z_1$. More generally the group C_0 acts on the (non-linear) space Z_1 by

$$(u, j) \mapsto j^u; \qquad j^u(x) = j(x) + u(x + j(x)) + (x + j(x))u^\dagger + u(x + j(x))u^\dagger; \tag{32}$$

in particular we have

$$j^{u_1 * u_2} = (j^{u_1})^{u_2}. \tag{33}$$

This action can be understood in terms of the perturbation theory of quantum stochastic flows [11]; it turns out that in the case of bounded structure maps, if j generates the flow J then j^u generates the flow J^u where

$$J_t^u(x) = U^{(j)}(t) J_t(x) U^{(j)*}(t) \tag{34}$$

and $U^{(j)}$ is the unique unitary solution of

$$dU^{(j)} = U^{(j)} J \otimes id(u), \qquad U^{(j)}(0) = 1.$$

Notice this is not of form (12) and requires a separate existence theorem for its solution [11]. When this theory is applied to inner flows, it is found that the group structure (31) gives rise to a natural generalisation of the Weyl (CCR) commutation relations [11].

Recently [8] quantum stochastic flows have been linked to cyclic cohomology. The Cuntz algebra $\mathscr{C}(\mathscr{A})$ [7] over the *-algebra $\mathscr{A}$ may be characterised as the *-algebra freely generated by elements $x \in \mathscr{A}$ and $q(x)$ where q is a linear *-map satisfying

$$q(xy) = q(x)y + xq(y) - q(x)q(y). \tag{35}$$

The connection with (25) is evident. Traces on $\mathscr{C}(\mathscr{A})$ correspond to cyclic cocycles over $\mathscr{A}$ [7]. Thus generators of flows can be used to construct cyclic cocycles [8]. The cocycles found in this way are rather meagre in number because of the paucity of traces on Ito algebras. $\mathbb{Z}_2$-graded Fermionic theories are more promising [14].

6. References

1. L. Accardi and I. Volovich, *The stochastic limit of quantum field theory*, Rome II preprint (1994).
2. D. B. Applebaum, *Stochastic dilations of the Bloch equation in Boson and Fermion noise*, J. Phys. A **19** (1986) 937–959.
3. A. Barchielli and G. Lupieri, *Quantum stochastic calculus, operation valued stochastic processes and continual measurements in quantum mechanics*, J. Math. Phys. **26** (1985).
4. C. Barnett, R. F. Streater and I. Wilde, *The Itô–Clifford Integral*, J. Funct. Anal. **48** (1982) 172–212.
5. V. P. Belvakin and P. Staszewski, *Nondemolition measurement of a free quantum particle*, Phys. Rev. A **45** (1992) 1347–1356.
6. W. Bradshaw, *Stochastic cocycles as a characterisation of quantum flows*, Bull. Sci. Math. France **116** (1992) 1–34.
7. A. Connes and J. Cuntz, *Quasi homomorphismes, cohomologie cyclique et positivité*, Commun. Math. Phys. **114** (1988) 515–526.
8. P. Beazley Cohen and R. L. Hudson, *Generators of quantum stochastic flows, Cuntz morphisms and cyclic cohomology*, Nottingham preprint (1994).
9. D. E. Evans and E. Christensen, *Cohomology of operator algebras and quantum dynamical semigroups*, J. London Math. Soc. (2) **20** (1979) 358–368.
10. M. P. Evans, *Existence of quantum diffusions*, Prob. Theory and Related Fields **81** (1989) 473–483.
11. M. P. Evans and R. L. Hudson, *Perturbations of quantum diffusions*, Jour. London Math. Soc. (2) **41** (1990) 373–384.
12. R. L. Hudson, Quantum diffusions and cohomology of algebras, pp. 479–485, *Proceedings of 1st World Congress of Bernoulli Society*, Tashkent, 1986, Vol. 1, ed. Yu Prohorov *et al.*, VNU (1987).
13. R. L. Hudson, Quantum stochastic flows, pp. 512–525, *Probability and Mathematical Statistics*, Proceedings, Vilnius, 1989, Vol. 1 (1990).
14. R. L. Hudson, *Fermion flows and supersymmetry*, Int. Jour. Theoret. Phys. **32** (1993) 2413–2422.
15. R. L. Hudson and J. M. Lindsay, *On characterising quantum stochastic evolutions*, Math. Proc. Camb. Phil. Soc. **102** (1987) 363–369.
16. R. L. Hudson and K. R. Parthasarathy, *Quantum Ito's formula and stochastic evolutions*, Commun. Math. Phys. **93** (1984) 301–323.
17. R. L. Hudson and K. R. Parthasarathy, *Unification of Boson and Fermion quantum stochastic calculus*, Commun. Math. Phys. **104** (1986) 457–470.
18. R. L. Hudson and K. R. Parthasarathy, *Stochastic dilations of uniformly continuous completely positive semigroups*, Acta. Math. Applicandae **2** (1984) 353–378.
19. R. L. Hudson and P. Robinson, *Quantum diffusions and the noncommutative torus*, Lett. Math. Phys. **15** (1988) 47–53.
20. R. L. Hudson and P. Shepperson, Stochastic dilations of quantum dynamical semigroups using one-dimensional quantum stochastic calculus, pp. 216–218, in *Quantum Probability V*, ed. L. Accardi et al., Springer LNM **1442** (1990).
21. J. L. Journée, *Structure des cocycles Markoviens sur l'espace de Fock*, Prob. Theor. Rel. Fields **75** (1987) 291–316.
22. G. Lindblad, *On the generators of quantum dynamical semigroups*, Commun. Math. Phys. **48** (1976) 119–130.

23. A. S. Lue, *Non-abelian cohomology of associative algebras*, Quart. Jour. Math. **19** (1968) 159–180.
24. H. Maassen and P. Robinson, *Quantum stochastic calculus and the dynamical Stark effect*, Rep. Math. Phys. **30** (1991) 185–203.
25. P.-A. Meyer, *Quantum probability for probabilists*, Springer LNM (1993).
26. K. R. Parthasarathy, *An introduction to quantum stochastic calculus*, Birkhäuser, Basel (1992).
27. P. Robinson, Quantum diffusions on the rotation algebra and the quantum Hall effect, pp. 326–333, in *Quantum Probability V*, ed. L. Accardi *et al.*, Springer LNM **1442** (1990).
28. I. E. Segal, *Tensor algebras over Hilbert spaces I*, Trans. Amer. Math. Soc. **81** (1956) 106–134.
29. I. R. Senitzky, *Dissipation in quantum mechanics. The harmonic oscillator*, Phys. Rev. **119** (1960) 670–679.
30. W. v. Waldenfels, Spontaneous light emission described by a quantum stochastic differential equation, pp. 516–534, in *Quantum Probability III*, ed. L. Accardi *et al.*, Springer LNM **1136** (1985).
31. N. Wiener, *The homogeneous chaos*, Amer. J. Math. **60** (1930) 897–936.

DEPARTMENT OF MATHEMATICS, UNIVERSITY OF NOTTINGHAM, UNIVERSITY PARK, NOTTINGHAM NG7 2RD, U.K.

E-mail address: rlh@maths.nott.ac.uk

Proceedings of Symposia in Pure Mathematics
Volume **59**, 1996

ENDOMORPHISMS OF $\mathcal{B}(\mathcal{H})$

OLA BRATTELI, PALLE E.T. JORGENSEN, AND GEOFFREY L. PRICE

ABSTRACT. The unital endomorphisms of $\mathcal{B}(\mathcal{H})$ of (Powers) index n are classified by certain $U(n)$-orbits in the set of non-degenerate representations of the Cuntz algebra $\mathcal{O}_n$ on $\mathcal{H}$. Using this, the corresponding conjugacy classes are identified, and a set of labels is given. For the subset of the endomorphisms which are shifts, this set of labels is $P/\sim$ where P is a set of pure states on the UHF- algebra M_{n^∞}, and $\sim$ is a non-smooth equivalence on P. Several subsets of P, giving concrete examples of non-conjugate shifts, are worked out in detail, including sets of product states, and a set of nearest neighbor states.

DEDICATED TO PROFESSOR DEREK ROBINSON
ON THE OCCASION OF HIS SIXTIETH BIRTHDAY.

0. INTRODUCTION

Recently the study of endomorphisms of von Neumann algebras has received increased attention, both in connection with the Jones index for subfactors and its applications [Jon], and also in connection with duality for compact groups [Wor] and super-selection sectors in algebraic quantum field theory. Two other articles (by W. Arveson and by R. Powers) in these proceedings deal with *semigroups* of endomorphisms of the type I_∞- factor. Here we restrict to the case of *single* endomorphisms of $\mathcal{B}(\mathcal{H})$. Potentially it is expected that the theory for $\mathcal{B}(\mathcal{H})$ may possibly be extended or modified to apply also to other factors, but so far only a few relatively isolated results (although still some very important ones) are known for endomorphisms of factors other than $\mathcal{B}(\mathcal{H})$. We report here on recent and new developments in the study of $\mathrm{End}(\mathcal{B}(\mathcal{H}))$. The methods used draw among other things

---

1991 *Mathematics Subject Classification.* 46L10, 46L50, 47A58, 47C15, 81S99.

Key words and phrases. Endomorphisms, shift, Cuntz-algebra, pure states, infinite tensor product, Hilbert space.

Research supported in part by the Norwegian Research Council for Ola Bratteli and Palle Jorgensen, by the U.S. National Science Foundation for Palle Jorgensen, and by the U.S. National Security Agency for Geoffrey Price.

This paper is the symposium contribution of P. E. T. Jorgensen.

on seminal ideas of von Neumann, and also on ideas of Powers from his pioneering work on the states on the CAR (canonical anticommutation relation)-algebra, and, more generally, states on the UHF (uniformly hyperfinite) C^*- algebras.

The work on End(M) for the case when M is a von Neumann factor of type II_1 (especially the hyperfinite case) is ongoing. It will not be treated here, but we refer to [Pow2], [Po-Pr], [EW], [Cho], and [ENWY].

1. Main Results

Let $\mathcal{B}(\mathcal{H})$ be the C^*-algebra of bounded linear operators on the separable, infinite dimensional Hilbert space $\mathcal{H}$. If $\alpha : \mathcal{B}(\mathcal{H}) \to \mathcal{B}(\mathcal{H})$ is a unital endomorphism, we say that α is ergodic if $\{X \in \mathcal{B}(\mathcal{H}) \mid \alpha(X) = X\} = \mathbb{C}1$, and that α is a shift if $\bigcap_{n=1}^{\infty} \alpha^n(\mathcal{B}(\mathcal{H})) = \mathbb{C}1$. The (Powers) index $n \in \{1, 2, \dots, \infty\}$ of α is defined as the n such that $\alpha(\mathcal{B}(\mathcal{H}))' \cap \mathcal{B}(\mathcal{H})$ is isomorphic to the factor of type I_n, [Pow2]. Thus $n = 1$ if and only if α is an automorphism. Let $\mathrm{End}_n(\mathcal{B}(\mathcal{H}))$ (respectively $\mathrm{Erg}_n(\mathcal{B}(\mathcal{H}))$, $\mathrm{Shift}_n(\mathcal{B}(\mathcal{H}))$) denote the set of unital endomorphisms (respectively ergodic unital endomorphisms, shifts) of $\mathcal{B}(\mathcal{H})$ of index n. We say that two elements $\alpha, \beta \in \mathrm{End}(\mathcal{B}(\mathcal{H}))$ are conjugate if there is an automorphism $\gamma \in \mathrm{Aut}(\mathcal{B}(\mathcal{H})) = \mathrm{End}_1(\mathcal{B}(\mathcal{H}))$ such that $\alpha = \gamma \circ \beta \circ \gamma^{-1}$, and α, β are approximately conjugate if for all $\epsilon > 0$ there is a $\gamma \in \mathrm{Aut}(\mathcal{B}(\mathcal{H}))$ such that $\|\alpha - \gamma \circ \beta \circ \gamma^{-1}\| < \epsilon$. It is easy to see that any two approximately conjugate endomorphisms α, β must have the same index n.

In [Pow2, Theorem 2.3] it was proved that if α, β are shifts of index $n \geq 2$ each allowing a pure, normal invariant state on $\mathcal{B}(\mathcal{H})$, then α and β are conjugate. The problem of whether there exist shifts without invariant vector states was left open in [Pow2], but we will both construct explicit classes of examples of shifts of order n without invariant vector- states in Sections 5–8, and prove a classification theorem.

Our construction of these special shift-conjugacy classes, and our analysis of their ergodic theoretic, and clustering type properties, are based on fundamental ideas of von Neumann, especially his 1938 *Compositio* paper [vNeu], and their extension by Guichardet [Gui] (notably [Gui3]). The imprint on our paper from von Neumann's legacy is most visible in our construction of explicit examples in Sections 6, 7, and 8 below.

In the study of $\mathrm{End}(\mathcal{B}(\mathcal{H}))$ we will make extensive use of ideas developed by von Neumann and other pioneers in operator algebras and in quantum theory, [vNeu], [Seg1–2], [Pow1], [ArWoo] (see also the beginning of Remarks 8.2).

Theorem 1.1. (see [Lac1, Theorem 4.5]) *Assume* $n \in \{2, 3, 4, \dots, \infty\}$. *Then the set of conjugacy classes in* $\mathrm{Shift}_n(\mathcal{B}(\mathcal{H}))$ *can be equipped with a natural Borel structure which is not countably separated. The same applies to* $\mathrm{End}_n(\mathcal{B}(\mathcal{H}))$ *and* $\mathrm{Erg}_n(\mathcal{B}(\mathcal{H}))$. *In particular there exist elements in* $\mathrm{Shift}_n(\mathcal{B}(\mathcal{H}))$ *which do not allow invariant vector states.*

This theorem will be proved in Section 5 (the Borel structure portion is new). In Section 5 we will give a complete labeling of the conjugacy classes in $\mathrm{Shift}_n(\mathcal{B}(\mathcal{H}))$ by $P/\sim$, where P is a subset of the pure state space of the UHF algebra M_{n^∞}, and $\sim$ is a certain equivalence relation on P. In Sections 5, 6, 7, and 8 we will look at some special elements in $P/\sim$. On the way to the proof, we will gain further insight into the shifts allowing invariant vector states.

In [Lac1], M. Laca continues the program initiated by Powers of analyzing the conjugacy classes of discrete shifts on $\mathcal{B}(\mathcal{H})$. The central theme of his approach, as it is here, is to exploit the correspondence between endomorphisms and representations of the Cuntz algebras which implement the endomorphisms. In his paper Laca succeeds in establishing the existence of uncountably many conjugacy classes of shifts of each index [Lac1, Remark 4.6.2]. He also obtains [Lac1, Theorem 4.5] a characterization of the conjugacy classes of shifts which identifies them with an equivalence class structure of a certain family of pure states on the subalgebra UHF_n of the Cuntz algebra $\mathcal{O}_n$. This result appears in a slightly different guise as our Theorem 1.1, which is included for the purposes of exposition. In [Pow2, Theorem 2.4] it was proved that any two shifts of $\mathcal{B}(\mathcal{H})$ with the same index are outer conjugate. Another version of this result is:

Theorem 1.2. (see [Pow2] and [Lac1, Proposition 2.3]) *Let α, β be two endomorphisms of $\mathcal{B}(\mathcal{H})$ of the same index $n \in \{1, 2, \dots, \infty\}$. Then there is a unitary $U \in \mathcal{B}(\mathcal{H})$ such that*

$$\alpha(X) = U\beta(X)U^*$$

for all $X \in \mathcal{B}(\mathcal{H})$.

Defining $\gamma(X) = UXU^*$, this relation can also be expressed as

$$\begin{aligned}\alpha(X) &= U\beta(U^*UXU^*U)U^* \\ &= \gamma\beta\gamma^{-1}(UXU^*)\end{aligned}$$

which is the form of outer conjugacy considered in [Pow2]. We will see that one cannot in general find a unitary U such that $\alpha(X) = \beta(UXU^*)$. This is proved in Section 3. Finally, using Voiculescu's non- commutative Weyl–von Neumann theorem [Voi1, Wor], we can establish

Theorem 1.3. *Let α, β be two endomorphisms of $\mathcal{B}(\mathcal{H})$ of the same index n, $2 \leq n < \infty$. Then α and β are approximately conjugate; i.e., there is a sequence $\gamma_k \in \mathrm{Aut}(\mathcal{B}(\mathcal{H}))$ such that*

$$\|\alpha - \gamma_k \circ \beta \circ \gamma_k^{-1}\| \to 0.$$

The sequence γ_k may furthermore be chosen such that $\alpha(X) - (\gamma_k \circ \beta \circ \gamma_k^{-1})(X)$ is compact for each $X \in \mathcal{B}(\mathcal{H})$, $k \in \mathbb{N}$.

We remark that when $n = 1$, it is well known that an automorphism α of $\mathcal{B}(\mathcal{H})$ is implemented by a unitary operator U, unique up to a scalar phase factor, and thus $\mathrm{Aut}(\mathcal{B}(\mathcal{H}))$ is indexed by the set $\mathrm{Rep}(C(\mathbb{T}), \mathcal{H})$ of non-degenerate representations of $C(\mathbb{T})$ on $\mathcal{H}$, modulo the canonical action of the circle group $\mathbb{T}$. These representations are well known from spectral theory, [Ped]. Thus $\mathrm{Shift}_1(\mathcal{B}(\mathcal{H}))$ and $\mathrm{Erg}_1(\mathcal{B}(\mathcal{H}))$ are empty, and $\mathrm{End}_1(\mathcal{B}(\mathcal{H}))$ is countably separated in its natural Borel structure. Theorem 1.2 is trivially true in the case $n = 1$ (just put $U = U_\alpha U_\beta^*$ where $\alpha = \mathrm{Ad}(U_\alpha)$, $\beta = \mathrm{Ad}(U_\beta)$), while Theorem 1.3 is false.

This work was essentially completed before we became aware of Laca's results. As mentioned above, some of our work overlaps with that in [Lac1], and we indicate below where this occurs. Our approach to the subject differs in several aspects, however. A major goal of our work, for example, is to develop techniques and concepts which differentiate between those endomorphisms which admit normal invariant states and those which do not (all endomorphisms have invariant states, however, see Remark 7.6). Since Powers already showed that for each index there is only one conjugacy class of *shifts* allowing invariant normal pure states (see Theorem 4.2, below), any method giving other conjugacy classes of course gives shifts without invariant vector states. (There does, however, exist a plethora of conjugacy classes of non-ergodic endomorphisms of a given index n each allowing (several) invariant vector states; just take discrete direct sums of the representations of $\mathcal{O}_n$ defined by Cuntz's states as in Section 4.) A special feature of our approach is that we obtain many examples of shifts not allowing invariant vector states by perturbing shifts allowing such states by various perturbation techniques (see Sections 6 and 7). Our constructions in Sections 5–7 are based primarily on consideration of (infinite) *product states* on UHF_n, whereas our construction in Section 8 uses instead certain *nearest neighbor states* on UHF_n. Both constructions lead to shifts which do not have invariant vector states, but, more importantly, the shifts on nearest neighbor states are *not* conjugate to those from Sections 6–7.

In Section 9, we construct explicitly extensions of endomorphisms of $\mathcal{B}(\mathcal{H})$ to automorphisms of $\mathcal{B}(\mathcal{H} \otimes \mathcal{H})$.

We finally point out the connection between our results and the results of Arveson on one-parameter semigroups of *- endomorphisms (see [Arv1–2]). If one translates Arveson's concepts, which are tailor-made for the semigroup $\mathbb{R}_+$, to the semigroup $\mathbb{N} \cup \{0\}$, then his *spectral C^*-algebra* for a shift of index n is nothing but the Toeplitz-Cuntz algebra $\mathcal{E}_n$, which in turn is an extension of $\mathcal{O}_n$ by the compact operators when n is finite, and $\mathcal{E}_\infty = \mathcal{O}_\infty$ [Eva]. Otherwise Arveson's Fock space methods have a different flavor from our infinite tensor product methods.

The Toeplitz-Cuntz algebras also play a role in the recent work in Dinh [Din], as well as [Lac1–2] and [Sta].

Another important difference between the present study and that of Arveson and Powers on continuous semigroups of endomorphisms is that they consider the classification of semigroups up to cocycle conjugacy, rather than conjugacy. In the discrete case cocycle conjugacy reduces to outer conjugacy, and Theorem 1.2 tells us that there is only one outer conjugacy class for each value of the index. In the case of continuous semigroups, however, there are at least countably many conjugacy classes for each value of the (appropriately defined) index, [Arv3], [Pow3].

2. Cuntz Algebras and Cuntz States

The Cuntz algebra $\mathcal{O}_n$ is uniquely defined as the C^*- algebra generated by $n = 2, 3, \ldots$ isometries $s_1, \ldots, s_n$ satisfying

$$s_i^* s_j = \delta_{ij} 1, \qquad \sum_{j=1}^{n} s_j s_j^* = 1, \tag{2.1}$$

[Cun]. There is a canonical representation of the n- dimensional unitary group

$U(n)$ in the automorphism group of $\mathcal{O}_n$ defined by

$$\tau_g(s_i) = \sum_{j=1}^{n} g_{ji} s_j \tag{2.2}$$

for $g = [g_{ij}]_{i,j=1}^n \in U(n)$.

Let π_1, π_2 be two non-degenerate representations of $\mathcal{O}_n$ on a Hilbert space $\mathcal{H}$, and put

$$S_i = \pi_1(s_i), \qquad T_i = \pi_2(s_i). \tag{2.3}$$

Then there exists a unitary operator $M = [m_{ij}] \in M_n(\mathcal{B}(\mathcal{H}))$ and a unitary operator $U \in \mathcal{B}(\mathcal{H})$ such that

$$T_i = \sum_{j=1}^{n} S_j m_{ji} = U S_i. \tag{2.4}$$

The operators M and U are given uniquely by

$$m_{ji} = S_j^* T_i, \qquad U = \sum_{j=1}^{n} T_j S_j^* \tag{2.5}$$

and we have the relations

$$m_{ji} = S_j^* U S_i, \qquad U = \sum_{j,i=1}^{n} S_j m_{ji} S_i^*. \tag{2.6}$$

Conversely, if $\{S_i\}$ is a realization of $\mathcal{O}_n$ on $\mathcal{H}$, and $[m_{ij}]$ is a unitary element in $M_n(\mathcal{B}(\mathcal{H}))$, then $\{T_i\}$ defined by (2.4) is a realization of $\mathcal{O}_n$ on $\mathcal{H}$. The same remark applies to a single unitary operator $U \in \mathcal{B}(\mathcal{H})$, and the other equation in (2.4). We will give explicit formulas for the transfer operators (2.6) in Sections 7–8 below for elements in $\mathrm{End}_n(\mathcal{B}(\mathcal{H}))$ from distinct conjugacy classes.

The C^*-algebra $\mathcal{O}_n$ is simple and antiliminal when $n > 1$, [Cun]. We define, naturally, $\mathcal{O}_1$ as the universal C^*-algebra generated by one unitary element, i.e., $\mathcal{O}_1 = C(\mathbb{T})$, and $\mathcal{O}_\infty$ as the algebra generated by isometries $s_1, s_2, \ldots$ satisfying merely the first relation in (2.1). Then $\mathcal{O}_\infty$ is still simple and antiliminal [Cun], while $\mathcal{O}_1$ of course is abelian.

With a slight abuse of terminology, we will say that π is a non-degenerate representation of $\mathcal{O}_\infty$ if π is a representation with $\sum_{i=1}^{\infty} \pi(s_i s_i^*) = 1$, where the sum is in the strong operator topology. With this convention, all the statements in the second paragraph of this section are still valid for $n = \infty$, and the infinite sums converge in the strong operator topology.

Let UHF_n be the fixed point subalgebra of $\mathcal{O}_n$ under the canonical action of the center of $U(n)$. Thus UHF_n is the closure of the linear span of operators of the form

$$s_{i_1} s_{i_2} \cdots s_{i_k} s_{j_k}^* s_{j_{k-1}}^* \cdots s_{j_1}^*$$

over $k = 0, 1, 2, \dots$. If $n < \infty$, then UHF_n is the UHF-algebra M_{n^∞}, which is the uniform closure of finite linear combinations of operators of the form $A_1 \otimes A_2 \otimes A_3 \otimes \cdots$, where each A_i acts on a fixed n-dimensional Hilbert space (i.e., $A_i \in M_n$) and all but finitely many of the A_i's are the identity. If $n = \infty$, then UHF_∞ is the (non- simple) AF algebra described as follows: Let $\mathcal{H}$ be a fixed infinite-dimensional separable Hilbert space. For each $k \in \mathbb{N}$, let $\mathcal{C}_k$ be the C^*-algebra of compact operators on $\bigotimes_1^k \mathcal{H}$, viewed as the C^*- algebra generated by all linear combinations of elements of the form $A_1 \otimes A_2 \otimes \cdots \otimes A_k \otimes I \otimes I \otimes \cdots$, where $A_i \in C(\mathcal{H})$. Then UHF_∞ is the C^*-algebra generated by the $\mathcal{C}_k$'s for $k \in \mathbb{N}$, and the identity. For more details on UHF_∞, see also [Cun], [Eva], [Br- Rob, Example 5.3.27] or [Lac1–2].

Let D_n denote the canonical diagonal subalgebra of UHF_n; that is, D_n is the abelian C^*-algebra obtained as the closure of the linear span of

$$s_{i_1} s_{i_2} \cdots s_{i_k} s_{i_k}^* s_{i_{k-1}}^* \cdots s_{i_1}^*.$$

Then D_n is maximal abelian in UHF_n. If $2 \le n < \infty$ then D_n is canonically isomorphic to $C(\prod_{k=0}^\infty \mathbb{Z}_n)$, where $\mathbb{Z}_n = \{1, \dots, n\}$ equipped with the discrete topology. If $n = \infty$, D_n is the abelian C^*-algebra spanned by $1 \otimes 1 \otimes 1 \otimes \cdots, C_0(\mathbb{N}) \otimes 1 \otimes 1 \otimes \cdots, C_0(\mathbb{N} \times \mathbb{N}) \otimes 1 \otimes \cdots$. (For details on this, see [Br-Rob; Example 5.3.27].)

If $n < \infty$, we defined the canonical endomorphism ψ of $\mathcal{O}_n$ by

$$\psi(x) = \sum_{k=1}^n s_k x s_k^*.$$

Then $\psi|_{\mathrm{UHF}_n}$ is the one-sided shift.

If $\eta_1, \dots, \eta_n \in \mathbb{C}$ with $\sum_{i=1}^n |\eta_i|^2 = 1$ the associated Cuntz state is the pure state ω_η on $\mathcal{O}_n$ defined by

$$\omega_\eta(s_{i_1} \cdots s_{i_k} s_{j_1}^* \cdots s_{j_\ell}^*) = \eta_{i_1} \cdots \eta_{i_k} \bar{\eta}_{j_1} \cdots \bar{\eta}_{j_\ell}$$

(this definition also goes through with obvious modifications for $n = \infty$ and $n = 1$). When $2 \le n < \infty$, $\omega_\eta|_{\mathrm{UHF}_n}$ is the infinite product on $\bigotimes_0^\infty M_n$ of the pure states on M_n defined by the vector $\eta = (\eta_1, \dots, \eta_n)$. When $n = +\infty$, $\omega_\eta|_{\mathrm{UHF}_\infty}$ is similarly the state on UHF_∞, described as before, defined by the unit vector $\eta \otimes \eta \otimes \eta \otimes \cdots$, [Cun], [ACE], [BEGJ] and [Voi2].

3. Endomorphisms

Theorem 3.1. ([Arv1], [Lac1; Theorem 2.1, Proposition 2.2]) *Let φ be a unital endomorphism of $\mathcal{B}(\mathcal{H})$ of Powers index $n \in \{1, 2, 3, \dots, +\infty\}$.*

It follows that there exists a non-degenerate representation of $\mathcal{O}_n$ on $\mathcal{H}$ such that

$$\varphi(X) = \sum_{i=1}^n S_i X S_i^* \tag{3.1}$$

where S_i is the representative of s_i. Conversely, any non-degenerate representation of $\mathcal{O}_n$ on $\mathcal{H}$ defines an endomorphism of index n by (3.1). The representation is unique up to the canonical action of $U(n)$.

Proof. Since $\varphi(\mathcal{B}(\mathcal{H}))$ is a unital subalgebra of $\mathcal{B}(\mathcal{H})$, isomorphic to $\mathcal{B}(\mathcal{H})$, we have a tensor product decomposition $\mathcal{H} = \mathcal{H}_0 \otimes \mathcal{K}$ of $\mathcal{H}$ such that $\varphi(\mathcal{B}(\mathcal{H}))$ identifies with $\mathcal{B}(\mathcal{H}_0)\otimes 1$, and then $\varphi(\mathcal{B}(\mathcal{H}))' \cap \mathcal{B}(\mathcal{H}) \cong 1 \otimes \mathcal{B}(\mathcal{K})$, [Dix]. Thus, $\mathrm{Index}(\varphi) = \mathrm{Dim}(\mathcal{K})$.

Let $(E_{ij})_{i,j=1}^n$ be a complete set of matrix units for $\varphi(\mathcal{B}(\mathcal{H}))'$. It follows that

$$E_{ii}\varphi(\mathcal{B}(\mathcal{H})) = \varphi(\mathcal{B}(\mathcal{H}))E_{ii} \cong \mathcal{B}(\mathcal{H}_0) \cong \mathcal{B}(\mathcal{H})$$

for $i = 1, 2, \ldots, n$, and $X \to \varphi(X)E_{ii}$ is a $*$-isomorphism between $\mathcal{B}(\mathcal{H})$ and $\mathcal{B}(E_{ii}\mathcal{H})$. By Wigner's theorem (which is Theorem 3.1 in the case $n = 1$) there is a unitary operator S_i from $\mathcal{H}$ onto $E_{ii}\mathcal{H}$ such that

$$\varphi(X)E_{ii} = S_i X S_i^*.$$

But then

$$\varphi(X) = \varphi(X)\sum_{i=1}^n E_{ii} = \sum_{i=1}^n \varphi(X)E_{ii} = \sum_{i=1}^n S_i X S_i^*.$$

We have

$$S_i^* S_i = 1, \qquad \sum_{i=1}^n S_i S_i^* = \sum_{i=1}^n E_{ii} = 1$$

so the S_i satisfy the Cuntz relations (2.1). Conversely, if S_i satisfy the Cuntz relations, then φ defined by (3.1) is an endomorphism such that $\varphi(\mathcal{B}(\mathcal{H}))' \cap \mathcal{B}(\mathcal{H})$ is spanned by $S_i S_j^*$, and consequently $\varphi(\mathcal{B}(\mathcal{H}))' \cap \mathcal{B}(\mathcal{H}) \cong M_n$ and φ has index n.

Let T_i, $i = 1, \ldots, n$ be another non-degenerate realization of $\mathcal{O}_n$ that implements φ:

$$\varphi(X) = \sum_{i=1}^n T_i X T_i^* = \sum_{i=1}^n S_i X S_i^*.$$

Multiply the last relation to the left by S_j^* and to the right by T_i to obtain

$$S_j^* T_i X = X S_j^* T_i.$$

Since this is true for any $X \in \mathcal{B}(\mathcal{H})$,

$$S_j^* T_i = h_{ji} 1$$

where $h_{ji} \in \mathbb{C}$. But then $h = [h_{ji}] \in U(n)$ and

$$\pi_2 = \pi_1 \circ \tau_h$$

where τ is the canonical action of $U(n)$ on $\mathcal{O}_n$, and π_1, π_2 are the representations determined by S, T, respectively.

Definition 3.2. For $n = 1, 2, \dots, \infty$, let

$$\mathrm{Rep}(\mathcal{O}_n, \mathcal{H})$$

denote the set of all non-degenerate representations of $\mathcal{O}_n$ on $\mathcal{H}$, and

$$\mathrm{Irr}(\mathcal{O}_n, \mathcal{H})$$

the set of all irreducible representations of $\mathcal{O}_n$ on $\mathcal{H}$, and

$$\mathrm{Rep}_s(\mathcal{O}_n, \mathcal{H})$$

the set of representations of $\mathcal{O}_n$ on $\mathcal{H}$ such that UHF_n is weakly dense in $\mathcal{B}(\mathcal{H})$. Of course the two latter sets are empty if $n = 1$.

The canonical action of $U(n)$ on $\mathcal{O}_n$ gives rise to an action of $U(n)$ on each of these spaces. Also, the unitary group $U(\mathcal{H})$ on $\mathcal{H}$ acts on each of the three spaces by $\pi(\cdot) \to U\pi(\cdot)U^*$ for $U \in U(\mathcal{H})$, $\pi \in \mathrm{Rep}(\mathcal{O}_n, \mathcal{H})$. The following corollary of Theorem 3.1 is immediate.

Theorem 3.3. *Let $\pi \to \varphi(\pi)$ be the surjective map from* $\mathrm{Rep}(\mathcal{O}_n, \mathcal{H})$ *onto* $\mathrm{End}_n(\mathcal{B}(\mathcal{H}))$ *defined in Theorem 3.1. Then:*

(3.2) $\varphi(\pi) \in \mathrm{Erg}_n(\mathcal{B}(\mathcal{H}))$ *if and only if* $\pi \in \mathrm{Irr}(\mathcal{O}_n, \mathcal{H})$
(3.3) $\varphi(\pi) \in \mathrm{Shift}_n(\mathcal{B}(\mathcal{H}))$ *if and only if* $\pi \in \mathrm{Rep}_s(\mathcal{O}_n, \mathcal{H})$
(3.4) ([Lac, Proposition 2.4]) $\varphi(\pi_1)$ *and* $\varphi(\pi_2)$ *are conjugate if and only if there is a* $g \in U(n)$ *and a* $U \in U(\mathcal{H})$ *such that*

$$\pi_2(\cdot) = U\pi_1(\tau_g(\cdot))U^*.$$

In short, the conjugacy classes in $\mathrm{End}_n(\mathcal{B}(\mathcal{H}))$ *correspond to the orbits in* $\mathrm{Rep}(\mathcal{O}_n, \mathcal{H})$ *under the joint actions of* $U(n)$ *and* $U(\mathcal{H})$.

Proof. To prove (3.2), it suffices to show that ([Lac, Proposition 3.1])

$$\pi(\mathcal{O}_n)' = \{X \in \mathcal{B}(\mathcal{H}) \mid \varphi(\pi)(X) = X\} \equiv \mathcal{B}(\mathcal{H})^{\varphi}. \tag{3.5}$$

But if $X \in \pi(\mathcal{O}_n)'$, then

$$\varphi(X) = \sum_{i=1}^{n} S_i X S_i^* = \sum_{i=1}^{n} S_i S_i^* X = 1 \cdot X = X$$

where $S_i = \pi(s_i)$, so $X \in \mathcal{B}(\mathcal{H})^{\varphi}$, and $\pi(\mathcal{O}_n)' \subseteq \mathcal{B}(\mathcal{H})^{\varphi}$. Conversely, if $X \in \mathcal{B}(\mathcal{H})^{\varphi}$, then $\sum_{i=1}^{n} S_i X S_i^* = X$. Multiplying to the left by S_j^* we obtain

$$XS_j^* = S_j^* X$$

and since $X^* \in \mathcal{B}(\mathcal{H})^{\varphi}$, we also derive

$$S_j X = X S_j.$$

Hence $X \in \pi(\mathcal{O}_n)'$ and so

$$\mathcal{B}(\mathcal{H})^{\varphi} \subseteq \pi(\mathcal{O}_n)'.$$

This establishes (3.5) and therefore (3.2).

To prove (3.3) we will more generally establish that ([Lac, Proposition 3.1])

$$\bigcap_k \varphi^k(\mathcal{B}(\mathcal{H})) = (\pi(\mathrm{UHF}_n))' \cap \mathcal{B}(\mathcal{H}). \tag{3.6}$$

This again will follow from

$$\varphi^k(\mathcal{B}(\mathcal{H}))' = \operatorname{lin\,span}\{S_{i_1} \cdots S_{i_k} S_{j_k}^* \cdots S_{j_1}^*\}. \tag{3.7}$$

But as

$$\varphi^k(X) = \sum_{i_1,\dots,i_k=1}^{n} S_{i_1} \cdots S_{i_k} X S_{i_k}^* \cdots S_i^*$$

and $(i_1,\dots,i_k) \to S_{i_1} \cdots S_{i_k}$ is a non-degenerate representation of $\mathcal{O}_{n^k}$, it suffices to prove (3.7) for $k = 1$. But as

$$\begin{aligned} S_i S_j^* \varphi(X) &= S_i S_j^* \sum_k S_k X S_k^* \\ &= S_i X S_j^* = \sum_k S_k X S_k^* S_i S_j^* \\ &= \varphi(X) S_i S_j^* \end{aligned}$$

we have

$$\operatorname{lin\,span}\{S_i S_j^*\} \subseteq \varphi(\mathcal{B}(\mathcal{H}))'.$$

Conversely, a general element $X \in \mathcal{B}(\mathcal{H})$ may be written

$$X = \sum_{ij} S_i X_{ij} S_j^* \qquad \text{where} \qquad X_{ij} = S_i^* X S_j$$

and, if $X \in \varphi(\mathcal{B}(\mathcal{H}))'$, then

$$\sum_{ij} S_i X_{ij} S_j^* \sum_k S_k Y S_k^* = \sum_k S_k Y S_k^* \sum_{ij} S_i X_{ij} S_j^*$$

for all $Y \in \mathcal{B}(\mathcal{H})$; that is,

$$\sum_{ij} S_i X_{ij} Y S_j^* = \sum_{ij} S_i Y X_{ij} S_j^*$$

for all $Y \in \mathcal{B}(\mathcal{H})$. Thus X_{ij} must be scalar multiples of 1, and X is a linear combination of $S_i S_j^*$. This establishes (3.7), and hence (3.6) and (3.3). (Of course, if $n = \infty$, linear span means the weak closure of the linear span.)

To prove (3.4), put $S_i = \pi_1(s_i)$, $T_i = \pi_2(s_i)$. If $\varphi(\pi_1)$, $\varphi(\pi_2)$ are conjugate, there exists a $\gamma = \mathrm{Ad}(U) \in \mathrm{Aut}(\mathcal{B}(\mathcal{H}))$ such that

$$\varphi(\pi_2) = \gamma\varphi(\pi_1)\gamma^{-1}$$

i.e.,

$$\sum_i T_i X T_i^* = U\left(\sum_i S_i U^* X U S_i^*\right)U^* = \sum_i (US_iU^*)X(US_iU^*)^*$$

for all $X \in \mathcal{B}(\mathcal{H})$. Since US_iU^* satisfy the Cuntz relations it follows from the uniqueness part of Theorem 3.1 that there exists a $g = [g_{ij}] \in U(n)$ such that

$$T_i = \sum_{j=1}^{n} g_{ji} U S_j U^*,$$

i.e.,

$$\pi_2(\cdot) = U(\pi_1 \circ \tau_g(\cdot))U^*.$$

The converse is established by doing the steps in converse order.

This finishes the proof of Theorem 3.3. $\square$

Proof of Theorem 1.2. By Theorem 3.1 there exist two realizations S, T of $\mathcal{O}_n$ on $\mathcal{H}$ such that

$$\alpha(X) = \sum_i S_i X S_i^*, \qquad \beta(X) = \sum_i T_i X T_i^*.$$

By (2.4) there is a unitary U such that

$$S_i = UT_i$$

for all i. But then

$$\alpha(X) = U\beta(X)U^*$$

for all $X \in \mathcal{B}(\mathcal{H})$. $\square$

Proof of Theorem 1.3. By Theorem 3.1 there exist two realizations S, T of $\mathcal{O}_n$ on $\mathcal{H}$ such that

$$\alpha(X) = \sum_{i=1}^{n} S_i X S_i^*, \qquad \beta(X) = \sum_{i=1}^{n} T_i X T_i^*.$$

Let $\epsilon > 0$. As $\mathcal{O}_n$ is a simple, antiliminal C^*- algebra it follows from Voiculescu's non-commutative Weyl– von Neumann theorem ([Voi1, Corollary 1.4] and [Wor]) that there exists a unitary U on $\mathcal{H}$ such that

$$S_i - UT_iU^*$$

are compact for $i = 1, \ldots, n$, and

$$\|S_i - UT_iU^*\| < \epsilon/2n.$$

Let $\gamma(X) = UXU^*$. Then

$$\begin{aligned}\alpha(X) - \gamma\beta\gamma^{-1}(X) &= \sum_{i=1}^{n} (S_i X S_i^* - UT_iU^* X (UT_iU^*)^*) \\ &= \sum_{i=1}^{n} (S_i - UT_iU^*) X S_i^* + \sum_{i=1}^{n} UT_iU^* X (S_i - UT_iU^*)^*.\end{aligned}$$

Thus $\alpha(X) - \gamma\beta\gamma^{-1}(X)$ is compact and

$$\begin{aligned}\|\alpha(X) - \gamma\beta\gamma^{-1}(X)\| &\leq 2n \cdot 1 \cdot \|X\| \epsilon/2n \\ &= \epsilon\|X\|.\end{aligned}$$

This proves Theorem 1.3. □

4. Shifts and Invariant States

Let α be an endomorphism of $\mathcal{B}(\mathcal{H})$. The next theorem gives a characterization of the normal α-invariant pure states on $\mathcal{B}(\mathcal{H})$.

Theorem 4.1. *Let α be a unital endomorphism of $\mathcal{B}(\mathcal{H})$ of index $n = 1, 2, \ldots, \infty$, and let π be a corresponding non-degenerate representation of $\mathcal{O}_n$. Let $S_i = \pi(s_i)$, $i = 1, \ldots, n$. Let ξ be a unit vector in $\mathcal{H}$, and let $\omega(X) = \langle \xi, \pi(X)\xi \rangle$ be the corresponding state on $\mathcal{O}_n$. The following conditions are equivalent:*

(4.1) *$\langle \xi, \alpha(X)\xi \rangle = \langle \xi, X\xi \rangle$ for all $X \in \mathcal{B}(\mathcal{H})$.*
(4.2) *ξ is a joint eigenvector for S_i^* for $i = 1, 2, \ldots, n$.*
(4.3) *ω is a Cuntz state on $\mathcal{O}_n$.*

Furthermore, the corresponding eigenvalues in (4.2) are $\bar{\eta}_i$:

$$S_i^* \xi = \bar{\eta}_i \xi \tag{4.4}$$

for $i = 1, 2, \ldots, n$, if and only if $\sum_{i=1}^{n} |\eta_i|^2 = 1$ and $\omega = \omega_\eta$.

Proof. (4.2) ⇔ (4.3) and the final remark are straightforward.

(4.2) ⇒ (4.1): If $S_i^* \xi = \bar{\eta}_i \xi$, then

$$\begin{aligned}\sum_{i=1}^{n} |\eta_i|^2 &= \sum_{i=1}^{n} \langle \bar{\eta}_i \xi, \bar{\eta}_i \xi \rangle = \sum_{i=1}^{n} \langle S_i^* \xi, S_i^* \xi \rangle \\ &= \sum_{i=1}^{n} \langle \xi, S_i S_i^* \xi \rangle = \langle \xi, \xi \rangle = 1.\end{aligned}$$

Furthermore

$$\langle \xi, \alpha(X)\xi \rangle = \sum_{i=1}^{n} \langle S_i^* \xi, X S_i^* \xi \rangle = \sum_{i=1}^{n} |\eta_i|^2 \langle \xi, X\xi \rangle = \langle \xi, X\xi \rangle.$$

(4.1) $\Rightarrow$ (4.2): Assume that $\langle \xi, \alpha(X)\xi \rangle = \langle \xi, X\xi \rangle$. We have

$$\langle \xi, \alpha(X)\xi \rangle = \sum_{i=1}^{n} \langle S_i^* \xi, X S_i^* \xi \rangle.$$

But $X \to \langle S_i^* \xi, X S_i^* \xi \rangle$ is a positive linear functional on $\mathcal{B}(\mathcal{H})$ of norm

$$\langle S_i^* \xi, S_i^* \xi \rangle = \langle \xi, S_i S_i^* \xi \rangle.$$

The sum of these norms is

$$\sum_{i=1}^{n} \langle \xi, S_i S_i^* \xi \rangle = \left\langle \xi, \left(\sum_{i=1}^{n} S_i S_i^* \right) \xi \right\rangle = \langle \xi, \xi \rangle = 1.$$

As the sum of these positive functionals is $\langle \xi, \cdot \xi \rangle$ and $\langle \xi, \cdot \xi \rangle$ is pure, it follows that

$$\langle S_i^* \xi, X S_i^* \xi \rangle = \| S_i^* \xi \|^2 \langle \xi, X \xi \rangle$$

for all $X \in \mathcal{B}(\mathcal{H})$, but then

$$S_i^* \xi = \bar{\eta}_i \xi$$

where $\eta_i \in \mathbb{C}$ is such that $|\eta_i| = \| S_i^* \xi \|$. □

Using Theorem 4.1 we can prove the following result, which is implicit in the proof of Theorem 2.3 of [Pow2], see also [Sta] for related results.

Theorem 4.2. *Suppose that α, β are ergodic unital endomorphisms of $\mathcal{B}(\mathcal{H})$, both of index $n \in \{2, 3, \dots\}$ and assume that both α and β allow a pure normal invariant state.*

It follows that α and β are conjugate, and both of them are shifts.

Proof. Let π_α, π_β be the representations of $\mathcal{O}_n$ corresponding to α, β, respectively. The ergodicity of α, β implies that π_α, π_β are irreducible, by Theorem 3.3. Let ξ_α, ξ_β be unit vectors in $\mathcal{H}$ such that $\langle \xi_\alpha, \alpha(\cdot)\xi_\alpha \rangle = \langle \xi_\alpha, \cdot \xi_\alpha \rangle$ and $\langle \xi_\beta, \beta(\cdot)\xi_\beta \rangle = \langle \xi_\beta, \cdot \xi_\beta \rangle$. By Theorem 4.1 the corresponding two states on $\mathcal{O}_n$ are Cuntz states; i.e., there exist unit vectors $\eta^\alpha = (\eta_1^\alpha, \dots, \eta_n^\alpha)$ and $\eta^\beta = (\eta_1^\beta, \dots, \eta_n^\beta)$ in $\ell^2(\{1, 2, \dots, n\})$ such that

$$\langle \xi_\alpha, \pi_\alpha(x)\xi_\alpha \rangle = \omega_{\eta^\alpha}(x) \qquad \text{and} \qquad \langle \xi_\beta, \pi_\beta(x)\xi_\beta \rangle = \omega_{\eta^\beta}(x)$$

for $x \in \mathcal{O}_n$. Now, choose $g = [g_{ij}] \in U(n)$ so that $\eta^\alpha = g^T \eta^\beta$, where g^T is the transpose of g. But since

$$\omega_\eta(\tau_g(s_i)) = \omega_\eta \left(\sum_j g_{ji} s_j \right) = \sum_j g_{ji} \eta_j = \omega_{g^T \eta}(s_i)$$

etc., one has

$$\omega_\eta \circ \tau_g = \omega_{g^T \eta}$$

for any $g \in U(n)$ and any unit vector $\eta \in \ell^2(\{1, 2, \dots\})$. In particular

$$\omega_{\eta^\beta} \circ \tau_g = \omega_{\eta^\alpha}. \tag{4.5}$$

As π_α and π_β are irreducible, it follows that ξ_α, is cyclic for π_α, and ξ_β is cyclic for π_β, and hence the relation (4.5) entails that π_α and $\pi_\beta \circ \tau_g$ are unitarily equivalent. By (3.4), α and β are conjugate. To show that α and β are shifts is equivalent to showing that $\pi_\alpha(\mathrm{UHF}_n)$ and $\pi_\beta(\mathrm{UHF}_n)$ are weakly dense in $\mathcal{B}(\mathcal{H})$. But π_α, π_β are unitarily equivalent to the representation defined by the Cuntz's states ω_{η^α}, ω_{η^β}, and these are irreducible in restriction to UHF_n by (8.8)–(8.9) below. □

5. Classification of Conjugacy Classes of Shifts

In this section we will prove Theorem 1.1, and find an explicit set of labels for the conjugacy classes in $\text{Shift}_n(\mathcal{B}(\mathcal{H}))$. In Sections 6, 7, and 8 we will consider some more explicit points in the label space. Assume that $n \in \{2, 3, \dots\}$. The case $n = \infty$ is somewhat more complicated and was treated in detail in [Lac]. The results are similar in that case, and we will restrict to finite n in the rest of this section. Consider unital shifts of Powers index n on $\mathcal{B}(\mathcal{H})$. By Theorem 3.3, these correspond to the set $\text{Rep}_s(\mathcal{O}_n, \mathcal{H})$ of representations π of $\mathcal{O}_n$ on $\mathcal{H}$ such that $\pi(\text{UHF}_n)$ is weakly dense in $\mathcal{B}(\mathcal{H})$. These representations identify with the cyclic representation defined by any vector state, defined by a unit vector in $\mathcal{H}$. We will characterize abstractly the corresponding states on $\mathcal{O}_n$, or, rather, the restriction of those states to UHF_n. So let P denote the set of pure states ω on UHF_n such that ω has a pure extension ω' to $\mathcal{O}_n$ with the property that, if $(\mathcal{H}_{\omega'}, \pi_{\omega'}, \Omega_{\omega'})$ is the corresponding representation, then $\pi_{\omega'}(\text{UHF}_n)'' = \mathcal{B}(\mathcal{H}_{\omega'})$. Let:

$$\begin{aligned}
\sigma(\cdot) &= \sum_i S_i \cdot S_i^* \text{ be the canonical endomorphism of } \text{UHF}_n \\
&\quad (= \text{the one-sided shift on } M_{n^\infty}) \\
A_m &= \underbrace{M_n \otimes \cdots \otimes M_n}_{m} \otimes 1 \otimes 1 \otimes \cdots \subseteq \text{UHF}_n \\
A_m^c &= \underbrace{1 \otimes \cdots \otimes 1}_{m} \otimes M_n \otimes M_n \otimes \cdots \subseteq \text{UHF}_n \\
&= \text{relative commutant of } A_m.
\end{aligned}$$

Then $\sigma(A_m^c) \subseteq A_{m+1}^c$ and $\sigma : A_m^c \to A_{m+1}^c$ is an isomorphism.

Lemma 5.1. *If ω is a pure state on* UHF_n *then $\omega \circ \sigma$ is a type I factor state of multiplicity $\leq n$.*

Proof.

$$\pi_\omega(\sigma(\text{UHF}_n))' = \pi_\omega(A_1) \cong M_n,$$

and the representation Hilbert space of $\omega \circ \sigma$ identifies with $\overline{\pi_\omega(\sigma(\text{UHF}_n))\Omega_\omega}$.

Lemma 5.2. *Let ω be a pure state on* UHF_n. *The following conditions are equivalent:*

(1) $\omega \in P$
(2) *For all $\epsilon > 0$ there is an $m \in \mathbb{N}$ such that*

$$\left\| (\omega \circ \sigma - \omega)|_{A_m^c} \right\| < \epsilon$$

(3) $\lim_{m\to\infty} \|\omega \circ \sigma^{m+1} - \omega \circ \sigma^m\| = 0$

Proof. Since σ^m maps UHF_n isometrically onto A_m^c, the equivalence of (2) and (3) is immediate. Since ω and $\omega \circ \sigma$ both are factor states by Lemma 5.1, it follows from [Pow1, Theorem 2.7] that (2) is equivalent to the representations π_ω and $\pi_{\omega\circ\sigma}$ being quasi- equivalent.

(1) $\Rightarrow$ (2). If $\omega \in P$, then

$$\omega \circ \sigma(x) = \sum_{i=1}^{n} \langle S_i^* \Omega_\omega, \pi_\omega(x) S_i^* \Omega_\omega \rangle$$

for $x \in \mathrm{UHF}_n$, where S_i^* are the representatives of s_i^* in the extension of π_ω to a representation of $\mathcal{O}_n$ on $\mathcal{H}_\omega$. But this shows that $\omega \circ \sigma$ is a normal state in the representation π_ω, and, as ω and $\omega \circ \sigma$ are factor states, they are quasi-equivalent.

(2) $\Rightarrow$ (1). If ω and $\omega \circ \sigma$ are quasi-equivalent, then the endomorphism $\pi_\omega(x) \rightarrow \pi_\omega(\sigma(x))$, $x \in \mathrm{UHF}_n$, extends by continuity to an endomorphism of $\mathcal{B}(\mathcal{H}_\omega)$ which we also call σ. But as $\pi_\omega(A_1) \subseteq \pi_\omega(\sigma(\mathrm{UHF}_n))'$, we have $\pi_\omega(A_1) \subseteq \sigma(\mathcal{B}(\mathcal{H}_\omega))'$. Realizing the elements in UHF_n as $n \times n$ matrices with entries in A_1^c, using that $A_1 \cong M_n$, one easily deduces the converse implication, and hence σ has Powers index n, and there exists by Theorem 3.1 a non-degenerate representation π of $\mathcal{O}_n$ on $\mathcal{H}_\omega$ such that

$$\sigma(X) = \sum_{i=1}^{n} S_i X S_i^*$$

where $S_i = \pi(s_i)$. But then $\sigma(\mathcal{B}(\mathcal{H}))'$ is spanned linearly by $S_i S_j^*$, $i, j = 1 \cdots n$, and, as $\sigma(\mathcal{B}(\mathcal{H}))' = \pi_\omega(A_1)$, $\pi_\omega(A_1)$ is just the linear span of $S_i S_j^*$, $i, j = 1 \cdots n$. Now, modifying the S_i's with an element in $U(n)$ if necessary, we may arrange it so that

$$S_i S_j^* = \pi_\omega(s_i s_j^*)$$

and this determines the S_i's up to a fixed phase factor. If

$$e_{ij}^{(k)} = \sigma^k(s_i s_j^*) \qquad \text{and} \qquad E_{ij}^{(k)} = \sigma^k(S_i S_j^*)$$

then

$$E_{ij}^{(k)} = \sigma^k(\pi_\omega(s_i s_j^*)) = \pi_\omega(\sigma^k(s_i s_j^*)) = \pi_\omega(e_{ij}^{(k)})$$

for $k = 1, 2, \ldots$, and thus we see that π_ω extends to a representation π of $\mathcal{O}_n$ by setting

$$\pi(s_i) = S_i.$$

Thus $\omega \in P$. $\square$

Lemma 5.3. *Two elements $\omega, \omega' \in P$ define unitarily equivalent representations of* UHF_n*, if and only, for $\forall \epsilon > 0$, $\exists\, m$ such that*

$$\| (\omega - \omega')|_{A_m^c} \| < \epsilon.$$

Proof. [Pow1, Theorem 2.7] again.

Lemma 5.4. *Assume that $\omega, \omega' \in P$. The following conditions are equivalent:*

(1) *ω and ω' define conjugate endomorphisms of $\mathcal{B}(\mathcal{H})$.*

(2) *There is a $g \in U(n)$ such that, for all $\epsilon > 0$, there is an $m \in \mathbb{N}$ with*

$$\left\| (\omega - \omega' \circ \tau_g)|_{A_m^c} \right\| < \epsilon$$

where $\tau_g = \bigotimes_{k=1}^{\infty} \operatorname{Ad} g$.

(3) *There is a $g \in U(n)$ such that*

$$\lim_{m\to\infty} \|\omega \circ \sigma^m - \omega' \circ \tau_g \circ \sigma^m\| = 0.$$

(4) *There is a $g \in U(n)$ and a unitary $U \in \mathrm{UHF}_n$ such that*

$$\omega(\cdot) = \omega'(U\tau_g(\cdot)U^*).$$

Proof. The equivalence of the three first conditions follows from Lemma 5.3 and Theorem 3.3. Since condition (2) means that the two pure states ω and $\omega' \circ \tau_g$ define unitarily equivalent representations, condition (4) follows from Kadison's transitivity theorem, [KR], and conversely (4) implies that ω and $\omega' \circ \tau_g$ define unitary equivalent representations. The proof is completed. □

We are now ready to prove Theorem 1.1, and to even give an explicit parametrization of the conjugacy classes in $\mathrm{Shift}_n(\mathcal{B}(\mathcal{H}))$. Let as before P be the set of pure states on UHF_n such that

$$\lim_{m\to\infty} \|\omega \circ \sigma^{m+1} - \omega \circ \sigma^m\| = 0$$

(this characterization is equivalent to the one given above). Define two states $\omega, \omega' \in P$ to be equivalent, $\omega \sim \omega'$, if they lie in the same orbit in P under the joint action of $U(n)$, and of $U(\mathrm{UHF}_n)$ = the unitary group of UHF_n. Then it follows from Lemma 5.2, Lemma 5.4, and Theorem 3.3, that there is a bijection between $P/\sim$ and the set of conjugacy classes of endomorphisms of $\mathcal{B}(\mathcal{H})$. Since $\mathcal{O}_n$ is type III, and $U(n)$ is compact, it follows, by the same reasoning as in Glimm's theorem (see [Gli] and [Ped]), that $P/\sim$ is not a standard Borel space. This is also implied by the fact that the orbits in $\mathrm{End}_n(\mathcal{B}(\mathcal{H}))$ under conjugacy all are norm dense by Theorem 1.3.

The slightly different proof in the case $n = \infty$ can be found in [Lac2].

Example 5.5. ([Lac1]) Let ξ_m, ξ'_m be unit vectors in $\mathbb{C}^n$, and let $\omega_m = \langle \xi_m, \cdot \xi_m \rangle$ and $\omega'_m = \langle \xi'_m, \cdot \xi'_m \rangle$ be the corresponding pure states on M_n. Consider the infinite tensor product states $\omega = \bigotimes_{m=1}^{\infty} \omega_m$ and $\omega' = \bigotimes_{m=1}^{\infty} \omega'_m$ on $\mathrm{UHF}_n = \bigotimes_{m=1}^{\infty} M_n$. These are pure states, and by Lemma 5.3 they induce unitarily equivalent representations if and only if

$$\lim_{m\to\infty} \|(\omega - \omega') \circ \sigma^m\| = 0. \tag{5.1}$$

It is well known, [Gui], that this condition can be expressed in the following equivalent ways:

$$\sum_{m=1}^{\infty} \|\omega_n - \omega_n'\|^2 < \infty, \tag{5.2}$$

$$\sum_{m=1}^{\infty} (1 - |\langle \xi_m, \xi_m' \rangle|) < \infty \tag{5.3}$$

$$\lim_{k\to\infty} \prod_{m=k}^{\infty} |\langle \xi_m, \xi_m' \rangle| = 1. \tag{5.4}$$

These conditions are non-commutative versions of the conditions for equivalence of infinite product measures on $\prod_1^\infty \mathbb{Z}_n$ given by Kakutani in 1948, [Kak]. Similar conditions for quasi-equivalence of quasi-free states, which are closely related to product states, have been given in [Po-St], [Ara1], [Ara2], [Dae].

If, furthermore, the phases of the vectors ξ_m' are chosen optimally with respect to ξ_m, i.e., such that $\langle \xi_m, \xi_m' \rangle \in \mathbb{R}_+$, then (5.2) is equivalent to

$$\sum_{m=1}^{\infty} \|\xi_m - \xi_m'\|^2 < \infty. \tag{5.5}$$

Note for example that the equivalence of (5.5) and (5.3) follows from

$$\|\xi_m - \xi_m'\|^2 = 2(1 - \mathrm{Re}\langle \xi_m, \xi_m' \rangle).$$

Using this, and Lemma 5.2, we see that $\omega = \bigotimes_{m=1}^{\infty} \omega_m$ is in P if and only if

$$\sum_{m=1}^{\infty} \|\omega_m - \omega_{m+1}\|^2 < \infty; \tag{5.6}$$

or, equivalently

$$\sum_{m=1}^{\infty} (1 - |\langle \xi_m, \xi_{m+1} \rangle|) < \infty,$$

or,

$$\lim_{k\to\infty} \prod_{m=k}^{\infty} |\langle \xi_m, \xi_{m+1} \rangle| = 1,$$

or, if the phases of ξ_m are chosen inductively such that $\langle \xi_m, \xi_{m+1} \rangle \in \mathbb{R}_+$,

$$\sum_{m=1}^{\infty} \|\xi_m - \xi_{m+1}\|^2 < \infty.$$

The (5.6) conditions are taken up again in Lemma 6.5 below. In Section 6 we will consider a condition (6.2) which is stronger than (5.6).

Finally, assume that $\omega = \bigotimes_{m=1}^{\infty} \omega_m$ and $\omega' = \bigotimes_{m=1}^{\infty} \omega'_m$ are both in P. By Lemma 5.4, and the remarks above, ω and ω' define non-conjugate shifts if and only if, for all $g \in U(n)$

$$\sum_{m=1}^{\infty} \|\omega_m - \omega'_m \circ Ad(g)\|^2 = +\infty \tag{5.7}$$

or, equivalently

$$\sum_{m=1}^{\infty} (1 - |\langle \xi_m, g\xi'_m \rangle|) = +\infty; \tag{5.8}$$

or the other two similar conditions. In this way we may analyze equivalence classes among the product states in P. See Section 6 for more details.

Example 5.6. As stated in Theorem 1.1, and clarified in Theorem 1.3 and the remarks after Lemma 5.4, there does not exist a smooth labeling of all the conjugacy classes in $\mathrm{Shift}_n(\mathcal{B}(\mathcal{H}))$. We will now give a complete labeling of a class of shifts which will be described in more detail in Section 7, but again this labeling cannot be taken to be smooth. Actually the conjugacy classes of the shifts obtained in this fashion agree exactly with those obtained in Example 5.5, and these classes contain the classes described in more detail in Sections 6 and 7 as subclasses. Let e_i, $i = 0, \dots, n-1$ be the orthonormal basis of $\mathbb{C}^n$ defined by (7.14), and define $\bigotimes_{m=0}^{\infty} \mathbb{C}^n \simeq \mathcal{L}^2(\Omega, \mu)$ as in the introduction to Section 7, so that μ is normalized Haar measure on $\Omega = \prod_{m=0}^{\infty} \mathbb{Z}_n$. In particular $\mathcal{L}^2(\Omega, \mu)$ contains the vector

$$\mathbb{1} = \bigotimes_{m=0}^{\infty} e_0 = e_0 \otimes e_0 \otimes \cdots$$

The following result describes a class of shifts which arise from product states on UHF_n, and they will be studied and characterized further in Sections 6–7, with view to their harmonic analysis. Our condition (5.16) below for *conjugacy* is closely related to [Sta; Theorem 3.6] and [Lac1; Theorem 4.3]; and we are grateful to M. Laca for bringing the reference [Sta] to our attention.

Theorem 5.7. *Let (U_p) be a sequence of unitaries on $\mathbb{C}^n$ satisfying*

$$\sum_{p=0}^{\infty} \|e_0 - U_p e_0\|^2 < \infty \tag{5.9}$$

and let $T_i = S_i \Gamma(U)$ where S_i is defined by (7.12) and $\Gamma(U) = \bigotimes_{p=0}^{\infty} U_p$ by (7.17). We have

$$e_0 \otimes e_0 \otimes e_0 \otimes \cdots = \mathbb{1} \in \mathcal{L}^2(\Omega, \mu). \tag{5.10}$$

The state ω_U defined on $M_{n^\infty} = \bigotimes_{m=0}^{\infty} M_n$ by

$$\begin{aligned}\omega_U(e_{i_1 j_1} \otimes e_{i_2 j_2} \otimes \cdots \otimes e_{i_m j_m} \otimes 1 \otimes 1 \otimes \cdots) \\ = \langle \mathbb{1}, T_{i_1} T_{i_2} \cdots T_{i_m} T_{j_m}^* \cdots T_{j_1}^* \mathbb{1} \rangle,\end{aligned} \tag{5.11}$$

where e_{ij}, $i, j = 1 \ldots, n$ is a set of matrix units for $M_n = B(\mathbb{C}^n)$, is a product state

$$\omega_U = \bigotimes_{m=0}^{\infty} \omega_{U,m} \tag{5.12}$$

where

$$\omega_{U,m} = \langle \xi_m, \cdot \xi_m \rangle, \tag{5.13}$$

$$\xi_0 = e_0 \tag{5.14}$$

$$\xi_m = U_0^* \cdots U_{m-1}^* e_0 \tag{5.15}$$

for $m = 1, 2, \ldots$. Hence, if (V_p) is another sequence of unitaries on $\mathbb{C}^n$ satisfying conditions (5.9)– (5.10), then the shift associated to $S_i\Gamma(V)$ is conjugate to the shift associated to $S_i\Gamma(U)$ if and only if there is a unitary $W \in U(n)$ such that

$$\sum_{m=0}^{\infty} (1 - |\langle V_0^* V_1^* \cdots V_m^* e_0, W U_0^* U_1^* \cdots U_m^* e_0 \rangle|) < +\infty. \tag{5.16}$$

Proof. Note first that the condition (5.9) is equivalent to $\bigotimes_{p=0}^{\infty} U_p e_0$ being a well-defined vector in $\bigotimes_{p=0}^{\infty} \mathbb{C}^n = \mathcal{L}^2(\Omega, \mu)$ (and (5.9) is implied by the condition $\sum_p \|1 - U_p\| < \infty$ considered in Section 7). Thus

$$\Gamma(U) = \bigotimes_{p=0}^{\infty} U_p \tag{5.17}$$

is a well-defined unitary operator on $\mathcal{L}^2(\Omega, \mu)$, so $T_i = S_i\Gamma(U)$ are well defined, and

$$\alpha \circ \gamma_U(A) = \sum_{i=1}^{n} T_i A T_i^* \tag{5.18}$$

for all $A \in \mathcal{B}(\mathcal{H})$, where α and γ_U are defined in Theorem 7.3.

It follows from (8.5) and (8.6) that

$$S_i^*(\eta_0 \otimes \eta_1 \otimes \eta_2 \otimes \cdots) = n^{-1/2} \eta_0(i)(\eta_1 \otimes \eta_2 \otimes \cdots) \tag{5.19}$$

$$S_i(\eta_0 \otimes \eta_1 \otimes \eta_2 \otimes \cdots) = n^{1/2}(\delta_i \otimes \eta_0 \otimes \eta_2 \otimes \cdots) \tag{5.20}$$

whenever $\eta_0 \otimes \eta_1 \otimes \eta_2 \otimes \cdots \in \mathcal{L}^2(\Omega, \mu)$. Using $T_i = S_i\Gamma(U)$, $T_i^* = \Gamma(U)^* S_i^*$, one then computes

(5.21)

$$T_i^*(\eta_0 \otimes \eta_1 \otimes \cdots) = n^{-1/2} \eta_0(i)(U_0^* \eta_1 \otimes U_1^* \eta_2 \otimes U_2^* \eta_3 \otimes \cdots)$$

$$T_i(\eta_0 \otimes \eta_1 \otimes \cdots) = n^{1/2}(\delta_i \otimes U_0 \eta_0 \otimes U_1 \eta_1 \otimes \cdots). \tag{5.22}$$

Iterating the formula for T_i^*, one computes

(5.23)
$$\begin{aligned} & T_{j_m}^* T_{j_{m-1}}^* \cdots T_{j_1}^* (e_0 \otimes e_0 \otimes e_0 \otimes \cdots) \\ &= n^{-m/2} (U_0^* e_0)(j_2)(U_0^* U_1^* e_0)(j_3) \\ &\quad \cdots (U_0^* U_1^* \cdots U_{m-2}^* e_0)(j_m)(U_0^* U_1^* \cdots U_{m-1}^* e_0 \otimes U_1^* U_2^* \cdots U_m^* e_0 \otimes \cdots) \end{aligned}$$

and hence

(5.24)
$$\begin{aligned} \langle \mathbb{1}, T_{i_1} \cdots T_{i_m} T_{j_m}^* \cdots T_{j_1}^* \mathbb{1} \rangle \\ &= n^{-m} \overline{U_0^* e_0(i_2)} \cdots \overline{(U_0^* U_1^* \cdots U_{m-2}^* e_0)(i_m)} \\ &\quad \cdots (U_0^* U_1^* \cdots U_{m-2}^* e_0)(j_m) \cdots (U_0^* e_0)(j_2) \\ &= n^{-m} \overline{\xi_0(i_1)}\, \overline{\xi_1(i_2)} \cdots \overline{\xi_{m-1}(i_m)} \\ &\quad \cdots \xi_{m-1}(j_m) \cdots \xi_0(j_1) \\ &= \omega_{U,0}(e_{i_1 j_1}) \omega_{U,1}(e_{i_2 j_2}) \cdots \omega_{U,m-1}(e_{i_m j_m}) \end{aligned}$$

if the ξ's and ω's are defined by (5.12)– (5.15).

Now, if we can show that $\mathbb{1}$ is a cyclic vector for the representation of UHF_n defined by the T's, the final conclusion of Theorem 7.4 follows from (5.7). But formuli (5.21)–(5.23) imply

(5.25)
$$\begin{aligned} & T_{i_1} T_{i_2} \cdots T_{i_m} T_{j_m}^* \cdots T_{j_1}^* (e_0 \otimes e_0 \otimes e_0 \otimes \cdots) \\ &= (U_0^* e_0)(j_2)(U_0^* U_1^* e_0)(j_2) \cdots (U_0^* U_1^* \cdots U_{m-2}^* e_0)(j_m) \\ & \cdots \delta_{i_1} \otimes U_0 \delta_{i_2} \otimes U_1 U_0 \delta_{i_3} \otimes \cdots \otimes U_{m-2} U_{m-3} \cdots U_0 \delta_{i_m} \otimes e_0 \otimes e_0 \otimes e_0 \otimes \cdots \end{aligned}$$

The linear combinations of these vectors for a fixed m are equal to the linear combinations of vectors of the form $\eta_0 \otimes \eta_1 \otimes \cdots \otimes \eta_{m-1} \otimes e_0 \otimes e_0 \otimes \cdots$ and hence $\mathbb{1}$ is cyclic for T. This ends the proof of Theorem 5.8. □

6. Construction of Shifts on $\mathcal{B}(\mathcal{H})$ With No Invariant States

We now consider a special case, and give an explicit construction of a family of shifts on $\mathcal{B}(\mathcal{H})$ which have no pure normal invariant states. This family may be constructed using the GNS representation theory corresponding to product states on UHF algebras of type n^∞. A shift α constructed in this manner will have Powers index n, i.e., $\alpha \in \mathrm{Shift}_n(\mathcal{B}(\mathcal{H}))$. This family of shifts was already considered in Example 5.5.

We begin by fixing an integer $n \geq 2$, and then we view $M_n(\mathbb{C})$ as the algebra of linear transformations on the n-dimensional vector space $\mathbb{C}^n$ over $\mathbb{C}$. For each $k \in \mathbb{N}$, let B_k be an isomorphic copy of $M_n(\mathbb{C})$, and in the usual way, we consider B_k to be embedded in the tensor product construction of the UHF algebra $\mathcal{A}$ of Glimm type n^∞, i.e., $\mathcal{A} = \bigotimes B_k$. We now construct a family of pure product states on $\mathcal{A}$ as follows. For each positive integer k, pick a unit vector h_k in $\mathbb{C}^n$, and let $e_k \in M_n(\mathbb{C})$ denote the corresponding rank one projection onto $\mathbb{C}h_k$. Throughout this section we impose the following conditions on the sequence of vectors $\{h_k\}$:

$$\lim_{k\to\infty} \|h_k - h\| = 0 \qquad \text{for some } h \in \mathbb{C}^n, \tag{6.1}$$

$$\sum_{k=1}^{\infty} \|h_k - h_{k+1}\| < \infty, \tag{6.2}$$

$$\prod_{k=m}^{\infty} |\langle h_k, h\rangle| = 0, \qquad \text{for all } m. \tag{6.3}$$

In fact, only *the last two* are really conditions, as (6.2) implies that (h_k) is Cauchy, and therefore (6.1) may be viewed as the definition of h. Using the first condition, there is an integer m such that $\langle h_k, h\rangle \neq 0$ for all $k \geq m$, so the third condition is equivalent to the divergence of the series $\sum_{k=m}^{\infty} -\ln\cos|\langle h_k, h\rangle|$. But one easily verifies that for sufficiently small x in $\mathbb{R}$, $x^2/2 \leq -\ln\cos(x) \leq x^2$, so the divergence of the series (and hence condition (6.3)) is equivalent to the following condition.

$$\sum_{k=1}^{\infty} \{\arccos(|\langle h_k, h\rangle|)\}^2 = \infty. \tag{6.3$'$}$$

Example. We provide an example of a sequence of vectors in $\mathbb{C}^2$ which satisfies (6.1), (6.2), (6.3′). Consider the sequence of real numbers

$$\{1, 1/2, 1/2, 1/3, 1/3, 1/3, \dots\} = \{\theta_k\},$$

i.e., each term $1/q$ appears q times in the sequence. Define h_k to be the vector $[\cos(\theta_k), \sin(\theta_k)]$. Then $\{h_k\}$ converges in norm to $h = [1, 0]$, so (6.1) is clearly satisfied. (An alternative example would be, $\theta_k = k^{-1/2}$, for $\forall k \in \mathbb{N}$.) To see that (6.2) holds, note that

$$\begin{aligned}
&\sum_{k=1}^{\infty} \|h_k - h_{k+1}\| \\
&\quad = \sum_{q=1}^{\infty} \|[\cos(q^{-1}), \sin(q^{-1})] - [\cos((q+1)^{-1}), \sin((q+1)^{-1})]\| \\
&\quad = \sum_{q=1}^{\infty} \{(\cos(q^{-1}) - \cos((q+1)^{-1})^2 + (\sin(q^{-1}) - \sin((q+1)^{-1}))^2\}^{1/2} \\
&\quad = \sum_{q=1}^{\infty} \{2 - 2(\cos(q^{-1} - (q+1)^{-1}))\}^{1/2} \\
&\quad = \sum_{q=1}^{\infty} 2^{1/2}\{1 - \cos((q(q+1))^{-1})\}^{1/2} \leq \sum_{q=1}^{\infty} 2^{1/2}\{1 - \cos^2((q(q+1))^{-1})\}^{1/2} \\
&\quad = \sum_{q=1}^{\infty} 2^{1/2} \sin((q(q+1))^{-1}) \leq \sum_{q=1}^{\infty} 2^{1/2}/(q(q+1)) = 2^{1/2} < \infty.
\end{aligned}$$

Finally, to see that (6.3′) holds, note that $|\langle h, h_k\rangle| = \cos(\theta_k)$, so

$$\sum_{k=1}^{\infty}\{\arccos(|\langle h, h_k\rangle|)\}^2 = \sum_{k=1}^{\infty}\theta_k^2 = \sum_{q=1}^{\infty} q(1/q)^2 = \sum_{q=1}^{\infty} 1/q = \infty.$$

For each positive integer k, let ρ_k be the pure state on B_k defined by $\rho_k(A) = \mathrm{tr}(e_k A)$, where tr is the non-normalized trace on B_k, and let ρ be the product state $\rho = \bigotimes \rho_k$ on $\mathcal{A}$. The state ρ is a pure state on $\mathcal{A}$, [Gui3, Corollary 2.2], so that in the corresponding GNS representation $(\pi_\rho, \mathcal{H}_\rho, \Omega_\rho)$ for ρ, the weak closure $\pi_\rho(\mathcal{A})''$ of $\pi_\rho(\mathcal{A})$ is isomorphic to $\mathcal{B}(\mathcal{H}_\rho)$.

Following the development by Guichardet on infinite tensor products of Hilbert spaces, [Gui3], (cf. [vNeu]) we record some important facts about $\mathcal{H}_\rho$ and $\mathcal{B}(\mathcal{H}_\rho)$. Consider all formal tensor products of vectors $x_1 \otimes x_2 \otimes \cdots$, where all but finitely many of the vectors x_k agree with the unit vectors h_k. Then there is a natural inner product which is defined on finite linear combinations of such vectors, satisfying

$$\left\langle \bigotimes x_k, \bigotimes y_k \right\rangle = \prod_{k=1}^{\infty} \langle x_k, y_k\rangle. \tag{6.4}$$

Note that all but finitely many of the inner products in the expression for the infinite product are 1. Then $\mathcal{H}_\rho$ is the Hilbert space completion, via the inner product above, of the set of finite linear combinations of vectors $\bigotimes x_k$, [Gui3, Section 1.1] (see also [vNeu, Section 3.11] and [Gui1–2]). Note that the cyclic unit vector Ω_ρ in the GNS representation for ρ is $\bigotimes h_k$.

Lemma 6.1. *Suppose $\{y_k : k \in \mathbb{N}\}$ is a sequence of unit vectors in $\mathbb{C}^n$ which satisfies $\sum_{k=1}^{\infty} \|y_k - h_k\| < \infty$. If for each $p \in \mathbb{N}$, H_p is the vector in $\mathcal{H}_\rho$ given by $H_p = y_1 \otimes \cdots \otimes y_p \otimes h_{p+1} \otimes h_{p+2} \otimes \cdots$, then $\{H_p\}$ is a Cauchy sequence.*

Proof. For positive integers $p < q$, $\|H_q - H_p\|_\rho \le \sum_{k=p+1}^{q} \|h_k - y_k\|$. □

Remark. As a result of the lemma it makes sense to represent the limit of such a Cauchy sequence by the symbol $\bigotimes y_k := \bigotimes_{k=1}^{\infty} y_k$ (cf. [Gui3, Proposition 1.1]).

Next we consider the algebra $C(\mathcal{H}_\rho)$ of compact operators in $\mathcal{B}(\mathcal{H}_\rho)$. Our presentation is implicit in the paper of Guichardet. For each $k \in \mathbb{N}$, select matrix units e_{ij}^k, $1 \le i, j \le n$, for B_k: i.e., for each k, $e_{ij}^k e_{pq}^k = \delta_{jp} e_{iq}^k$, and $\sum_{j=1}^{n} e_{jj}^k = I$. We impose the condition $e_{11}^k = e_k$ for each k. For each k, let $\{h_{k1}, \dots, h_{kn}\}$ be an orthonormal basis for $\mathbb{C}^n$ selected so that $h_{k1} = h_k$ and $h_{kj} = e_{jj}^k h_{kj}$, $j \in \{1, \dots, n\}$. Next let $\mathcal{I}$ be the set of all ordered sequences $P = \{p_1, p_2, \dots\}$ where $p_k \in \{1, 2, \dots, n\}$ for each k, and all but finitely many of the p_k are 1. We define δ_{PQ}, $P, Q \in \mathcal{I}$, to be 1 if $p_k = q_k$ for all k, and otherwise 0. We use the notation $\bigotimes h(P)$ to represent the unit vector $\bigotimes h_{kp_k}$ in $\mathcal{H}_\rho$. From the discussion above, linear combinations of the vectors $\bigotimes h(P)$ are dense in $\mathcal{H}_\rho$ and furthermore, $\langle \bigotimes h(P), \bigotimes h(Q)\rangle = \delta_{PQ}$, $P, Q \in \mathcal{I}$. The following result is clear.

Lemma 6.2. *The set $\{\bigotimes h(P) : P \in \mathcal{I}\}$ forms an orthonormal basis for $\mathcal{H}_\rho$.*

Next for $R, S \in \mathcal{I}$, we use the notation E_{RS} to represent the rank one operator in $\mathcal{B}(\mathcal{H}_\rho)$ which satisfies $E_{RS}\left(\bigotimes h(P)\right) = \delta_{PS}\left(\bigotimes h(R)\right)$. It is sometimes useful to write E_{RS} as $e^1_{r_1s_1} \otimes e^2_{r_2s_2} \otimes \cdots$. From the previous equation and the previous lemma, it follows that the operators E_{RS} form a complete set of matrix units for $C(\mathcal{H}_\rho)$ (= the compact operators), i.e, E_{RS}, satisfy the identities

$$E^*_{PQ} = E_{QP} \tag{6.5}$$

$$E_{PQ}E_{RS} = \delta_{QR}E_{PS}, \qquad P, Q, R, S \in \mathcal{I}, \tag{6.6}$$

and the set of finite linear combinations of the matrix units E_{PQ} is a uniformly dense subalgebra of $C(\mathcal{H}_\rho)$.

We next show that there is a natural way to make sense of the symbol $I \otimes E_{PQ}$ as a rank n operator in $C(\mathcal{H}_\rho)$, and then we use these operators to define a shift on $\mathcal{B}(\mathcal{H}_\rho)$ of index n. To begin, let $\bigotimes h(P)$, $P \in \mathcal{I}$, be any vector in the orthonormal basis for $\mathcal{H}_\rho$, and let m be a positive integer sufficiently large so that $p_k = 1$ for any $k \geq m$. Let v be any unit vector in $\mathbb{C}^n$. By condition (6.2), $\sum_{j=1}^\infty \|h_{m+j+1} - h_{m+j}\| < \infty$, so by Lemma 6.1, the symbol $v \otimes h_{1p_1} \otimes h_{2p_2} \otimes h_{3p_3} \otimes \cdots$ represents a unit vector in $\mathcal{H}_\rho$. Hence the symbol $I \otimes E_{PQ}$ represents a rank n operator in $C(\mathcal{H}_\rho)$ which maps, for any $v \in \mathbb{C}^n$, the vector $v \otimes h_{1q_1} \otimes h_{2q_2} \otimes h_{3q_3} \otimes \cdots$ to the vector $v \otimes h_{1p_1} \otimes h_{2p_2} \otimes h_{3p_3} \otimes \cdots$. Furthermore it is not difficult to show that the operators $I \otimes E_{PQ}$ satisfy the identities

$$(I \otimes E_{PQ})(I \otimes E_{RS}) = \delta_{QR}(I \otimes E_{PS}). \tag{6.7}$$

If $A \in B_1$, then clearly $\pi_\rho(A)$ and $I \otimes E_{PQ}$ commute, for all P, Q in $\mathcal{I}$. Hence $I \otimes E_{PQ} \in \pi_\rho(B_1)'$. On the other hand, consider the subalgebra $\bigotimes_{k=2}^\infty B_k$ of $\mathcal{A}$ generated by $B_2, B_3, \ldots$. From [Pow1, Lemma 2.3], $\pi_\rho(B_1)'$ coincides with $\pi_\rho(\bigotimes B_k)''$, which is a Type I subfactor of $\mathcal{B}(\mathcal{H}_\rho)$. Thus the set of finite linear combinations of compact operators of the form

$$\sum_{j=1}^n e^1_{jj} \otimes e^2_{r_2s_2} \otimes e^3_{r_3s_3} \otimes \cdots = \sum_{j=1}^n E_{R_jS_j}, \tag{6.8}$$

$R, S \in \mathcal{I}$, where for $R = \{r_1, r_2, r_3, \ldots\}$, R_j is the sequence $\{j, r_2, r_3, \ldots\}$, forms a weakly dense subalgebra of $\pi_\rho\left(\bigotimes_{k=2}^\infty B_k\right)''$. We summarize these results below.

Theorem 6.3. *For any $P, Q \in \mathcal{I}$, the symbol $I \otimes E_{PQ}$ represents a compact operator of rank n in $C(\mathcal{H}_\rho)$. The set of such operators forms a complete set of matrix units for the subalgebra of compact operators of the Type I subfactor $\pi_\rho(B_1)'$ of $\mathcal{B}(\mathcal{H}_\rho)$.*

The results of the preceding theorem enable us to define a shift α of index n on $\mathcal{B}(\mathcal{H}_\rho)$ which satisfies $\alpha(C(\mathcal{H}_\rho)) \subset C(\mathcal{H}_\rho)$. Fix $S = \{1, 1, 1, \ldots\}$ in $\mathcal{I}$. Since $I \otimes E_{SS}$ is a rank n projection in $C(\mathcal{H}_\rho)$, there exist partial isometries $W_1, \ldots, W_n$ in $\mathcal{B}(\mathcal{H}_\rho)$, each of rank one, satisfying

$$W_j^* W_j = E_{SS} \qquad \text{and} \qquad \sum_{j=1}^n W_j W_j^* = I \otimes E_{SS}.$$

Define operators $V_1, \dots, V_n$ in $\mathcal{B}(\mathcal{H}_\rho)$ by $V_j = \sum_{K\in\mathcal{I}}(I\otimes E_{KS})W_j(E_{SK})$ (cf. [Pow2, Theorem 2.4] it is straightforward to show that V_j's are isometries satisfying the Cuntz algebra relation $\sum_{j=1}^n V_j^* V_j = I$. We may thus define a shift α on $\mathcal{B}(\mathcal{H}_\rho)$ by setting, for $A \in \mathcal{B}(\mathcal{H}_\rho)$, $\alpha(A) = \sum_{j=1}^n V_j A V_j^*$. Note that for $P, Q \in \mathcal{I}$,

$$\begin{aligned}
\alpha(E_{PQ}) &= \sum_{j=1}^n V_j E_{PQ} V_j^* \\
&= \sum_{j=1}^n \left\{\sum_K (I\otimes E_{KS})W_j E_{SK}\right\} E_{PQ} \left\{\sum_L E_{LS} W_j^* (I\otimes E_{SL})\right\} \\
&= \sum_{j=1}^n (I\otimes E_{PS}) W_j^* E_{SS} W_j (I\otimes E_{SQ}) \\
&= \sum_{j=1}^n (I\otimes E_{PS}) W_j^* W_j (I\otimes E_{SQ}) \\
&= (I\otimes E_{PS})(I\otimes E_{SS})(I\otimes E_{SQ}) = I\otimes E_{PQ},
\end{aligned}$$

so that it is natural to use the notation $\alpha(A) = I\otimes A$ to denote this shift on $\mathcal{B}(\mathcal{H}_\rho)$. By Theorem 6.3 and the previous calculation, $\alpha(\mathcal{B}(\mathcal{H}_\rho))' = \pi_\rho(B_1)''$, so $\alpha \in \mathrm{End}_n(\mathcal{B}(\mathcal{H}_\rho))$, i.e., $[\mathcal{B}(\mathcal{H}_\rho) : \alpha(\mathcal{B}(\mathcal{H}_\rho))] = n^2$.

The following theorem gives a concrete realization of the representation of $\mathcal{O}_n$ defined in Theorem 3.1 and Lemma 5.2 in the present setting.

Theorem 6.4. *Let $V_1, \dots, V_n$ be the isometries defined as above. Then the mapping $\alpha(A) = \sum_{j=1}^n V_j A V_j^*$ is a shift endomorphism on $\mathcal{B}(\mathcal{H}_\rho)$ of index n satisfying the identities $\alpha(E_{PQ}) = I\otimes E_{PQ}$, $P, Q \in \mathcal{I}$.*

Proof. To finish the proof we must show that α is a shift. But it is not difficult to show that $\alpha^k(\mathcal{B}(\mathcal{H}_\rho))' \supset \pi_\rho(B_1\otimes\cdots\otimes B_k)$; and since $[\bigcup_k \pi_\rho(B_1\otimes\cdots\otimes B_k)]'' = \mathcal{B}(\mathcal{H}_\rho)$, it follows that $\bigcap_k \alpha^k(\mathcal{B}(\mathcal{H}_\rho))' = \mathbb{C}I$. □

Next we prove some preliminary results to be used in showing that there are no normal α-invariant pure states on $\mathcal{B}(\mathcal{H}_\rho)$. We could of course just refer to Lemma 5.4 and (5.7) for this, but we prefer to give an interesting direct argument. We state the following well- known result for convenience.

Lemma 6.5. *Let H, H' be unit vectors in $\mathcal{H}_\rho$, and let ω, ω' be the corresponding (pure normal) vector states on $\mathcal{B}(\mathcal{H}_\rho)$. Then $\|\omega - \omega'\| \le 2\|H - H'\|$.*

Proof. The result follows from the inequality

$$\begin{aligned}
|\omega(A) - \omega'(A)| &= |\langle AH, H\rangle - \langle AH', H'\rangle| \\
&\le |\langle AH, H\rangle - \langle AH, H'\rangle| + |\langle AH, H'\rangle - \langle AH', H'\rangle| \\
&\le \|AH\|\cdot\|H - H'\| + \|A(H - H')\|\cdot\|H\|.
\end{aligned}$$

□

For the following two results the notation $\mathcal{I}_m$, $m \in \mathbb{N}$, is used to denote the set of sequences $P \in \mathcal{I}$ whose entries are all 1 with the possible exception of the first $m-1$ entries. Observe that $\mathcal{I}_1 \subset \mathcal{I}_2 \subset \cdots$ and $\bigcup_m \mathcal{I}_m = \mathcal{I}$.

Lemma 6.6. *Let E be an orthogonal rank one projection in $\mathcal{B}(\mathcal{H}_\rho)$. Then for any $\epsilon > 0$, there is a finite linear combination E' of the rank one operators E_{PQ}, $P, Q \in \mathcal{I}$, which is an orthogonal projection satisfying $\|E - E'\| < \epsilon$.*

Proof. Since the E_{PQ}'s, $P, Q \in \mathcal{I}$, form a full set of matrix units for $C(\mathcal{H}_\rho)$, there are, for some m, sequences $P(1), P(2), \dots, P(m), Q(1), \dots, Q(m)$ of $\mathcal{I}_m$, and complex numbers c_j, $j = 1, \dots, m$, such that $\left\| E - \sum_{j=1}^{m} c_j E_{P(j)Q(j)} \right\| < \epsilon$. Hence the sum $\sum_{j=1}^{m} c_j E_{P(j)Q(j)}$ takes the form $A \otimes e_m \otimes e_{m+1} \otimes \cdots$, for some $A \in \pi_\rho(B_1 \otimes \cdots \otimes B_{m-1})$. Using standard functional calculus techniques [Gli, Lemma 1.6] we may assume that A is a projection in $\pi_\rho(B_1 \otimes \cdots \otimes B_{m-1})$ and the result follows. □

Remark. Note that if $\epsilon < 1$ then the projection constructed in the proof of the previous lemma must also be rank one.

The following lemma identifies an important *clustering property* which we take up again in Sections 7–8 below.

Lemma 6.7. *Let E' be any projection of the form $A \otimes e_m \otimes e_{m+1} \otimes \cdots$ as in the previous lemma. Let $H' \in \mathcal{H}_\rho$ be any unit vector obtained as a finite linear combination of vectors in the orthonormal basis $\{h(P) : P \in \mathcal{I}\}$ of $\mathcal{H}_\rho$. Let ω' be the vector state corresponding to H'. Then $\lim_{k\to\infty} \omega'(\alpha^k(E')) = 0$.*

Proof. Suppose for some $m \in \mathbb{N}$, H' is a finite linear combination of the vectors $\bigotimes h(R(1)), \dots, \bigotimes h(R(m))$, for some sequences $R(j)$ in $\mathcal{I}_m$. Then we may write H' in the form $\Phi \otimes h_m \otimes h_{m+1} \otimes h_{m+2} \otimes \cdots$, where Φ is a unit vector in the $(m-1)$-fold tensor product of $\mathbb{C}^n$. From the preceding theorem, $\alpha^k(E')$ has the form $I \otimes I \otimes \cdots \otimes I \otimes A \otimes e_m \otimes e_{m+1} \otimes \cdots$, where the first k tensors are I. Note that from the form of H' and of E' we have

$$\begin{aligned} \omega'(\alpha^k(E')) &\le \prod_{j=1}^{\infty} \langle e_{m+j} h_{m+j+k}, h_{m+j+k} \rangle \\ &= \prod_{j=1}^{\infty} |\langle h_{m+j}, h_{m+j+k} \rangle|^2 \end{aligned}$$

By condition (6.1), $\lim_{k\to\infty} h_{m+j+k} = h$, and by condition (6.3),

$$\prod_{j=1}^{\infty} |\langle h_{m+j}, h \rangle| = 0.$$

Applying these two conditions to the infinite product above, it is not hard to show that

$$\lim_{k\to\infty} \left\{ \prod_{j=1}^{\infty} |\langle h_{m+j}, h_{m+j+k} \rangle|^2 \right\} = 0, \tag{6.9}$$

and the result follows. □

Theorem 6.8. *Let α be a shift on $\mathcal{B}(\mathcal{H}_\rho)$ constructed as above. Then there are no pure normal α-invariant states on $\mathcal{B}(\mathcal{H}_\rho)$.*

Proof. Any pure normal state ω on $\mathcal{B}(\mathcal{H}_\rho)$ is a vector state $\omega = \langle H, \cdot H \rangle$, for some unit vector in $\mathcal{H}_\rho$. Given $\epsilon > 0$, there is a vector H' such that H' is a finite linear combination of the basis vectors $\bigotimes h(P)$, $P \in \mathcal{I}$, with $\|H - H'\| < \epsilon/3$. Let E be an orthogonal rank one projection in $C(\mathcal{H}_\rho)$, then by Lemma 6.6 there is a rank one projection E' which is a finite linear combination of the matrix units E_{PQ}, $P, Q \in \mathcal{I}$, such that $\|E - E'\| < \epsilon/3$. Let $\omega' = \langle H', \cdot H' \rangle$, then $\|\omega - \omega'\| < 2\epsilon/3$, by Lemma 6.5. Then since $\|\alpha^k(E) - \alpha^k(E')\| = \|E - E'\|$ we have, for all k, $|\omega(\alpha^k(E)) - \omega'(\alpha^k(E'))| < \epsilon$. But $\lim_{k\to\infty} \omega'(\alpha^k(E)) = 0$, by the previous lemma. Since ϵ is arbitrary we have $\lim_{k\to\infty} \omega(\alpha^k(E)) = 0$ also. Hence if ω were an α-invariant state then $\omega(E) = 0$ for all orthogonal rank one projections in $C(\mathcal{H}_\rho)$. But then $\omega|_{C(\mathcal{H}_\rho)} = 0$ which contradicts the normality of ω. This contradiction yields the result. □

Corollary 6.9. *Let α be a shift on $\mathcal{B}(\mathcal{H}_\rho)$ constructed as above. Then there are no normal α-invariant states on $\mathcal{B}(\mathcal{H}_\rho)$.*

Proof. Suppose ω is a finite linear combination $\sum_{j=1}^m a_j\omega_j$, $a_j \in \mathbb{R}^+$, of normal states $\omega_j = \langle H_j, \cdot H_j \rangle$. For any $\epsilon > 0$ one may, as in the proof of the theorem, choose unit vectors H_j', each of which is a finite linear combination of the basis vectors $\bigotimes h(P)$, $P \in \mathcal{I}$, and satisfying $\|H_j - H_j'\| < \epsilon/3$. Then if $\omega' = \sum_{j=1}^m a_j\omega_j'$, $\|\omega - \omega'\| < 2\epsilon/3$, and for E' chosen as above it is clear that $\lim_{k\to\infty} \omega'(\alpha^k(E')) = 0$, so that $\lim_k \sup\{|\omega(\alpha^k(E))|\} < \epsilon$, for all rank one projections E in $C(\mathcal{H}_\rho)$. Since ϵ is arbitrary we therefore obtain $\lim_{k\to\infty} \omega(\alpha^k(E)) = 0$, whence, as in the proof of the theorem, ω cannot be α-invariant. Finally, since any normal state ω of $\mathcal{B}(\mathcal{H}_\rho)$ may be approximated arbitrarily closely in norm by states which are finite linear combinations of vector states, we have $\lim_{k\to\infty} \omega(\alpha^k(E)) = 0$ for these states as well, so such a state cannot be α-invariant. □

7. Clustering Properties

Let $n \in \mathbb{N}$, $n > 1$, be given, and let $\mathcal{O}_n$ be the corresponding *Cuntz-algebra* on generators $(s_i)_{i=0}^{n-1}$ and relations, $s_i^* s_j = \delta_{ij}\mathbb{1}$, and $\sum_{i=0}^{n-1} s_i s_i^* = \mathbb{1}$. Let $\mathcal{H}$ be a separable (infinite-dimensional) complex Hilbert space. Then we saw that each element in $\mathrm{Rep}(\mathcal{O}_n, \mathcal{H})$ is specified by an assignment, $s_i \mapsto S_i$ of isometries of $\mathcal{H}$, subject to the Cuntz-relations,

$$S_i^* S_j = \delta_{ij} I \qquad \text{and} \qquad \sum_{i=0}^{n-1} S_i S_i^* = I \tag{7.1}$$

where I denotes the identity operator on $\mathcal{H}$. In Theorem 3.1, we identified the $U(n)$*-equivalence* (denoted $\sim$) on $\mathrm{Rep}(\mathcal{O}_n, \mathcal{H})$, and a (bijective) isomorphism

$$\mathrm{End}_n(\mathcal{B}(\mathcal{H})) \cong \mathrm{Rep}(\mathcal{O}_n, \mathcal{H})/\sim . \tag{7.2}$$

The element $\alpha \in \mathrm{End}_n(\mathcal{B}(\mathcal{H}))$ which corresponds to a given $(S_i) \in \mathrm{Rep}(\mathcal{O}_n, \mathcal{H})$ is

$$\alpha(A) = \sum_{i=0}^{n-1} S_i A S_i^* \tag{7.3}$$

defined for $\forall A \in \mathcal{B}(\mathcal{H})$. We also saw in Section 2 that α in (7.3) is a *shift* precisely when the operators

$$S_{i_1} \cdots S_{i_p} S_{j_p}^* \cdots S_{j_1}^* \tag{7.4}$$

act irreducibly on $\mathcal{H}$. (Note that the family in (7.4) is *indexed by* (variable) $p \in \mathbb{N}$, and double-multi-indices, $i_1, \ldots, i_p, j_1, \ldots, j_p$.)

For any two elements (S_i) and (T_j) in $\operatorname{Rep}(\mathcal{O}_n, \mathcal{H})$, it is clear from (7.1) that the matrix

$$(S_i^* T_j) \in \mathcal{M}_n(\mathcal{B}(\mathcal{H})) \tag{7.5}$$

is *unitary.* Note that the matrix entries, $M_{ij} = S_i^* T_j$ are generally just in $\mathcal{B}(\mathcal{H})$. It also follows (as noted in (2.4)–(2.5) above) that, *conversely*, if $(S_i) \in \operatorname{Rep}(\mathcal{O}_n, \mathcal{H})$, and $(M_{ij}) \in \mathcal{M}_n(\mathcal{B}(\mathcal{H}))$ is given unitary, *then* the operators T_j defined by

$$T_j = \sum_i S_i M_{ij} \tag{7.6}$$

satisfy the Cuntz-relations (7.1) and

$$S_i^* T_j = M_{ij}. \tag{7.7}$$

We think of the unitary operator-valued matrix (M_{ij}) as a *non-commutative Radon-Nikodym derivative* relating two elements in $\operatorname{Rep}(\mathcal{O}_n, \mathcal{H})$. By (7.2), it will therefore also be relating the corresponding elements in $\operatorname{End}_n(\mathcal{B}(\mathcal{H}))$.

We will show that there is a distinguished (up to unitary equivalence) element $(S_i) \in \operatorname{Rep}(\mathcal{O}_n, \mathcal{H})$ corresponding to a certain Haar measure (details below). It will be a shift, and we shall refer to it as the *Haar-shift.* It has a pure invariant state which is defined directly in terms of the constant function on Ω (where Ω is the infinite product group defined from $\mathbb{Z}_n$, see (7.11) below) and the Haar measure on this compact group Ω (see (7.11) below). Our purpose in the present section is to be able to read off from the Radon- Nikodym derivative (7.7) when some second element $(T_j) \in \operatorname{Rep}(\mathcal{O}_n, \mathcal{H})$ also has a pure invariant state. Recall, by (7.2–7.3), that

$$\beta(A) := \sum_j T_j A T_j^* \qquad (\text{for } A \in \mathcal{B}(\mathcal{H})) \tag{7.8}$$

is the element in $\operatorname{End}_n(\mathcal{B}(\mathcal{H}))$ which corresponds to the given (T_j); and that the possible existence of pure and invariant states refers then to the possible existence of unit-vectors $\xi \in \mathcal{H}$ such that

$$\langle \xi, A\xi \rangle = \langle \xi, \beta(A)\xi \rangle \qquad \text{for } \forall A \in \mathcal{B}(\mathcal{H}). \tag{7.9}$$

We saw in Theorem 4.1 that such a vector ξ exists if and only if there is a solution $c = (c_i) \in \ell_n^2$, with $\sum_{i=0}^{n-1} |c_i|^2 = 1$, to the *simultaneous eigenvalue problem,*

$$T_j^* \xi = c_j \xi \qquad \text{for } 0 \leq j < n. \tag{7.10}$$

Definition 7.1. Following [Jo-Pe], we now describe *the Haar-shift of index n*. We recall the residue group, $\mathbb{Z}_n := \mathbb{Z}/n\mathbb{Z} \simeq \{0, 1, \dots, n-1\}$, and the corresponding infinite Cartesian product group,

$$\Omega = (\mathbb{Z}_n)^{\mathbb{N}} = \prod_{p=1}^{\infty} \mathbb{Z}_n. \tag{7.11}$$

It is viewed as a compact abelian group under coordinate addition. The corresponding normalized Haar measure on $\Omega = \Omega_n$ will be denoted μ. It is the product measure corresponding to assigning equal weights n^{-1} at the n coordinates (of each factor). Points in Ω are denoted

$$x = (x_p)_{p=1}^{\infty} = (x_1, x_2, \dots),$$

and we have the right and left Bernoulli-shifts given respectively by

$$\sigma_i(x_1, x_2, \dots) = (i, x_1, x_2, \dots), \qquad \text{and} \qquad \sigma(x_1, x_2, \dots) = (x_2, x_3, \dots).$$

Clearly then, $\sigma \circ \sigma_i = \mathrm{id}_\Omega$ for all i, and furthermore,

$$\mu = \frac{1}{n} \sum_{i=0}^{n-1} \mu \circ \sigma_i^{-1}, \qquad \mu \circ \sigma_i = \frac{1}{n} \mu,$$

and therefore $\mu \circ \sigma^{-1} = \mu$.

It follows (see [Jo-Pe]) that we get a Cuntz-algebra system (S_i) on $\mathcal{H} = \mathcal{L}^2(\Omega, \mu)$ as follows: The operators S_i, and their adjoints S_i^*, will be acting on $\mathcal{H}$, and are given by,

$$S_i^* \xi = n^{-1/2} \xi \circ \sigma_i \qquad \text{for } \forall \xi \in \mathcal{H} = \mathcal{L}^2(\Omega, \mu). \tag{7.12}$$

The corresponding shift α from (7.3) will be called *the Haar-shift*. The vector state ω_0 on $\mathcal{B}(\mathcal{H})$, given by Haar-measure μ, and the constant function $\mathbb{1}$, is α-invariant: For $A \in \mathcal{B}(\mathcal{H})$, we therefore have,

$$\omega_0(A) = \langle \mathbb{1}, A\mathbb{1} \rangle_{\mathcal{L}^2} = \int_\Omega (A\mathbb{1})(x)\, d\mu(x), \tag{7.13a}$$

$$\omega_0(\alpha(A)) = \omega_0(A). \tag{7.13b}$$

We shall need the *character* on the group $\mathbb{Z}_n$, defined as follows: For $p \in \mathbb{Z}$, set

$$e(p) := \exp(i2\pi p/n), \tag{7.14}$$

and, for $j \in \mathbb{Z}_n$, $k \in \mathbb{Z}_n$, $e(jk)$ is given by this, with $p = jk$ and jk representing the product in the ring $\mathbb{Z}_n$. We shall write, $e_j(k) := \langle j, k \rangle = e(jk)$

Definition 7.2. To be able to describe our Radon-Nikodym derivative, we shall need a certain *unitary representation* acting on $\mathcal{H} = \mathcal{L}^2(\Omega, \mu)$.

Consider first the infinite product of identical copies of the group $U(n)$ of all n by n unitary matrices. Inside this product, we have the infinite-dimensional *subgroup* of elements $U = (U_p)_{p=1}^{\infty}$ subject to,

$$\sum_{p=1}^{\infty} \|I - U_p\| < \infty \tag{7.15}$$

where I is the $n \times n$ identity matrix $\begin{pmatrix} 1 & 0 & \cdots & 0 \\ 0 & 1 & & \\ \vdots & & \ddots & \\ 0 & & & 1 \end{pmatrix}$, and $\|\cdot\|$ is the C^*-norm on the $n \times n$ matrices. (In fact the weaker condition $\sum_p \|I - U_p\|^2 < \infty$ will suffice.) This subgroup will be denoted G_n, and it has a natural *unitary representation* on $\mathcal{L}^2(\Omega, \mu)$ which we proceed to describe.

Using (7.14), we note that the discrete group Λ, which is *dual* to $\Omega = \prod \mathbb{Z}_n$, is $\Lambda = \coprod \mathbb{Z}_n$ consisting of elements, $\lambda = (y_1, \dots, y_q, 0, 0, \dots)$ with at most a finite number of nonzero points y_j in $\mathbb{Z}_n$, followed by an *infinite* string of zeros. We get an *orthonormal basis* e_λ, indexed by $\lambda \in \Lambda$, and given by,

$$e_\lambda(x) := \prod_{p=1}^{q} e(y_p x_p) = \prod_{p=1}^{q} \langle y_p, x_p \rangle. \tag{7.16}$$

Note that we may also view (7.16) as an *infinite product*, but the factors after q will all be one. For $y \in \mathbb{Z}_n$, we further have the functions $e_y \in \ell_n^2 = \ell^2(\mathbb{Z}_n) \simeq \mathbb{C}^n$, given by, $e_y(x) := e(yx)$, see (7.14) above. This is again an orthonormal basis, now in ℓ_n^2, relative to the Haar measure on $\{0, \dots, n-1\}$, i.e., equal weights n^{-1}. Each unitary n by n matrix may then be identified with a unitary transformation on ℓ_n^2 relative to this basis. For an element, $U = (U_p)_{p=1}^{\infty}$ in G, we define $\Gamma(U)$ on the basis $\{e_\lambda\}_{\lambda \in \Lambda}$ as follows:

$$(\Gamma(U)e_\lambda)(x) := \prod_{p=1}^{\infty} (U_p e_{y_p})(x_p). \tag{7.17}$$

Using the argument from Section 6, we may then check that the right-hand side in (7.17) represents a well- defined element in $\mathcal{L}^2(\Omega, \mu)$, with the infinite product *convergent in mean-square.* We omit the simple argument which is based directly on the summability (7.15) defining the subgroup G. It is also clear that, $U \to \Gamma(U)$, is then a unitary representation of G acting on $\mathcal{L}^2(\Omega, \mu)$.

The construction of the unitary representation, $U \mapsto \Gamma(U)$ in (7.17) is parallel to the corresponding one, $U \mapsto \widetilde{\Gamma}(U)$ from [Gui2–3]; but with the $\widetilde{\Gamma}$ representation acting on von Neumann's Hilbert space $\mathcal{H}(h_p)$ associated with some (fixed) sequence $(h_p)_{p=1}^{\infty}$ (specified as in Section 6; see especially Lemma 6.6 and formula (6.4) for details): When $U = (U_p)_{p=1}^{\infty}$ in G is given, then $\widetilde{\Gamma}(U)$ is defined on the generic monomials in $\mathcal{H}(h_p)$ by the ansatz:

$$\widetilde{\Gamma}(U) \bigotimes_{p=1}^{\infty} \eta_p := \bigotimes_{p=1}^{\infty} U_p \eta_p .$$

We shall need below a specific unitary isomorphism defined in (7.24)

$$W : \mathcal{H}(h_p) \overset{\approx}{\to} \mathcal{L}^2(\Omega, \mu)$$

which *intertwines* the two representations, i.e., we have

$$\Gamma(U)W = W\widetilde{\Gamma}(U) \qquad \text{for } \forall U \in G.$$

We are now ready to state the main result of the present section.

Theorem 7.3. *Let $n \in \mathbb{N}$, $n > 1$, be given, and let Ω be the corresponding infinite product* (7.11)*. Let $\mathcal{H} = \mathcal{L}^2(\Omega, \mu)$, and let*

$$\alpha(A) = \sum_i S_i A S_i^*, \qquad A \in \mathcal{B}(\mathcal{H})$$

be the Haar-shift from (7.12)*. Let $U = (U_p)_{p=1}^{\infty} \in G_n$ be given such that, for all $a, b \in \ell_n^2$ of unit-norm, i.e., $\sum |a_i|^2 = \sum |b_i|^2 = 1$, we have*

$$\sum_{p=1}^{\infty} \left(\cos^{-1} |\langle a, U_1 \cdots U_p b\rangle|\right)^2 = \infty. \tag{7.18}$$

Let $\gamma = \gamma_U \in \operatorname{Aut}(\mathcal{B}(\mathcal{H}))$ be given by

$$\gamma_U(A) = \Gamma(U) A \Gamma(U)^* \qquad \text{for } \forall A \in \mathcal{B}(\mathcal{H}). \tag{7.19}$$

Then

$$\beta := \alpha \circ \gamma_U \tag{7.20}$$

is a shift of multiplicity n which has no pure normal invariant states. Moreover, we have

$$\beta(A) = \sum_j T_j A T_j^* \qquad \text{for } \forall A \in \mathcal{B}(\mathcal{H}) \tag{7.21}$$

where

$$T_j := S_j \Gamma(U) \qquad (\text{for } \forall j).$$

Proof. By Theorem 3.3, the endomorphism β in (7.21) may be defined alternatively from an element $(T_j') \in \operatorname{Rep}(\mathcal{O}_n, \mathcal{H})$ with corresponding Radon-Nikodym derivative,

$$S_j^* T_k' = n^{-1/2} e(jk) \Gamma(U); \tag{7.22}$$

and this representation is the one we identify (up to unitary equivalence) in Section 6 above, but acting in von Neumann's infinite-product Hilbert space. The result then follows from our Theorems 3.1 and 6.8 above. Let $H := \ell^2_n$, and let $(h_p)_{p=1}^\infty$ be a sequence of vectors in H such that $\|h_p\| = 1$, and $\exists h \in H$ such that,

(i) $\lim_{p\to\infty} h_p = h$,
(ii) $\sum_{p=1}^\infty \|h_p - h_{p+1}\| < \infty$ (recall that (i) is implied by (ii)), and
(iii) $\sum_p \left(\cos^{-1} |\langle h_p, h\rangle|\right)^2 = \infty$.

Then we saw in Theorems 6.4 and 6.8 that von Neumann's Hilbert space $\mathcal{H}(h_p)$ (specifics in [vNeu]) carries a shift $\tilde\beta(A) = I \otimes A$ which has *no* pure invariant states. If $v_i \in H$ is an orthonormal basis, and

$$\tilde T_i \xi := v_i \otimes \xi \qquad \text{for } \forall \xi \in \mathcal{H}(h_p), \tag{7.23}$$

then $(\tilde T_i) \in \operatorname{Rep}(\mathcal{O}_n, \mathcal{H}(h_p))$ and $\tilde\beta(A) = \sum_i \tilde T_i A \tilde T_i^*$ for $\forall A \in \mathcal{B}(\mathcal{H}(h_p))$. If $v \in H$ is chosen such that $V_p v = h_p$ for a sequence of unitaries $(V_p)_{p=1}^\infty$, then the unitaries, $U_p := V_p V_{p+1}^*$, satisfy, $h_p = U_p h_{p+1}$; and we have a correspondence between our conditions (i)–(iii) on the one hand, and the two conditions (7.15) and (7.18) for the sequence $(U_p)_{p=1}^\infty$ on the other hand. (Note that (7.18) is equivalent to $\sum_p (1 - |\langle a, U_1 \cdots U_p b\rangle|) = \infty$.) But, if $(T'_j) \in \operatorname{Rep}(\mathcal{O}_n, \mathcal{L}^2(\Omega, \mu))$ is given by (7.22), and $(\tilde T_j) \in \operatorname{Rep}(\mathcal{O}_n, \mathcal{H}(h_p))$ by (7.23), *then* we can show that they are intertwined by a unitary isomorphism, $W : \mathcal{H}(h_p) \to \mathcal{L}^2(\Omega, \mu)$. To describe W, pick, for each $p \in \mathbb{N}$, an orthonormal basis $(b_{j_p}^{(p)})$, indexed by $j_p \in \mathbb{Z}_n$, such that $b_0^{(p)} = h_p$; and, using Lemma 6.2, we get an associated orthonormal basis,

$$b(\lambda) := \bigotimes_{p=1}^{q} b_{j_p}^{(p)} \otimes \bigotimes_{i=q+1}^{\infty} h_i \tag{7.24}$$

for $\mathcal{H}(h_p)$. We then define our W by sending the basis element $b(\lambda)$ in (7.24) to $e_\lambda \in \mathcal{L}^2(\Omega, \mu)$, corresponding to the Λ- index given by $\lambda = (j_1, \dots, j_q, 0, 0, \dots)$; and it can easily be checked now that $W : \mathcal{H}(h_p) \xrightarrow{\approx} \mathcal{L}^2(\Omega, \mu)$ has the stated intertwining property, i.e., that $T'_i W = W \tilde T_i$ for $\forall i$. The proof is completed. □

Remarks 7.4. (i) The fact that β from (7.21–7.22) satisfies (7.20) follows from substitution of $T'_k = n^{-1/2} \sum_j e(jk) S_j \Gamma(U)$ into (7.21), (in fact also directly from Theorem 3.3 with $g = [n^{-1/2} e(jk)]_{jk}$):

$$\begin{aligned}
\beta(A) &= n^{-1} \sum_{j_1} \sum_{j_2} \sum_{k} e(kj_1)\overline{e(kj_2)} S_{j_1} \Gamma(U) A \Gamma(U)^* S_{j_2}^* \\
&= \sum_j S_j \gamma_U(A) S_j^* = \sum_j T_j A T_j^* \\
&= \alpha(\gamma_U(A)) \qquad \text{for } \forall A \in \mathcal{B}(\mathcal{L}^2(\Omega, \mu)).
\end{aligned}$$

(ii) Let $n \in \mathbb{N}$, $n > 1$, be given and let G_n denote the subgroup in $\prod_1^\infty U(n) = U(n)^{\mathbb{N}}$ defined by condition (7.15) above. Let α denote the Haar-shift of $\mathcal{B}(\mathcal{L}^2(\Omega_n))$.

Theorem 7.3 is then the assertion that $\{\alpha \circ \gamma_U : U \in G_n\}$ contains more than one conjugacy class, so it makes explicit the analysis from [Pow2, Theorem 2.3]. We showed that, when U in G_n satisfies (7.18), then $\alpha \circ \gamma_U$ represents a conjugacy class *different from* that of α.

(iii) In Example 5.6 and Theorem 5.8, we gave a complete abstract labeling of all the conjugacy classes of shifts considered in the present section. The labeling is the set of tensor products $\bigotimes_p U_p$, $\bigotimes_p V_p$ satisfying (5.9) (or stronger (7.15)) modulo the equivalence relation $\bigotimes_p U_p \sim \bigotimes_p V_p$ defined by (5.16). This labeling is non-smooth, as we may expect from Theorem 1.1, and there is a continuum of distinct conjugacy classes of this form. The classes we have considered in Examples 5.5 and 5.6 and Sections 6 and 7 (which are all the same except for the difference between (5.5) and (6.1), and between (5.9) and (7.15)) do not allow a complete smooth labeling. It would be interesting to understand how numerical labels separating conjugacy classes of n-shifts may possibly be assigned, like the clustering labels in (7.25) and (7.26) below. The situation is somewhat analogous to that in von Neumann factors. Von Neumann had a discrete labeling ($\mathrm{I}_n, n = 1, 2, \ldots, \infty$, II_1, II_∞, and III). In 1967 Powers introduced a real label λ to distinguish isomorphism classes III_λ in III, $0 \leq \lambda \leq 1$, and Connes and Takesaki introduced a non-smooth label, the flow of weights, to distinguish III_0 classes. A modest attempt of introducing some continuous labels is done in Remark 7.6. The set $P/\sim$ from Theorem 1.1 and Section 5 above (see especially details in Example 5.6) *is* in fact a complete labeling of the n-shift conjugacy classes. We showed also that *some of* the labels for n-shift conjugacy classes may be identified as points in our group G_n, but there are certainly other labels as well. We will encounter one of them in Section 8.

To stress the difference between the two conjugacy classes represented by the Haar-shift α, and by $\beta_U := \alpha \circ \gamma_U$, from Theorem 7.3 above, we include the following:

Corollary 7.5. *Let $n \in \mathbb{N}$, $n > 1$, be given. Let α be the corresponding Haar-shift, and let $U \in G_n$ be given subject to (7.18), and let $\beta_U := \alpha \circ \gamma_U$ be the corresponding transformed n- shift. Then we have, for all $A \in C(\mathcal{L}^2(\Omega_n, \mu))$ (= the compact operators) and all $\xi \in \mathcal{L}^2(\Omega_n, \mu)$, the two limits*

$$\lim_{k\to\infty} \langle \xi, \alpha^k(A)\xi \rangle = \omega_0(A)\|\xi\|^2 \tag{7.25}$$

and

$$\lim_{k\to\infty} \langle \xi, \beta_U^k(A)\xi \rangle = 0. \tag{7.26}$$

Proof. We have already noted that (7.26) is contained in the proof of our Lemma 6.7 and Theorem 7.3 above. We showed that the problem for $\mathcal{L}^2(\Omega_n, \mu)$ was equivalent to one in the von Neumann tensor product space $\mathcal{H}(h_p)$ for a certain sequence $(h_p)_{p=1}^\infty$ of vectors in ℓ_n^2; and we checked (7.26) in $\mathcal{H}(h_p)$ in Theorem 6.8 (and especially Lemma 6.7) by an approximation both in ξ and A.

Formula (7.25), on the other hand, may be checked directly from (7.8), and an iteration of the formula (7.12) for S_i^*. We recall from (7.13) that $\omega_0(\cdot)$ is calculated directly from the Haar- measure μ on Ω_n. We omit further details on (7.25), but refer instead to the paper [Jo- Pe]. □

Remark 7.6. Formula (7.25), and recent ideas from [Pow3], suggest the possibility of other conjugacy invariants for $\mathrm{Shift}_n(\mathcal{B}(\mathcal{H}))$. If α is a shift of index n on $\mathcal{B}(\mathcal{H})$, then any weak limit point of the sequence $(\alpha^m(A))$ as $m \to \infty$ has to be in $\bigcap_m \alpha^m(\mathcal{B}(\mathcal{H})) = \mathbb{C}1$ for all $A \in \mathcal{B}(\mathcal{H})$, and hence all weak limit points are scalar multiples of 1. Thus, if δ is any free ultrafilter on $\mathbb{N}$, we may define a state $\omega(\delta)$ on $\mathcal{B}(\mathcal{H})$ by

$$\omega(\delta)(A)1 = w - \lim_{N\to\delta} A_N$$

where

$$A_N = \frac{1}{N}\sum_{m=0}^{N-1} \alpha^m(A),$$

and then, of course,

$$\lim_{N\to\delta} \frac{1}{N}\sum_{m=0}^{N-1} \langle \xi, \alpha^m(A)\xi\rangle = \omega(\delta)(A)\|\xi\|^2.$$

As $\|\alpha(A_N) - A_N\| \le \frac{2}{N}\|A\| \to 0$ for $N \to \infty$, the state $\omega(\delta)$ is then α-invariant.

If there is a state ω on $\mathcal{B}(\mathcal{H})$ such that $\langle \xi, \alpha^m(A)\xi\rangle$ tends to $\omega(A)\|\xi\|^2$ in any stronger sense, for example

$$\lim_{m\to\infty} \langle \xi, \alpha^m(A)\xi\rangle = \omega(A)\|\xi\|^2 \tag{7.27}$$

or

$$\lim_{N\to\infty} \frac{1}{N}\sum_{m=0}^{N-1} \langle \xi, \alpha^m(A)\xi\rangle = \omega(A)\|\xi\|^2 \tag{7.28}$$

then $\omega(\delta) = \omega$, independently of δ.

Now if α has an invariant vector state, then this state is a Cuntz state in restriction to the representation π of $\mathcal{O}_n$ defining α, by Theorem 4.1. If ω denotes the normal extension of this state to $\mathcal{B}(\mathcal{H})$, then

$$\lim_{m\to\infty} \langle \xi, \alpha^m(A)\xi\rangle = \omega(A)\|\xi\|^2$$

for all $A \in C(\mathcal{H})$, by the same reasoning as in Corollary 7.5. (This is the absorption property of [Pow3].) But any state on $C(\mathcal{H})$ (= the compact operators), has a unique extension as a state to $\mathcal{B}(\mathcal{H})$, and hence $\omega(\delta) = \omega$ for any δ, also in this case. On the other hand, if α does not have invariant vector states, then $\omega(\delta)$ is necessarily non- normal.

Note also that if $x \in \mathrm{UHF}_n$, and $\xi \in \mathcal{H}$ with $\|\xi\| = 1$, then

$$\lim_{m\to\infty} |\langle \xi, \alpha^m(\pi(x))\xi\rangle - \langle \xi, \alpha^{m+1}(\pi(x))\xi\rangle| = 0$$

by Lemma 5.2, and hence, if $\omega_0(\delta)$ is defined on $\mathcal{B}(\mathcal{H})$ by

$$\omega_0(\delta)(A)I = w - \lim_{m\to\delta} \alpha^m(A),$$

then $\omega_0(\delta)$, restricted to the weakly dense subalgebra $\pi(\mathrm{UHF}_n)$ of $\mathcal{B}(\mathcal{H})$, is α-invariant, and clearly $\omega(\delta)$ is an extension of $\omega_0(\delta)$ from $\pi(\mathrm{UHF}_n)$ to $\mathcal{B}(\mathcal{H})$, i.e.,

$$\omega(\delta)(\pi(x))I = w - \lim_{m\to\delta} \alpha^m(\pi(x))$$

for $x \in \mathrm{UHF}_n$. If we put $\omega(\alpha,\delta) = \omega(\delta)$, and if $\gamma \in \mathrm{Aut}(\mathcal{B}(\mathcal{H}))$, it is easily verified that

$$\omega(\gamma\alpha\gamma^{-1},\delta) = \omega(\alpha,\delta)\circ\gamma^{-1}.$$

It is presently unclear how to get a conjugacy invariant out of this, and relate this invariant to $P/\sim$. On the other hand, we are able to verify the absorption property (7.27) for a class of shifts related to those considered in the previous section. (For more on the absorption property, see [Pow3; Definition 2.12].) In the following result we return to the notation of Section 6.

Theorem 7.7. *Suppose $\{h_k : k \in \mathbb{N}\}$ is a sequence of unit vectors in $\mathbb{C}^n$ satisfying the conditions*

(i) $\sum_{k=1}^{\infty} \|h_{k+1} - h_k\| < \infty$,
(ii) $\lim_{m\to\infty} \prod_{k=m}^{\infty} \langle h_k, h\rangle = 1$,

where $h = \lim_{k\to\infty} h_k$. Let $\rho = \bigotimes \rho_k$ be the pure product state on UHF_n where $\rho_k = \langle h_k, \cdot h_k\rangle$, where GNS representation $(\pi_\rho, \mathcal{H}_\rho, \Omega_\rho)$. Let ω be the symmetric pure product state $\omega = \bigotimes \omega_h$ on UHF_n, $\omega_h = \langle h, \cdot h\rangle$. Then $\lim_{k\to\infty}\langle \xi, \alpha^k(A)\xi\rangle = \omega(A)\|\xi\|^2$, for all $A \in \mathcal{B}(\mathcal{H}_\rho)$ and all $\xi \in \mathcal{H}_\rho$, where α is the shift given by $\alpha(A) = I \otimes A$ on $\mathcal{B}(\mathcal{H}_\rho)$.

Remark 7.8. Note that condition (i) is the same as (6.2) above. If one assumes that condition (i) holds, then (ii) is the negation of condition (6.3).

Proof. We first recall that only condition (6.2) was used in the proof of Theorem 6.4 so that there exists a shift α on $\mathcal{B}(\mathcal{H}_\rho)$ which satisfies $\alpha(A) = I \otimes A$ for all A in $\mathcal{B}(\mathcal{H}_\rho)$.

Next, since for sufficiently large m, $\prod_{k=m}^{\infty}\langle h_k, h\rangle$ exists and is nonzero, it follows ([vNeu, Lemma 2.4.2]) that $\prod_{k=m}^{\infty} |\langle h_k, h\rangle|$ exists and also $\sum_{k=1}^{\infty}|\theta_k| < \infty$, where $\theta_k \in (-\pi, \pi]$ is the argument of $\langle h_k, h\rangle$. Hence, since $|e^{i\theta} - 1| \le |\theta|$ for $\theta \in (-\pi,\pi]$,

$$\begin{aligned}\sum_{k=1}^{\infty} |1 - \langle h_k, h\rangle| &\le \sum_{k=1}^{\infty}\{|\langle h_k, h\rangle|\cdot|e^{i\theta_k} - 1| + ||\langle h_k, h\rangle| - 1|\}\\ &\le \sum_{k=1}^{\infty}\{|\theta_k| + ||\langle h_k, h\rangle| - 1|\}\\ &< \infty,\end{aligned}$$

so ([Gui3, Proposition 1.1]), $h\otimes h\otimes h\cdots$ represents a unit vector in the Hilbert space $\mathcal{H}_\rho$. For simplicity we write $H = h\otimes h\otimes\cdots$.

Now suppose that ξ is a unit vector in $\mathcal{H}_\rho$, then arguing as in Lemma 6.7 there is a positive integer m and a unit vector ξ' which is a finite linear combination of vectors among the orthonormal set $\{\bigotimes h(P) : P \in I_m\}$, and satisfying $\|\xi - \xi'\| < \epsilon/4$. Write $\xi' = \sum_{P\in I_m} a_P(\bigotimes h(P))$. The maximum number of nonzero terms in this sum is

$N = n^m$. Then using the fact that the vector H lies in $\mathcal{H}_\rho$, one may show that there exists a positive integer $M > m$ sufficiently large so that if for each $P \in I_m$ one obtains a new vector H_P from $\bigotimes h(P)$ by replacing the tail $\otimes h_{M+1} \otimes h_{M+2} \otimes \cdots$ of $\bigotimes h(P)$ with $\otimes h \otimes h \otimes \cdots$, then $\|\bigotimes h(P) - H_P\| < \epsilon/(4N)$. Then if ξ'' is the vector $\sum_{P \in I_m} a_P H_P$, one sees that ξ'' is a unit vector satisfying $\|\xi' - \xi''\| \le \sum_{P \in I_m} |a_P| \, \|\bigotimes h(P) - H_P\| \le N \cdot 1 \cdot \epsilon/(4N) = \epsilon/4$. Hence $\|\xi - \xi''\| < \epsilon/2$, and therefore, by Lemma 6.5, $|\langle \xi, \alpha^k(A)\xi\rangle - \langle \xi'', \alpha^k(A)\xi''\rangle| \le \epsilon\|A\|$, for all $A \in \mathcal{B}(\mathcal{H}_\rho)$ and all $k \in \mathbb{N}$.

But if k is chosen to be greater than M, note that $\langle \xi'', \alpha^k(A)\xi''\rangle = \langle H, AH\rangle = \omega(A)$. Since ϵ is arbitrary, we obtain $\lim_{k\to\infty}\langle \xi, \alpha^k(A)\xi\rangle = \omega(A)$. □

8. Nearest Neighbor States

In Sections 6 and 7, we constructed shifts on $\mathcal{B}(\mathcal{H})$ coming from product states on UHF_n. In this section, we will consider a state on UHF_n which is a prototype of what could be called a nearest neighbor state, since it couples nearest neighbors in the tensor product decomposition $\mathrm{UHF}_n = M_n \otimes M_n \otimes M_n \otimes \cdots$. We will study this shift by perturbing the shifts with invariant states considered in Section 4. To this end we need to describe the latter more explicitly. We assume $n \in \{2, 3, \dots\}$.

Let $\eta = (\eta_0, \eta_1, \dots, \eta_{n-1})$ be a sequence of complex numbers with

$$\sum_{k=0}^{n-1} |\eta_k|^2 = 1.$$

We also assume for the moment that $\eta_k \ne 0$ for $k = 0, \dots, n-1$. Let $\Omega = \prod_{k=0}^{\infty} \mathbb{Z}_n$, so that Ω is homeomorphic to the Cantor set. Equip Ω with the infinite product measure μ obtained from the measure on $\mathbb{Z}_n$ with weights $|\eta_0|^2, |\eta_1|^2, \dots, |\eta_{n-1}|^2$ on the n points. Define continuous open injections $\sigma_i : \Omega \to \Omega$ by

$$\sigma_i(x_0, x_1, x_2, \dots) = (i, x_0, x_1, \dots) \tag{8.1}$$

and define the shift $\sigma : \Omega \to \Omega$ by

$$\sigma(x_0, x_1, x_2, \dots) = (x_1, x_2, x_3, \dots). \tag{8.2}$$

The corresponding element in $\mathrm{Rep}(\mathcal{O}_n, \mathcal{L}^2(\mu))$ may be identified by: Define

$$S_i^* \xi = \bar{\eta}_i \xi \circ \sigma_i \tag{8.3}$$

$$S_i \xi = \bar{\eta}_i^{-1} \chi_{\sigma_i \Omega} \xi \circ \sigma, \tag{8.4}$$

or

$$(S_i^* \xi)(x_0, x_1, x_2, \dots) = \bar{\eta}_i \xi(i, x_0, x_1, \dots) \tag{8.5}$$

$$(S_i \xi)(x_0, x_1, x_2, \dots) = \bar{\eta}_i^{-1} \delta_{i x_0} \xi(x_1, x_2, \dots). \tag{8.6}$$

One checks, using the formula (see [Kak])

$$\int_\Omega \psi(x_0, x_1, \dots)\, d\mu(x_0, x_1, \dots) = \sum_{i=0}^{n-1} |\eta_i|^2 \int_\Omega \psi(i, x_1, x_2, \dots)\, d\mu(x_1, x_2, \dots),$$

that S_i^* is indeed the adjoint of S_i, and that S_i satisfy the Cuntz relations (2.1).

In fact notice that *distinct* weight distributions, $p = (p_i)_{i\in\mathbb{Z}_n}$, where $p_i := |\eta_i|^2 > 0$, give corresponding *orthogonal* (i.e., mutually singular) measures $\mu = \mu_p$ on $\Omega = \prod_0^\infty \mathbb{Z}_n$ by an application of Kakutani's theorem [Kak]. However the individual operators S_i in (8.6) depend both on the p_i's and on the phases $\eta_i|\eta_i|^{-1}$. Note also that the constant function $\xi = \mathbb{1}$ is a joint eigenvector for $S_1^*, \dots, S_n^*$ with eigenvalues $\bar\eta_1, \dots, \bar\eta_n$, and hence $\langle \mathbb{1}, \cdot\mathbb{1}\rangle$ defines the Cuntz state on $\mathcal{O}_n$ by Theorem 4.1. We have

$$(8.7)\qquad (S_{i_1}\cdots S_{i_k}\mathbb{1})(x_0,x_1,x_2,\dots) = \bar\eta_{i_1}^{-1}\delta_{i_1,x_0}\bar\eta_{i_2}^{-1}\delta_{i_2,x_1}\cdots\bar\eta_{i_k}^{-1}\delta_{i_k,x_{k-1}}$$

and hence $\mathbb{1}$ is a cyclic vector for the representation.

Note further that

$$(8.8)\quad \begin{aligned}&(S_{i_1}\cdots S_{i_k}S_{j_k}^*\cdots S_{j_1}^*\xi)(x_0,x_1,\dots)\\ &\qquad = \bar\eta_{i_1}^{-1}\delta_{i_1,x_0}\cdots\bar\eta_{i_k}^{-1}\delta_{i_k,x_{k-1}}\bar\eta_{j_k}\cdots\bar\eta_{j_1}\xi(j_1,\dots,j_k,x_k,x_{k+1},\dots)\end{aligned}$$

and hence

$$(8.9)\quad \begin{aligned}&(S_{i_1}\cdots S_{i_k}S_{i_k}^*\cdots S_{i_1}^*\xi)(x_0,x_1,\dots)\\ &\qquad = \delta_{i_1,x_0}\delta_{i_2,x_1}\cdots\delta_{i_k,x_{k-1}}\xi(i_1,\dots i_k,x_k,x_{k+1},\dots)\\ &\qquad = \delta_{i_1,x_0}\delta_{i_2,x_1}\cdots\delta_{i_k,x_{k-1}}\xi(x_0,x_1,x_2,\dots).\end{aligned}$$

It follows from (8.8) that UHF_n acts irreducibly on $\mathcal{L}^2(\Omega,\mu)$, confirming by Theorem 3.3 that the corresponding endomorphism of $\mathcal{B}(\mathcal{H})$ is a shift. It follows from (8.9) that $\pi(D_n)$ identifies with $C(\Omega)$ acting as multiplication operators on $\mathcal{L}^2(\Omega,\mu)$.

Since, as we have seen in the proof of Theorem 4.2, the canonical action of $U(n)$ acts transitively on the Cuntz states, one may obtain concrete realizations of the representation associated to a unit vector $(\eta_0,\dots,\eta_{n-1})$ in $\mathbb{C}^n$, where some of the components are zero, by applying canonical actions on states where all the components are nonzero.

For simplicity, let us specialize to the case

$$(8.10)\qquad \eta_i = n^{-1/2};\qquad i = 0,\dots,n-1,$$

so

$$(8.11)\qquad S_i^*\xi = n^{-1/2}\xi\circ\sigma_i$$

$$(8.12)\qquad S_i\xi = n^{1/2}\chi_{\sigma_i\Omega}\xi\circ\sigma$$

for $\xi \in \mathcal{L}^2(\Omega,\mu)$.

In this case $\mathcal{L}^2(\Omega,\mu)$ has an orthonormal basis consisting of all finite products

$$(8.13)\qquad e_j(x) = \langle j_0,x_0\rangle\langle j_1,x_1\rangle\cdots\langle j_k,x_k\rangle$$

for $j = (j_0, j_1, \dots, j_k, 0, 0, \dots) \in \hat{\Omega}$, and $x = (x_0, x_1, \dots) \in \Omega$, where

$$\langle j, x \rangle = \exp(2\pi i\, jx/n) \tag{8.14}$$

for $j, x \in \mathbb{Z}_n$.

We will now make a realization $T_1, \dots, T_n$ of $\mathcal{O}_n$ on $\mathcal{L}^2(\Omega, \mu)$ which defines a shift without pure normal invariant states. Any such realization has the form

$$T_i = \sum_{j=1}^{n} S_j m_{ji}$$

by (2.4), where $[m_{ji}]$ is a unitary matrix in $M_n(\mathcal{B}(\mathcal{L}^2(\Omega, \mu)))$. We take $[m_{ji}]$ to be a diagonal matrix with m_{ii} being the multiplication operator on $\mathcal{L}^2(\Omega, \mu)$ defined by the function

$$m_{ii}(x_0, x_1, x_2, \dots) = \langle i, x_0 \rangle. \tag{8.15}$$

In formulas (2.5) and (2.6) above, we introduced, for general pairs $(S_i), (T_i)$ in $\mathrm{Rep}(\mathcal{O}_n, \mathcal{H})$, the *unitary transfer operator* U which relates them. Recall that, for a general such pair, U is given by,

$$U = \sum_j T_j S_j^* = \sum_i \sum_j S_i m_{ij} S_j^*;$$

and, for the present concrete pair, a calculation yields,

$$(U\xi)(x_0, x_1, \dots) = \langle x_0, x_1 \rangle \xi(x_0, x_1, \dots) \tag{8.16}$$

for $\forall \xi \in \mathcal{L}^2(\Omega, \mu)$ and for $\forall x = (x_0, x_1, \dots) \in \Omega$.

We are now ready to give the new shift associated with *nearest neighbor states.* As we note in Remark 8.3 below, this shift is *not* conjugate to any one of those from Sections 6–7. Recall they were constructed from infinite product states.

Theorem 8.1. *Let* $(S_i) \in \mathrm{Rep}(\mathcal{O}_n, \mathcal{L}^2(\mu))$ *be given by* (8.12), *and let* α *be the corresponding Haar shift. Let* $T_i \in \mathrm{Rep}(\mathcal{O}_n, \mathcal{L}^2(\mu))$ *be given by,* $T_i = S_i m_{ii}$, *with the functions* $m_{ii}(\cdot)$ *on* Ω *defined in* (8.15)*; and let,* $\beta(A) := \sum_i T_i A T_i^*$, *(for* $\forall A \in \mathcal{B}(\mathcal{L}^2(\mu))$*) be the corresponding endomorphism.*

Then β *is a shift of Powers index* n, *and* β *does not allow invariant vector states. The corresponding state* ω *of* UHF_n *is given by*

$$\begin{aligned}\omega(e^{(1)}_{i_1 j_1} \otimes e^{(2)}_{i_2 j_2} \otimes \cdots \otimes e^{(k)}_{i_k j_k}) &= \langle \mathbb{1}, T_{i_1} T_{i_2} \cdots T_{i_k} T^*_{j_k} \cdots T^*_{j_1} \mathbb{1} \rangle \\ &= \frac{1}{n^k} \delta_{i_k j_k} \langle i_1, i_2 \rangle \langle i_2, i_3 \rangle \cdots \langle i_{k-1}, i_k \rangle \\ &\quad \cdot \overline{\langle j_1, j_2 \rangle}\, \overline{\langle j_2, j_3 \rangle} \cdots \overline{\langle j_{k-1}, j_k \rangle}.\end{aligned}$$

Proof. We have

$$T_i^* = \bar{m}_{ii} S_i^*$$

so by (8.11) and (8.13),

$$(8.17) \qquad (T_i^*\xi)(x_0, x_1, \dots) = \overline{\langle i, x_0\rangle} n^{-1/2}\xi(i, x_0, x_1, \dots)$$

for all $\xi \in \mathcal{L}^2(\Omega, \mu)$. Assume now (ad absurdum) that ξ is a joint eigenvector of the T_i^*'s:

$$(8.18) \qquad T_i^*\xi = \lambda_i\xi$$

where $\lambda_i \in \mathbb{C}$ and $\sum_{i=0}^{n-1} |\lambda_i|^2 = 1$. Combining with (8.17) we have

$$(8.19) \qquad \lambda_i\xi(x_0, x_1, x_2, \dots) = \overline{\langle i, x_0\rangle} n^{-1/2}\xi(i, x_0, x_1, \dots)$$

for $i = 0, \dots, n-1$; that is,

$$(8.20) \qquad \xi(y_0, y_1, y_2, \dots) = \lambda_{y_0} n^{1/2}\langle y_0, y_1\rangle\xi(y_1, y_2, y_3, \dots).$$

By recursion,

$$(8.21) \qquad \xi(y_0, y_1, y_2, \dots) \\ = n^{m/2}\lambda_{y_0}\lambda_{y_1}\cdots\lambda_{y_{m-1}}\langle y_0, y_1\rangle\langle y_1, y_2\rangle\cdots\langle y_{m-1}, y_m\rangle\xi(y_m, y_{m+1}, \dots).$$

By the axiom of choice, there exist non-zero functions ξ satisfying (8.19): One divides all strings $(y_0, y_1, \dots)$ into equivalence classes characterized by having the same tail up to translations, and then one assigns an arbitrary value of ξ to one element in each equivalence class and uses the recursion (8.19) to compute the value of ξ on the other elements in the class. We will now, however, argue that (8.19) has no nonzero solution $\xi \in \mathcal{L}^2(\Omega, \mu)$. We will show this by demonstrating that if $\xi \in \mathcal{L}^2(\Omega, \mu)$ and ξ satisfies (8.19), then ξ is orthogonal to all the vectors in the orthonormal basis (8.13) for $\mathcal{L}^2(\Omega, \mu)$. The proof uses the Fourier transform on the compact abelian group Ω, and the corresponding basis: If $e_j(x)$ is the element given by (8.13), choose $m > k+1$ in (8.19) to obtain

$$(8.22) \qquad \widetilde{\xi}(j_0, j_1, \dots, j_k, 0, 0, \dots) =: \langle e_j, \xi\rangle = \int_\Omega \overline{e_j(y)}\xi(y)\, d\mu(y)$$
$$= n^{-m/2}\sum_{y_0=1}^{n}\cdots\sum_{y_{m-1}=1}^{n}\lambda_{y_0}\lambda_{y_1}\cdots\lambda_{y_{m-1}}$$
$$\cdot\overline{\langle j_0, y_0\rangle}\,\overline{\langle j_1, y_1\rangle}\cdots\overline{\langle j_k, y_k\rangle}\cdot\langle y_0, y_1\rangle\langle y_1, y_2\rangle\cdots\langle y_{m-2}, y_{m-1}\rangle$$
$$\cdot\int_\Omega \langle y_{m-1}, y_m\rangle\xi(y_m, y_{m+1}, \dots)\, d\mu(y_m, y_{m+1}, \dots)$$
$$= n^{-m/2}\sum_{y_0=1}^{n}\cdots\sum_{y_{m-1}=1}^{n}\lambda_{y_0}\lambda_{y_1}\cdots\lambda_{y_{m-1}}\cdot\overline{\langle j_0, y_0\rangle}\,\overline{\langle j_1, y_1\rangle}\cdots\overline{\langle j_k, y_k\rangle}$$
$$\cdot\langle y_0, y_1\rangle\langle y_1, y_2\rangle\cdots\langle y_{m-2}, y_{m-1}\rangle\cdot\widetilde{\xi}(-y_{m-1}, 0, 0, 0, \dots).$$

In the case $k = 0$, $m = 1$ an analogous formula takes the form

(8.23)
$$\begin{aligned}\widetilde{\xi}(j_0, 0, 0, \dots) &= n^{-1/2} \sum_{y_0=1}^{n} \lambda_{y_0} \overline{\langle j_0, y_0 \rangle} \int \langle y_0, y_1 \rangle \xi(y_1, y_2, \dots)\, d\mu(y_1, y_2, \dots) \\ &= n^{-1/2} \sum_{y_0=1}^{n} \lambda_{y_0} \overline{\langle j_0, y_0 \rangle}\, \widetilde{\xi}(-y_0, 0, \dots).\end{aligned}$$

It follows, with $\widetilde{\xi}(j) = \widetilde{\xi}(j, 0, 0, \dots)$, that

(8.24)
$$\begin{aligned}\sum_{j \in \mathbb{Z}_n} |\widetilde{\xi}(j)|^2 &= n^{-1} \sum_{j,y,z \in \mathbb{Z}_n} \overline{\lambda}_y \lambda_z \langle j, y \rangle \overline{\langle j, z \rangle}\, \overline{\widetilde{\xi}(-y)}\, \widetilde{\xi}(-z) \\ &= n^{-1} \sum_{y,z \in \mathbb{Z}_n} n\delta(y - z) \overline{\lambda}_y \lambda_z \overline{\widetilde{\xi}(-y)}\, \widetilde{\xi}(-z) \\ &= \sum_{y \in \mathbb{Z}_n} |\lambda_y|^2\, |\widetilde{\xi}(-y)|^2,\end{aligned}$$

so

(8.25)
$$\sum_{j \in \mathbb{Z}_n} |\widetilde{\xi}(j)|^2 = \sum_{j \in \mathbb{Z}_n} |\lambda_j|^2\, |\widetilde{\xi}(-j)|^2.$$

Since $\sum_{j \in \mathbb{Z}_n} |\lambda_j|^2 = 1$, it follows immediately that if at least two of the λ_j are nonzero, then $\widetilde{\xi}(j, 0, 0, \dots) = 0$ for all $j \in \mathbb{Z}_n$. But the recursion relation (8.22) then implies that $\widetilde{\xi}(j_0, j_1, \dots, j_k, 0, 0, \dots) = 0$ for each $(j_0, j_1, j_2, \dots, j_k, 0, 0, \dots) \in \hat{\Omega}$. It follows that

(8.26)
$$\xi = 0 \qquad \text{in } \mathcal{L}^2(\Omega, \mu)$$

and (8.18) has no nonzero solution.

If all λ_j except one are zero, e.g., $(\lambda_j) = (1, 0, \dots, 0)$, then it follows directly from (8.21) that

$$\xi(y_0, y_1, y_2, \dots) = 0$$

unless $(y_0, y_1, y_2, \dots) = (0, 0, 0, \dots)$. But this single point has Haar measure zero, so again $\xi = 0$ in $\mathcal{L}^2(\Omega, \mu)$.

This completes the proof that (8.18) cannot have a nontrivial solution. This means, by Theorem 4.1, that the endomorphsim $\beta(A) := \sum_i T_i A T_i^*$, $A \in \mathcal{B}(\mathcal{H})$, *cannot* have an invariant vector state.

To complete the proof of Theorem 8.1 we have to show that β really is a shift (using Theorem 3.3), and to compute the corresponding state on UHF_n. But using (8.17) and the corresponding formula

$$(T_i \xi)(x_0, x_1, \dots) = \langle x_0, x_1 \rangle n^{1/2} \delta_{i x_0} \xi(x_1, x_2, \dots)$$

one computes

$$\begin{aligned}
&(T_{i_1}\cdots T_{i_k}T_{j_k}^*\cdots T_{j_1}^*\xi)(x_0,x_1,\dots)\\
&=\delta_{i_1,x_0}\delta_{i_2,x_1}\cdots\delta_{i_k,x_{k-1}}\langle x_0,x_1\rangle\langle x_1,x_2\rangle\cdots\langle x_{k-1},x_k\rangle\\
&\qquad\cdot\overline{\langle j_1,j_2\rangle}\,\overline{\langle j_2,j_3\rangle}\cdots\overline{\langle j_{k-1},j_k\rangle}\,\overline{\langle j_k,x_k\rangle}\,\xi(j_1,\dots,j_k,x_k,x_{k+1},\dots)\\
&=\delta_{i_1,x_0}\delta_{i_2,x_1}\cdots\delta_{i_k,x_{k-1}}\langle i_1,i_2\rangle\langle i_2,i_3\rangle\cdots\langle i_{k-1},i_k\rangle\langle i_k,x_k\rangle\overline{\langle j_1,j_2\rangle}\,\overline{\langle j_2,j_3\rangle}\\
&\qquad\cdots\overline{\langle j_{k-1},j_k\rangle}\,\overline{\langle j_k,x_k\rangle}\,\xi(j_1,\dots,j_k,x_k,x_{k+1},\dots).
\end{aligned}$$

It follows immediately from this formula that the representation of UHF_n on $\mathcal{L}^2(\Omega,\mu)$ defined by the T_i's is irreducible, and thus by Theorem 3.3 β is a shift. Furthermore

$$\begin{aligned}
&\omega(e^{(1)}_{i_1j_1}\otimes e^{(2)}_{i_2j_2}\otimes\cdots\otimes e^{(k)}_{i_kj_k})\\
&=\langle \mathbb{1},T_{i_1}\cdots T_{i_k}T_{j_k}^*\cdots T_{j_1}^*\mathbb{1}\rangle\\
&=\frac{1}{n^{k+1}}\sum_{x_0}\cdots\sum_{x_k}\delta_{i_1,x_0}\cdots\delta_{i_k,x_{k-1}}\cdot\langle i_1,i_2\rangle\langle i_2,i_3\rangle\cdots\langle i_{k-1},i_k\rangle\langle i_k,x_k\rangle\\
&\qquad\cdot\overline{\langle j_1,j_2\rangle}\,\overline{\langle j_2,j_3\rangle}\cdots\overline{\langle j_{k-1},j_k\rangle}\,\overline{\langle j_k,x_k\rangle}\\
&=\frac{1}{n^{k+1}}\langle i_1,i_2\rangle\cdots\overline{\langle j_{k-1},j_k\rangle}\sum_{x_k}\langle x_k,i_k-j_k\rangle\\
&=\frac{1}{n^k}\delta_{i_kj_k}\langle i_1,i_2\rangle\langle i_2,i_3\rangle\cdots\langle i_{k-1},i_k\rangle\cdot\overline{\langle j_1,j_2\rangle}\,\overline{\langle j_2,j_3\rangle}\cdots\overline{\langle j_{k-1},j_k\rangle}.
\end{aligned}$$

This ends the proof of Theorem 8.1 □

Remarks 8.2. As already remarked after (8.21), the equation (8.19) always has a continuum of function solutions which are not measurable, and thus are not in $\mathcal{L}^2(\Omega,\mu)$ or define states on $\mathcal{O}_n$ in any reasonable sense. Also note that (8.19) has the formal infinite product "solution"

$$\xi(y_0,y_1,y_2,\dots)=\prod_{k=0}^{\infty}n^{1/2}\lambda_{y_k}\langle y_k,y_{k+1}\rangle.$$

One way of stating Theorem 8.1 is that these infinite products do not converge to a non-zero vector in $\mathcal{L}^2(\Omega,\mu)$. We will in the following remark on special cases of (8.19) where "solutions" exist which are not in $\mathcal{L}^2(\Omega,\mu)$, and also supply some related operator theoretic observations. Since, for the general case of (8.18), or (7.10) above, $\mathcal{L}^2$-solutions correspond to pure normal invariant states, the non $\mathcal{L}^2$ "solutions" correspond to states on $\mathcal{O}_n$ which are *not normal* in the given Haar respresentation, and the "solutions" give us a clue to what these states are, namely the Cuntz states defined by the appropriate λ's. This lies at the heart of why one uses C^*- algebras rather than merely Hilbert spaces in various contexts: States which cannot be realized by vectors in the Hilbert space, can be realized as states on an appropriate C^*-algebra. In the analysis of quantum systems with infinitely many degrees of freedom, examples of this abound (sometimes under the name of the van Hove phenomenon); see [Br-Rob, p. 224], [Hov], and [Seg2].

(i) In the special case where the vector (λ_i) in (8.18) is $(1, 0, \dots, 0)$, we noted that a possible "eigenvector" ξ must then necessarily be a constant times something like the delta mass at $0 = (0, 0, \dots)$ in the group Ω. If ξ shall define the appropriate state on $\mathcal{O}_n$, it should rather be the "square root" of the Dirac delta mass. This solution is not in $\mathcal{L}^2(\Omega, \mu)$, of course, unless the constant is zero. Specifically, the assertion about ξ in this special case is that $\xi(x_0, x_1, \dots) = 0$ unless $x_0 = x_1 = \dots = 0$.

(ii) The most interesting special case of (8.18) turns out to be the *equi-distribution*, $\lambda_i = n^{-1/2}$ (for $\forall i$). In that case, the recursive formula (8.20) [for a possible $\mathcal{L}^2$-solution ξ to (8.18)] then takes the following geometric form: Using (8.2), we may define the *isometric* operator S on $\mathcal{L}^2(\Omega, \mu)$ by $S\xi := \xi \circ \sigma$, and (8.18)–(8.20) become the single condition,

$$\xi = US\xi \tag{8.27}$$

where U is the unitary transfer operator from (2.6) and (8.16).

For this, moreover, the details for the (8.22) calculation simplify as follows. The present argument is again based on the Ω-Λ duality and the corresponding Fourier transform. Let $\lambda_i = n^{-1/2}$; we supply the recursion. For $j \approx (j, 0, 0, \dots) \in \Lambda := \hat{\Omega}$, we get:

(8.28)
$$\begin{aligned}
\widetilde{\xi}(j) &= \int_\Omega \overline{\langle x_0, j\rangle}\, \xi(x_0, \dots)\, d\mu(x_0, \dots) \\
&= n^{-2} \sum_{y_0} \sum_{y_1} \overline{\langle y_0, j\rangle}\, \langle y_0, y_1\rangle \int_\Omega \langle y_1, x_2\rangle \xi(x_2, \dots)\, d\mu(x_2 \cdots) \\
&= n^{-1} \sum_{y_1} \delta(j - y_1) \int_\Omega \langle y_1, x_2\rangle \xi(x_2, \dots)\, d\mu(x_2 \cdots) \\
&= n^{-1} \int_\Omega \overline{\langle -j, x_2\rangle}\, \xi(x_2, \dots)\, d\mu(x_2 \cdots) \\
&= n^{-1} \widetilde{\xi}(-j)
\end{aligned}$$

valid for $\forall j \in \mathbb{Z}_n := \mathbb{Z}/n\mathbb{Z}$, and with all summations being over $\mathbb{Z}_n$, again with $\mathbb{Z}_n$ viewed as an additive group. Replacing j by $-j$ yields, $\widetilde{\xi}(j) = 0$, for $\forall j \in \mathbb{Z}_n$; or, more specifically, $\widetilde{\xi}(j, 0, 0, \dots) = 0$, for $\forall j \in \mathbb{Z}_n$, and (8.26) follows as before.

By a calculation quite analogous to the one above, we get, $\forall (i_0, \dots, i_k, 0, \dots) \in \Lambda = \hat{\Omega} = \coprod_0^\infty \mathbb{Z}_n$, that

$$\widetilde{\xi}(i_0, \dots, i_k, 0, 0, \dots) = n^{-1} \overline{\langle i_0, i_1\rangle}\, \widetilde{\xi}(i_2 - i_0, i_3, \dots, i_k, 0, 0, \dots). \tag{8.29}$$

But then, by induction, $\widetilde{\xi}$ must vanish identically on $\Lambda = \hat{\Omega} = \coprod_0^\infty \mathbb{Z}_n$.

(iii) A more operator theoretic way to see that $US\xi = \xi$ implies $\xi = 0$ is this: If $\xi \in \mathcal{L}^2(\Omega, \mu)$ is arbitrary, one computes as above,

$$\begin{aligned}
\langle e_j, (US)^2 \xi\rangle &= ((US)^2\xi)^\sim (j_0, j_1, \dots j_k, 0, 0, \dots) \\
&= n^{-1}\, \overline{\langle j_0, j_1\rangle}\, \widetilde{\xi}(j_2 - j_0, j_3, \dots, j_k, 0, 0, \dots).
\end{aligned}$$

Because of the n^{-1} factor, it follows by iteration that

$$|\langle e_j, (US)^{2m}\xi\rangle| \leq n^{-m}\|\xi\|.$$

We will now show that (8.27) has no solution by proving that the unitary part of the *Wold decomposition* of $T = US$ is zero. Recall from [Nik] that if T is *any isometry* on $\mathcal{H} = \mathcal{L}^2(\Omega, \mu)$, i.e., $T^*T = 1$, then $\mathcal{H}$ has a decomposition, $\mathcal{H} = \mathcal{H}_1 \oplus \mathcal{H}_2$ into T-invariant subspaces such that $T|_{\mathcal{H}_1}$ is unitary, and $V = T|_{\mathcal{H}_2}$ is a shift. That V is a *shift* means that $\lim_{n\to\infty} V^n\xi = 0$ for any $\xi \in \mathcal{H}_2$. Put $\mathcal{L} = \mathcal{H}_2 \ominus V\mathcal{H}_2 = \mathcal{H} \ominus T\mathcal{H}$. (If (ξ_i) is an orthonormal basis for $\mathcal{L}$, and one defines $\xi_{ij} = V^j\xi_i = T^j\xi_i$, then (ξ_{ij}) is an orthonormal basis for $\mathcal{H}_2$. Thus $\mathcal{H}_2 = \bigoplus_{m=0}^{\infty} V^m\mathcal{L}$, and V decomposes into a direct sum of $\dim\mathcal{L}$ copies of the *Hilbert shift*, defined by, $\xi_{ij} \mapsto \xi_{i,j+1}$, for fixed i, and $j = 0, 1, \ldots$; $\dim\mathcal{L}$ is called the *multiplicity* of the shift.) The subspaces $\mathcal{H}_1$ and $\mathcal{H}_2$ may be identified through the formuli

$$\mathcal{H}_1 = \bigcap_m T^m\mathcal{H} = \bigcap_m \{\xi \in \mathcal{H} : \|T^{*m}\xi\| = \|\xi\|\}$$

and

$$\mathcal{H}_2 = \mathcal{H}_1^{\perp} = \bigoplus_{m=0}^{\infty} T^m\mathcal{L}.$$

Returning to our specific isometry $T = US$, we have to show that $\mathcal{H}_1 = 0$: Let $\xi \in \mathcal{H}_1$. Then $\xi \in T^{2m}\mathcal{H}$, so for each m there exists a $\xi_m \in \mathcal{H}$ with $\xi = T^{2m}\xi_m$. But then

$$\|\xi_m\| = \|T^{2m}\xi_m\| = \|\xi\|,$$

and hence

$$\begin{aligned}|\langle e_j, \xi\rangle| &= |\langle e_j, T^{2m}\xi_m\rangle| \\ &\leq n^{-m}\|\xi_m\| = n^{-m}\|\xi\|.\end{aligned}$$

Letting $m \to \infty$, we see that

$$\langle e_j, \xi\rangle = 0,$$

and, since $j \in \Lambda$ is arbitrary, the desired conclusion, $\xi = 0$, follows. We conclude that $T = US$ is a *shift* on $\mathcal{L}^2(\mu)$, equivalently a completely non-unitary isometry. This seems of independent interest as the isometry S is *not* a shift, recall $S^*\mathbb{1} = \mathbb{1}$. An inspection also reveals that the shift T has the multiplicity $(n-1)\cdot\infty$ where n is the index of the Haar shift. Of course, the infinite product,

$$\xi(x_0, x_1, x_2, \ldots) := \prod_{k=0}^{\infty} \langle x_k, x_{k+1}\rangle$$

$$\left(\text{or, more formally, } \prod_{k=0}^{\infty} e^{i2\pi x_k x_{k+1}/n} = e^{i(2\pi/n)\sum_{k=0}^{\infty} x_k x_{k+1}}\right)$$

is a "formal" solution to (8.27); but our present considerations imply that this infinite product is indeed purely formal, and *not* convergent in $\mathcal{L}^2(\Omega, \mu)$. Specifically (ad absurdum), convergence in $\mathcal{L}^2(\mu)$ would put the limit-function, $\xi(x)$ (for $x \in \Omega$), in $\bigcap_{k=1}^{\infty} T^k(\mathcal{L}^2(\mu))$. But this intersection *is* the unitary term in the Wold-decomposition, and we proved that it must be zero.

Note furthermore that our stronger conclusion, based on this Wold decomposition argument, is the assertion that there can be no sequence $(\xi_k)_{k=1}^{\infty}$ in $\mathcal{L}^2(\mu)$ such that the limit, $\lim_{k\to\infty} T^k \xi_k$, exists in $\mathcal{L}^2(\mu)$ and is non-zero.

(iv) With the notation

$$e_{ij} := e_{i_1 j_1}^{(1)} \otimes \cdots \otimes e_{i_k j_k}^{(k)} \otimes 1 \otimes \cdots \in \mathrm{UHF}_n,$$

and

$$i = (i_1, \dots, i_k, 0, 0, \dots) \in \Lambda_k = \coprod_1^k \mathbb{Z}_n \subseteq \Lambda = \coprod_1^\infty \mathbb{Z}_n = \widehat{\left(\prod_1^\infty \mathbb{Z}_n\right)},$$

the formula for the state ω in Theorem 8.1 above is

$$\omega(e_{ij}) = n^{-k} \delta(i_k - j_k) \xi(i) \overline{\xi(j)}$$

where the function $\xi(\cdot)$ is defined (as in (iii) above) on Λ by

$$\xi(i) = \prod_{p=1}^{\infty} \langle i_p, i_{p+1} \rangle, \tag{8.30}$$

and, for positive definite functions $F_k(\cdot,\cdot)$ on $\Lambda_k \times \Lambda_k$, satisfying the consistency relations

$$F_k(i,j) = \sum_{r \in \mathbb{Z}_n} \langle r, i_k - j_k \rangle F_{k+1}(ir, jr),$$

there are corresponding states ω_F on UHF_n given by the more general formula

$$\omega_F(e_{ij}) := F(i,j) \xi(i) \overline{\xi(j)} \tag{8.31}$$

for $(i,j) \in \Lambda \times \Lambda$ having the same length. When F is so chosen, the object is to identify operators T_i, depending on F, and satisfying the Cuntz-relations, such that

$$\omega_F(e_{ij}) = \langle \mathbb{1}, T_{i_1} \cdots T_{i_k} T_{j_k}^* \cdots T_{j_1}^* \mathbb{1} \rangle_{\mathcal{L}^2}$$

is given by the expression on the right hand side in (8.31) and $\mathbb{1}$ denotes the constant unit function on Ω. Specifically we may get such states ω_F in the set P from Section 5 as follows: Let $\omega = \bigotimes_1^\infty \omega_k$ (each ω_k is a state on M_n) be a product state in P as described in Example 5.5, and for $i = (i_1, \dots, i_k, 0, 0, \dots)$, $j = (j_1, \dots, j_k, 0, 0, \dots)$ in Λ, let

$$F_{\omega,k}(i,j) := \prod_{m=1}^{k} \omega_m(e_{i_m j_m}^{(m)}) \cdot \sum_{r \in \mathbb{Z}_n} \langle r, i_k - j_k \rangle \omega_{k+1}(e_{rr}^{(k+1)}).$$

Then it can be shown that the corresponding state ω_{F_ω} in (8.31) is in P (details in a sequel paper), and we get an associated element in $\mathrm{Rep}_s(\mathcal{O}_n, \mathcal{L}^2)$. Furthermore, we may choose the product state $\bigotimes_k \omega_k$ in P such that the corresponding *shift* β on $\mathcal{B}(\mathcal{L}^2)$, i.e., $\beta(A) = \sum_i T_i A T_i^*$, is *non-conjugate* to the one from Theorem 8.1, and also not to those from Sections 6–7. Note that the function $\xi(\cdot)$ in (8.31) is well defined on the subgroup Λ of Ω, but, as noted in (iii) above, it is not sufficiently *almost periodic* to extend naturally to the compactification Ω.

Remark 8.3. It can be proved that if ω is the state on UHF_n defined in Theorem 8.1, and ω' is any infinite tensor product of pure states on UHF_n, then

$$\|\omega \circ \sigma^m - \omega' \circ \sigma^m\| = 2$$

for all $m \in \mathbb{N}$. If $\omega' \in P$, it follows from Lemma 5.4 that the corresponding shifts of $\mathcal{B}(\mathcal{H})$ are non- conjugate, and hence the shift considered in this section is not conjugate to any one of those discussed in Sections 6 and 7. The proof will be deferred to a subsequent paper where nearest neighbor states and similar states will be treated more systematically.

9. Extending Unital Endomorphisms to Automorphisms

In [Arv2], [AK] it was proved that a continuous one-parameter semigroup of unital *-endomorphisms of $\mathcal{B}(\mathcal{H})$ has an extension to a group of *-automorphisms of $\mathcal{B}(\mathcal{H} \otimes \mathcal{H})$ when $\mathcal{B}(\mathcal{H})$ is embedded as $1 \otimes \mathcal{B}(\mathcal{H})$. Using techniques from [PR], let us establish a similar (but simpler) result for a single endomorphism.

Theorem 9.1. *Let α be a unital *- endomorphism of $\mathcal{B}(\mathcal{H})$ of index n, and embed $\mathcal{B}(\mathcal{H})$ into $\mathcal{B}(\mathcal{H}\otimes\mathcal{H})$ as $1\otimes\mathcal{B}(\mathcal{H})$. Then α has an extension β to $\mathcal{B}(\mathcal{H}\otimes\mathcal{H})$ such that β is a *-automorphism. Furthermore, if α is a shift, and H_0 is the Hilbert space of dimension n, and $M_n = \mathcal{B}(H_0)$, then $\mathcal{B}(\mathcal{H} \otimes \mathcal{H})$ contains $M_{n^\infty} = \bigotimes_{k=-\infty}^{\infty} M_n$ as a weakly dense C^*-subalgebra in such a manner that the restriction of β to M_{n^∞} is just the two-sided right shift, and $\bigotimes_{k=0}^{\infty} M_n \subseteq \bigotimes_{k=-\infty}^{\infty} M_n$ is weakly dense in $1 \otimes \mathcal{B}(\mathcal{H})$.*

Proof. Since $\alpha(\mathcal{B}(\mathcal{H}))' \cong \mathcal{B}(H_n)$, we have a tensor product decomposition

$$\mathcal{H} = H_n \otimes \mathcal{K}$$

such that

$$\alpha(\mathcal{B}(\mathcal{H})) = 1_{H_n} \otimes \mathcal{B}(\mathcal{K}).$$

Let $\alpha' : \mathcal{B}(\mathcal{H}) \to \mathcal{B}(\mathcal{K})$ be the corresponding *-isomorphism such that

$$\alpha(A) = 1 \otimes \alpha'(A)$$

for $A \in \mathcal{B}(\mathcal{H})$. Choose a particular unit vector in H_0, and let $H' = \bigotimes_{k=-\infty}^{-1} H_0$ be the corresponding von Neumann reduced tensor product. (For the first part of Theorem 9.1, we do not need any structure on H' other than it is a separable infinite dimensional Hilbert space.) First, let β' be *any* *- isomorphism $\mathcal{B}(H') \to \mathcal{B}(H' \otimes H_0)$ and define

$$\beta : \mathcal{B}(H' \otimes \mathcal{H}) \to \mathcal{B}(H' \otimes \mathcal{H})$$

by

$$\beta(B \otimes A) = \beta'(B) \otimes \alpha'(A)$$

for $B \in \mathcal{B}(H')$, $A \in \mathcal{B}(\mathcal{H})$ and the *last* tensor product is according to the decomposition

$$H' \otimes \mathcal{H} = (H' \otimes H_0) \otimes \mathcal{K}.$$

Then β is a *-automorphism extending α. For the last part of the theorem we define β', more specifically, as the *-isomorphism, $\mathcal{B}(\bigotimes_{-\infty}^{-1} H_0) \to \mathcal{B}(\bigotimes_{-\infty}^{0} H_0)$, implemented by the right-shift, $U : \bigotimes_{-\infty}^{-1} H_0 \to \bigotimes_{-\infty}^{0} H_0$, defined by

$$U(\cdots \psi_{-3} \otimes \psi_{-2} \otimes \psi_{-1}) = \cdots \underset{-2}{\psi_{-3}} \otimes \underset{-1}{\psi_{-2}} \otimes \underset{0}{\psi_{-1}}.$$

Now, if $N_0 = \alpha(\mathcal{B}(\mathcal{H}))' \cap \mathcal{B}(\mathcal{H})$ and, inductively

$$N_{k+1} = \alpha(N_k), \qquad k = 0, 1, \ldots$$

then by [Pow2, Lemma 2.1], or the ideas surrounding (3.6) in the proof of Theorem 3.3, it follows that the N_k's are mutually commuting I_n factors, with $\{\bigcup_k N_k\}' = \mathbb{C}(1)$. Putting N_{-k} equal to the bounded operators of the $-k$'th tensor factor in $\bigotimes_{-\infty}^{-1} H_0$ (tensor 1 on the remaining factors), it follows that all the N_k's mutually commute, $\beta(N_k) = N_{k+1}$ for $k \in \mathbb{Z}$, and the C^*- algebra generated by the N_k's is weakly dense in $\mathcal{B}(H' \otimes \mathcal{H})$. Theorem 9.1 follows. □

Remark 9.2. If α is a shift, and β and $\mathcal{B}(H' \otimes \mathcal{H})$ are constructed according to the recipe above, then all elements in the weakly dense *- subalgebra $\bigcup_{k=-1}^{-\infty} (\bigotimes_k^{\infty} M_n)$ of $\mathcal{B}(H' \otimes \mathcal{H})$ will ultimately be mapped into $1_{H'} \otimes \mathcal{B}(\mathcal{H})$. Thus any asymptotic property of α (such as having an absorbing state), readily translates into a similar property for β.

Acknowledgments. The present paper came about in connection with the second named author's (Palle Jorgensen) visit to the University of Oslo, with support from *the Norwegian Research Council*, and the visit by the last two authors (Palle Jorgensen and Geoffrey L. Price) to the Massachusetts Institute of Technology in connection with *the 1994 von Neumann Symposium*, and with support from the *U.S. National Science Foundation.* Hospitality from the respective hosts (O. Bratteli and E. Størmer in Oslo, and I.E. Segal at MIT) is greatly appreciated. In addition to the two funding agencies already mentioned, this work was also supported by a *U.S.-Western Europe NSF grant*, by a *NATO grant*, and (for Palle Jorgensen) a *University of Iowa Faculty Scholarship Award.* We acknowledge helpful conversations with G.A. Elliott, R.T. Powers, E. Størmer, and S. Pedersen. The paper was finished while Ola Bratteli visited *The Fields Institute for Research in Mathematical Sciences* with Norwegian Research Council funding. He is grateful for patience from the host G.A. Elliott there. Finally we wish to thank M. Laca for generously making many valuable suggestions to a preliminary version of our paper, and for allowing us to quote from [Lac].

References

[Ara1] H. Araki, *On quasi-free states of CAR and Bogoliubov automorphism*, Publ. Res. Inst. Math. Sci. **6** (1970), 385–442.

[Ara2] H. Araki, *On quasi-free states of the canonical commutation relations* II, Publ. Res. Inst. Math. Sci. **7** (1971), 121–152.

[ACE] H. Araki, A.L. Carey, and D.E. Evans, *On $\mathcal{O}_{n+1}$*, J. Operator Theory **12** (1984), 247–264.

[Ar-Woo] H. Araki and E.J. Woods, *Complete Boolean algebras of type I factors*, Publ. RIMS **2** (1966), 157–242.

[Arv1] W.B. Arveson, *Continuous analogues of Fock space* I, Mem. Amer. Math. Soc. **80** (1989), no. 409.

[Arv2] ———, *Continuous analogues of Fock space* IV*: Essential states*, Acta Math. **164** (1990), 265–300.

[Arv3] ———, *E_0-semigroups in quantum field theory*, These proceedings..

[AK] W. Arveson and A. Kishimoto, *A note on extensions of semigroups of *-endomorphisms*, Proc. Amer. Math. Soc. **116** (1992), 769–774.

[BEGJ] O. Bratteli, D.E. Evans, F.M. Goodman, and P.E.T. Jorgensen, *A dichotomy for derivations on $\mathcal{O}_n$*, Publ. RIMS **22** (1986), 103–107.

[Br-Rob] O. Bratteli and D.W. Robinson, *Operator algebras and quantum statistical mechanics*, vol. II, Spring-Verlag, Berlin–New York, 1981.

[Bra] O. Bratteli, *Inductive limits of finite dimensional C^*-algebras*, Trans. Amer. Math. Soc. **171** (1972), 195–234.

[Cho] M. Choda, *Shifts on the hyperfinite II_1 factor*, J. Operator Theory **17** (1987), 223–235.

[Cob] L.A. Coburn, *The C^*-algebra generated by an isometry*, Bull. Amer. Math. Soc. **73** (1967), 722–736.

[Cun] J. Cuntz, *Simple C^*-algebras generated by isometries*, Commun. Math. Phys. **57** (1977), 173–185.

[Dae] A. van Daele, *Quasi-equivalence of quasi-free states on the Weyl algebra*, Commun. Math. Phys. **21** (1971), 171–191.

[Din] H.T. Dinh, *On discrete semigroups of *-endomorphisms of type I factors*, Internat. J. Math. **3** (1992), 609–628.

[Dix] J. Dixmier, *Les algèbres d'opérateurs dans l'espaces Hilbertien*, 2nd ed., Gauthier-Villars, Paris, 1969.

[ENWY] M. Enomoto, M. Nagisa, Y. Watatani, and H. Yoshida, *Relative commutant algebras of Powers' binary shifts on the hyperfinite II_1 factor*, Math. Scand. **68** (1991), 115–130.

[EW] M. Enomoto and Y. Watatani, *Endomorphisms of type II_1 factors*, Preprint (1994).

[Eva] D.E. Evans, *On $\mathcal{O}_n$*, Publ. Res. Inst. Math. Sci. **16** (1980), 915–927.

[Gli] J. Glimm, *On a certain class of operator algebras*, Trans. Amer. Math. Soc. **95** (1960), 318–340.

[Gui1] A. Guichardet, *Tensor products of C^*-algebras*, Math. Inst. Aarhus University, Lecture Notes **12** (1969).

[Gui2] ———, *Symmetric Hilbert spaces and related topics*, LNM, vol. 261, Spring-Verlag, Berlin–New York, 1972.

[Gui3] ———, *Produits tensoriels infinis et représentations des relations d'anticommutation*, Ann. Ec. Norm. Sup. **83** (1966), 1–52.

[Hov] L. van Hove, *Les difficulties de divergence pour un modéle particulier de champ quantifié*, Physica **18** (1952), 145–159.

[Jon] V.F.R. Jones, *Hecke algebra representations of braid groups and link polynomials*, Ann. Math. **126** (1987), 335–388.

[KR] R.V. Kadison and J.R. Ringrose, *Fundamentals of the theory of operator algebras*, Vol. II, Academic Press, New York, 1983.

[Kak] S. Kakutani, *On equivalence of infinite product measures*, Ann. Math. **49** (1948), 214–224.

[Lac1] M. Laca, *Endomorphisms of $\mathcal{B}(\mathcal{H})$ and Cuntz-Algebras*, Preprint (1991).

[Lac2] M. Laca, *Gauge invariant states on $\mathcal{O}_\infty$*, Preprint.

[Jo-Pe] P.E.T. Jorgensen and S. Pedersen, *Harmonic analysis of fractal limit-measures induced by representations of a certain C^*-algebra*, J. Funct. Anal. (to appear).

[vNeu] J. von Neumann, *On infinite direct products*, Compositio Math. **6** (1938), 1–77.

[Nik] N.K. Nikolskii, *Treatise on the shift operator*, Grundlehren, vol. 273, Springer-Verlag, Berlin–New York, 1986.

[Ped] G.K. Pedersen, *C^*-algebras and their automorphism groups*, Academic Press and London Math. Soc., London, 1979.

[Pow1] R.T. Powers, *Representations of uniformly hyperfinite algebras and their associated von Neumann rings*, Ann. Math. **86** (1967), 138–171.

[Pow2] ———, *An index theory for semigroups of $*$-endomorphisms of $\mathcal{B}(\mathcal{H})$ and type II_1 factors*, Canad. J. Math. **40** (1988), 86–114.

[Pow3] ———, *New examples of continuous spatial semigroups of $*$- endomorphisms of $\mathcal{B}(\mathcal{H})$*, These proceedings..

[Po-Pr] R.T. Powers and G.L. Price, *Cocycle conjugacy classes of shifts on the hyperfinite II_1 factor*, J. Funct. Anal. **121** (1994), 275–295.

[Po-St] R.T. Powers and E. Størmer, *Free states of the canonical anti-commutation relations*, Commun. Math. Phys. **16** (1970), 1–33.

[PR] R.T. Powers and D.W. Robinson, *An index for continuous semigroups of *-endomor phisms of $\mathcal{B}(\mathcal{H})$*, J. Funct. Anal. **84** (1989), 85–96.

[Seg1] I.E. Segal, *The structure of a class of representations of the unitary group on a Hilbert space*, Proc. Amer. Math. Soc. **8** (1957), 197–203.

[Seg2] I.E. Segal, *Mathematical problems of relativistic physics*, American Mathematical Society, Providence, RI, 1963.

[Sta] P.J. Stacey, *Product shifts on $\mathcal{B}(\mathcal{H})$*, Proc. Amer. Math. Soc. **113** (1991), 955–963.

[Voi1] D. Voiculescu, *A non-commutative Weyl–von Neumann theorem*, Rev. Roumaine Math. Pures Appl. **21** (1976), 97–113.

[Voi2] D. Voiculescu, *Symmetries of some reduced free product C^*- algebras*, LNM 1132, Operator Algebras and Their Connection to Topology and Ergodic Theory, Springer-Verlag, Berlin, 1985, pp. 556–588.

[Wor] S.L. Woronowicz, *Compact matrix pseudogroups*, Commun. Math. Phys. **111** (1987), 613–665.

Mathematics Institute, University of Oslo, PB 1053 – Blindern, N-0316 Oslo, Norway

E-mail address: bratteli@math.uio.no

Department of Mathematics, University of Iowa, Iowa City, IA 52242, U.S.A.

E-mail address: jorgen@math.uiowa.edu

Department of Mathematics 9E, U.S. Naval Academy, Annapolis, MD 21402, U.S.A.

E-mail address: GLP@sma.usna.navy.mil

Proceedings of Symposia in Pure Mathematics
Volume **59**, 1996

Absolutely Continuous Spectrum in Random Schrödinger Operators

Abel Klein

ABSTRACT. The spectrum of the Anderson Hamiltonian $H_\lambda = -\Delta + \lambda V$ on the Bethe Lattice is absolutely continuous inside the spectrum of the Laplacian, if the disorder λ is sufficiently small. More precisely, given any closed interval I contained in the interior of the spectrum of the (centered) Laplacian Δ on the Bethe lattice, for small disorder H_λ has purely absolutely continuous spectrum in I with probability one (i.e., $\sigma_{ac}(H_\lambda) \cap I = I$ and $\sigma_{pp}(H_\lambda) \cap I = \sigma_{sc}(H_\lambda) \cap I = \emptyset$ with probability one). The proof is discussed and regularity properties are proven for the spectral measures restricted to such intervals of absolute continuity.

1. Introduction

The Anderson Hamiltonian [**6**] is the random Schrödinger operator

$$H_\lambda = \tfrac{1}{2}\Delta + \lambda V \quad \text{on} \quad \ell^2(\mathbb{Z}^d). \tag{1.1}$$

Here the (centered) Laplacian Δ is defined by

$$(\Delta u)(x) = \sum_y u(y)\,, \tag{1.2}$$

where the sum runs over all nearest neighbors of x, and V is a random potential, with $V(x)$, $x \in \mathbb{Z}^d$, being independent, identically distributed random variables with common probability distribution μ. The real parameter λ is called the *disorder*.

The spectrum of the Hamiltonian H_λ is given by

$$\sigma(H_\lambda) = \sigma(\tfrac{1}{2}\Delta) + \lambda \operatorname{supp} \mu = [-d, d] + \lambda \operatorname{supp} \mu \tag{1.3}$$

with probability one, due to ergodic considerations [**30, 9**]. For each choice of V the spectrum of H_λ can be decomposed into pure point spectrum, $\sigma_{pp}(H_\lambda)$, absolutely continuous spectrum, $\sigma_{ac}(H_\lambda)$, and singular continuous spectrum, $\sigma_{sc}(H_\lambda)$. Ergodicity also gives the existence of sets $\Sigma_{\lambda,pp}$, $\Sigma_{\lambda,ac}$, $\Sigma_{\lambda,sc} \subset \mathbb{R}$ such that

1991 *Mathematics Subject Classification.* Primary 82B44; Secondary 81Q10.
Key words and phrases. Random Schrödinger operators, Anderson model, extended states, absolutely continuous spectrum, localization.
The author was supported in part by NSF Grant DMS-9208029.

$\sigma_{pp}(H_\lambda) = \Sigma_{\lambda,pp}$, $\sigma_{ac}(H_\lambda) = \Sigma_{\lambda,ac}$ and $\sigma_{sc}(H_\lambda) = \Sigma_{\lambda,sc}$ with probability one [**23, 9**].

In the physics literature ([**6, 28, 35, 1, 2, 25**] and others) the following picture is given: In one and two dimensions, as long as the potential is random (i.e., $\lambda \neq 0$), the model shows *exponential localization* (i.e., pure point spectrum with exponentially decaying eigenfunctions). In three and more dimensions both localized and *extended states* (i.e., absolutely continuous spectrum) are expected for small disorder, with the energies of extended and localized states being separated by a *mobility edge.*

In one dimension there are now mathematical proofs of exponential localization for any disorder (e.g., [**15, 23, 8, 11**] and others). In the multidimensional case exponential localization has been proven for large disorder or low energy (e.g., [**14, 13, 10, 32, 11, 5, 4, 18, 16**] and others); for small disorder there exist energies $E^{\pm}_{\lambda,loc}$, with $|E^{\pm}_{\lambda,loc}| > d$ and $\lim_{\lambda\to 0} E^{\pm}_{\lambda,loc} = \pm d$, such that H_λ has pure point spectrum in $(-\infty, E^{-}_{\lambda,loc}) \cup (E^{+}_{\lambda,loc}, \infty)$ [**33, 12, 4, 34**]. Only localization in two dimensions for small disorder is still an open problem.

But up to now there is no mathematical proof of the occurrence of absolutely continuous spectrum in the Anderson model on $\mathbb{Z}^d$.

The Bethe lattice (or Cayley tree), $\mathbb{B}$, is an infinite connected graph with no closed loops and a fixed degree (number of nearest neighbors) at each vertex (site or point). The degree is called the coordination number and the connectivity, K, is one less the coordination number ($K \geq 2$, so $\mathbb{B}$ is not the line $\mathbb{R}$) . The distance between two sites x and y will be denoted by $d(x, y)$ and is equal to the length of the shortest path connecting x and y. The Anderson Hamiltonian on the Bethe lattice is given by (1.1) with $\mathbb{B}$ substituted for $\mathbb{Z}^d$. It was first studied by Abou-Chacra, Anderson and Thouless [**1**]; their self-consistent approximation for the study of localization becomes exact in the Bethe lattice. The resulting equations were further studied by Abou-Chacra and Thouless [**2**], who argued that, on the Bethe lattice, for sufficiently low disorder there should be an energy at which localization breaks down, which converges to $\frac{K+1}{2}$ in the zero disorder limit. An outline of expected results was given by Kunz and Souillard [**24**]. Miller and Derrida [**26**] performed a weak disorder expansion inside the spectrum of the zero disorder Hamiltonian, and computed perturbatively the density of states and conducting properties corresponding to extended states. They also found the existence of an energy, which converges (from outside) to the edge $\sqrt{K}$ of the spectrum of $\frac{1}{2}\Delta$ in the zero disorder limit, above which the density of states and conducting properties vanish to all orders in perturbation theory. The Bethe lattice model was also discussed by Mirlin and Fyodorov [**27**].

Ergodicity considerations still apply (see Acosta and Klein [**3**, Appendix]), so the first equality in (1.3) and the following statements are still valid in the Bethe lattice, where we have $\sigma(\Delta) = [-2\sqrt{K}, 2\sqrt{K}]$. In particular,

$$\sigma(H_\lambda) = [-\sqrt{K}, \sqrt{K}] + \lambda \operatorname{supp} \mu \tag{1.4}$$

with probability one.

The nature of the spectrum for energies E with $\sqrt{K} \leq |E| \leq \frac{K+1}{2}$, at low disorder, does not seem to have been properly discussed in the physics literature. The Abou-Chacra and Thouless calculations indicate that we do not have localization in these intervals for small disorder; Miller and Derrida's weak disorder expansions suggest that the corresponding states cannot be "too extended", since they should not have conducting properties. So there should be continuous spectrum, but of which type? It could be absolutely continuous, but it would be of a different nature then the one inside the free spectrum. It is plausible that near $E = \pm\frac{K+1}{2}$ there is a transition from pure point spectum to either singular continuous spectrum or absolutely continuous spectrum "without conducting properties", and that near $E = \pm\sqrt{K}$ there is another transition where the spectrum changes to absolutely continuous "with conducting properties". If so, there is a *mobility interval* instead of a *mobility edge*!

Localization for large disorder or low energies has only recently been proven by Aizenman and Molchanov [**5**], and Aizenman [**4**] proved localization for energies beyond $\frac{K+1}{2}$ at weak disorder, confirming half of Abou-Chacra and Thouless' prediction [**2**].

On the Bethe lattice we proved that the Anderson Hamiltonian has "extended states" for small disorder [**19, 20**]. More precisely, given any closed interval I contained in the interior of the spectrum of $\frac{1}{2}\Delta$ on the Bethe lattice, we proved that for small disorder H_λ has purely absolutely continuous spectrum in I with probability one, and its integrated density of states is continuously differentiable on the interval. These results agree with Miller and Derrida's conclusions [**26**]. We have also studied the spreading of wave packets evolving under the Anderson Hamiltonian on the Bethe Lattice: for small disorder we showed that the averaged mean square distance travelled by a particle in a time t grows as t^2 for large t [**21**].

The precise results requires hypotheses about the single site potential probability distribution μ: the localization results require μ to have a bounded density with respect to Lebesgue measure, while the extended states results call for μ to have a characteristic function $h(t) = \int e^{-itv} d\mu(v)$ which is twice differentiable with bounded first and second derivatives on $(0, \infty)$. (True for any any probability distribution μ with a finite second moment (e.g., uniform, Gaussian or Bernoulli distributions) and for the Cauchy distribution).

In this article we show that by requiring more differentiability of the characteristic function $h(t)$ we can get absolutely continuous spectrum with certain regularity properties, which will be expressed in terms of the spectral measures $\nu_{\lambda,x}$, $x \in \mathbb{B}$, given by

$$d\nu_{\lambda,x}(E) = \langle \delta_x, dP_\lambda(E)\delta_x \rangle ,$$

where $dP_\lambda(E)$ is the spectral measure of the operator H_λ.

DEFINITION. Let $p \in \mathbb{N}$. A probability measure μ will be said to be p-admissible if its characteristic function $h(t)$ is $2p$-times differentiable on $(0, \infty)$, with all $2p$ derivatives bounded on $(0, \infty)$. (True for any any probability distribution μ with $\int |v|^{2p} d\mu(v) < \infty$).

THEOREM 1.1. *Let H_λ be the Anderson Hamiltonian on the Bethe lattice with μ p-admissible, $p \in \mathbb{N}$. Then for any $E \in (0, \sqrt{K})$ there exists $\lambda(E) > 0$, such that given any λ with $|\lambda| < \lambda(E)$, with probability one the spectrum of H_λ in $[-E, E]$ is purely absolutely continuous (i.e., $\Sigma_{\lambda,ac} \cap [-E, E] = [-E, E]$ and $\Sigma_{\lambda,pp} \cap [-E, E] = \Sigma_{\lambda,sc} \cap [-E, E] = \emptyset$), the spectral measures $\nu_{\lambda,x}$ being absolutely continuous in this interval with $\frac{d\nu_{\lambda,x}}{dE} \in L^{2p}([-E, E])$ for all $x \in \mathbb{B}$.*

The $p = 1$ case was proved in [**19, 20**].

The Green's function of H_λ is given by

$$G_\lambda(x,y;z) = \left\langle \delta_x, (H_\lambda - z)^{-1}\delta_y \right\rangle \tag{1.5}$$

for $x, y \in \mathbb{B}$ and $z = E + i\eta$ with $E \in \mathbb{R}$, $\eta > 0$. For any $x \in \mathbb{B}$ and any potential V, $G_\lambda(x,x;E+i\eta)$ is a continuous function of $(\lambda, E, \eta) \in \mathbb{R} \times \mathbb{R} \times (0,\infty)$, so $\mathbb{E}(|G_\lambda(x,x;E+i\eta)|^{2p})$ is also a continuous function of $(\lambda, E, \eta) \in \mathbb{R} \times \mathbb{R} \times (0,\infty)$ for any $p \in \mathbb{N}$. Theorem 1.1 will follow from the fact that we can let $\eta \downarrow 0$ inside the spectrum of $\frac{1}{2}\Delta$.

THEOREM 1.2. *Let H_λ be the Anderson Hamiltonian on the Bethe lattice with μ p-admissible, $p \in \mathbb{N}$. Then for any $E \in (0, \sqrt{K})$ there exists $\lambda(E) > 0$, such that for all $x \in \mathbb{B}$ the continuous function*

$$(\lambda, E', \eta) \in (-\lambda(E), \lambda(E)) \times [-E, E] \times (0,\infty) \longrightarrow \mathbb{E}(|G_\lambda(x,x;E'+i\eta)|^{2p}) \tag{1.6}$$

has a continuous extension to $(-\lambda(E), \lambda(E)) \times [-E, E] \times [0,\infty)$. In particular,

$$\sup_{\lambda;\, |\lambda| < \lambda(E)} \ \sup_{E';\, |E'| \le E} \ \sup_{\eta;\, 0<\eta} \mathbb{E}(|G_\lambda(x,x;E'+i\eta)|^{2p}) < \infty . \tag{1.7}$$

Theorem 1.1 follows from (1.7). Let $E \in (0, \sqrt{K})$ and $\lambda(E) > 0$ as in Theorem 1.2, so (1.7) holds. For $|\lambda| < \lambda(E)$ and any $x \in \mathbb{B}$ we use Fubini's Theorem and Fatou's Lemma to obtain

$$\mathbb{E}\left(\liminf_{\eta \downarrow 0} \int_{-E}^{E} |G_\lambda(x,x;E'+i\eta)|^{2p}\, dE' \right) \le \liminf_{\eta\downarrow 0} \int_{-E}^{E} \mathbb{E}(|G_\lambda(x,x;E'+i\eta)|^{2p})\, dE' < \infty . \tag{1.8}$$

Thus we must have

$$\liminf_{\eta\downarrow 0} \int_{-E}^{E} (Im\, G_\lambda(x,x;E'+i\eta))^{2p}\, dE' \le \liminf_{\eta\downarrow 0} \int_{-E}^{E} |G_\lambda(x,x;E'+i\eta)|^{2p}\, dE' < \infty \tag{1.9}$$

with probability one. Since $G_\lambda(x,x;E+i\eta)$ is the Stieltjes transform of the measure $\nu_{\lambda,x}$, it follows (see [**31**, Theorem 2.1]) that, with probability one, the spectral measures $\nu_{\lambda,x}$ are absolutely continuous in the interval $(-E,E)$ with $\frac{d\nu_{\lambda,x}}{dE} \in L^{2p}(-E,E)$ for all $x \in \mathbb{B}$. Theorem 1.1 now follows.

2. Formulas

Let H_λ be the Anderson Hamiltonian on the Bethe lattice with μ p-admissible. We fix an arbitrary site in $\mathbb{B}$ which we will call the origin and denote by 0. Given two nearest neighbors sites $x, y \in \mathbb{B}$, we will denote by $\mathbb{B}^{(x|y)}$ the lattice obtained by removing from $\mathbb{B}$ the branch emanating from x that passes through y; if we do not specify which branch was removed we will simply write $\mathbb{B}^{(x)}$. Each vertex in $\mathbb{B}^{(x)}$ has degree $K+1$, with the single exception of x which has degree K.

Given $\Lambda \subset \mathbb{B}$, we will use $H_{\lambda,\Lambda}$ to denote the operator H_λ restricted to $\ell^2(\Lambda)$ with Dirichlet boundary conditions. The Green's function corresponding to $H_{\lambda,\Lambda}$ will be denoted by

$$G_{\lambda,\Lambda}(x,y;z) \quad = \quad \left\langle \delta_x, (H_{\lambda,\Lambda} - z)^{-1}\delta_y \right\rangle \tag{2.1}$$

for $x, y \in \Lambda$ and $z = E + i\eta$ with $E \in \mathbb{R}$, $\eta > 0$. We will write H_λ, $H_\lambda^{(x|y)}$ and $H_\lambda^{(x)}$ for $H_{\lambda,\mathbb{B}}$, $H_{\lambda,\mathbb{B}^{(x|y)}}$ and $H_{\lambda,\mathbb{B}^{(x)}}$, respectively. Similarly, we will use $G_\lambda(x,y;z)$ for $G_{\lambda,\mathbb{B}}(x,y;z)$ and $G_\lambda(z)$, $G_\lambda^{(x|y)}(z)$, $G_\lambda^{(x)}(z)$ for $G_\lambda(0,0;z)$, $G_{\lambda,\mathbb{B}^{(x|y)}}(x,x;z)$, $G_{\lambda,\mathbb{B}^{(x)}}(x,x;z)$, respectively.

We start by stating some useful formulas (see [**20**] for more details). The next proposition is a consequence of the resolvent equation.

PROPOSITION 2.1. *For any $\lambda \in \mathbb{R}$, $E \in \mathbb{R}$ and $\eta > 0$ we have*

$$G_\lambda(z) \quad = \quad -\left(z - \lambda V(0) + \tfrac{1}{4} \sum_{x:\, d(x,0)=1} G_\lambda^{(x|0)}(z) \right)^{-1} \tag{2.2}$$

and, for any two nearest neighbors sites $x, y \in \mathbb{B}$,

$$G_\lambda^{(x|y)}(z) \quad = \quad -\left(z - \lambda V(x) + \tfrac{1}{4} \sum_{x':\, d(x',x)=1,\, x' \neq y} G_\lambda^{(x'|x)}(z) \right)^{-1}. \tag{2.3}$$

PROPOSITION 2.2. *For any $\lambda \in \mathbb{R}$, $E \in \mathbb{R}$ and $\eta > 0$ we have*

$$G_\lambda(z) \quad = \quad \frac{i}{\pi} \int_{\mathbb{R}^2} e^{i(z-\lambda V(0))\varphi^2} \exp\left\{ \tfrac{i}{4} \sum_{x:\, d(x,0)=1} G_\lambda^{(x|0)}(z)\, \varphi^2 \right\} d^2\varphi\,, \tag{2.4}$$

and, for any two nearest neighbors sites $x, y \in \mathbb{B}$,

$$e^{\frac{i}{4} G_\lambda^{(x|y)}(z)\varphi^2} \quad = \tag{2.5}$$

$$-\frac{1}{\pi} \int_{\mathbb{R}^2} e^{-i\varphi\cdot\varphi'}\, \partial \left\{ e^{i(z-\lambda V(x))\varphi'^2} \exp\left\{ \tfrac{i}{4} \sum_{x':\, d(x',x)=1,\, x'\neq y} G_\lambda^{(x'|x)}(z)\, \varphi'^2 \right\} \right\} d^2\varphi'\,,$$

where $\varphi^2 = \varphi \cdot \varphi$ and $\partial f(\varphi^2) = f'(\varphi^2)$.

PROOF. If we perform the integration in (2.4) and (2.5) we obtain (2.2) and (2.3). $\square$

DEFINITION. For any $\lambda \in \mathbb{R}$, $E \in \mathbb{R}$ and $\eta > 0$ let

$$\xi_{\lambda,z}(t,s) \quad = \quad \mathbb{E}\left(\exp\left\{ \frac{i}{4} \left(G_\lambda^{(0)}(z)\, t \;-\; \overline{G_\lambda^{(0)}(z)}\, s \right) \right\} \right) \tag{2.6}$$

$$\quad = \quad \mathbb{E}\left(\exp\left\{ \frac{1}{4} \left[i\mathcal{R}_\lambda^{(0)}(z)\,(t-s) \;-\; \mathcal{I}_\lambda^{(0)}(z)\,(t+s) \right] \right\} \right) \tag{2.7}$$

for $t, s > 0$, where $\mathcal{R}_\lambda^{(x|0)}(z) + i\mathcal{I}_\lambda^{(x|0)}(z)$ is the decomposition of $G_\lambda^{(x|0)}(z)$ into its real and imaginary parts.

If $\lambda = 0$ we can calculate $G_0^{(0)}(z)$ [**3**] obtaining

$$\xi_{0,z}(t,s) = e^{\frac{i}{2K}\{(-z+\sqrt{z^2-K})t - \overline{(-z+\sqrt{z^2-K})s}\}}, \tag{2.8}$$

where we always make the choice Im $\sqrt{\ } > 0$. If $|E| < \sqrt{K}$, we have the pointwise limit

$$\xi_{0,E}(t,s) \equiv \lim_{\eta \downarrow 0} \xi_{0,z}(t,s) = e^{\frac{1}{2K}\{-iE(t-s)-\sqrt{K-E^2}(t+s)\}}. \tag{2.9}$$

LEMMA 2.3. *For any $\lambda \in \mathbb{R}$, $E \in \mathbb{R}$ and $\eta > 0$ we have*

$$\mathbb{E}(|G_\lambda(z)|^{2p}) = \tag{2.10}$$
$$\frac{1}{\pi^{2p}} \int_{\mathbb{R}^{2p}\times\mathbb{R}^{2p}} e^{iE(\varphi_+^2-\varphi_-^2)-\eta(\varphi_+^2+\varphi_-^2)} h(\lambda(\varphi_+^2-\varphi_-^2))[\xi_{\lambda,z}(\varphi_+^2,\varphi_-^2)]^{K+1} d^{2p}\varphi_+ d^{2p}\varphi_-,$$

and

$$\xi_{\lambda,z}(\varphi_+^2,\varphi_-^2) = \frac{1}{\pi^{2p}} \int_{\mathbb{R}^{2p}\times\mathbb{R}^{2p}} e^{-i(\varphi_+\cdot\varphi'_+ - i\varphi_-\cdot\varphi'_-)} \times \tag{2.11}$$
$$\partial_+^p \partial_-^p \left\{ e^{iE({\varphi'_+}^2-{\varphi'_-}^2)-\eta({\varphi'_+}^2+{\varphi'_-}^2)} h(\lambda({\varphi'_+}^2-{\varphi'_-}^2))[\xi_{\lambda,z}({\varphi'_+}^2,{\varphi'_-}^2)]^K \right\} d^{2p}\varphi'_+ d^{2p}\varphi'_-,$$

with

$$\partial_\pm g(\varphi_+^2,\varphi_-^2) = \frac{\partial}{\partial\varphi_\pm^2} g(\varphi_+^2,\varphi_-^2). \tag{2.12}$$

PROOF. Equations (2.10) and (2.11) follow from (2.4) and (2.5) by taking the pth power of each side, multiplying it by its complex conjugate, and taking expectations of each side, using the independence of the potential at different sites. □

3. Nonlinear analysis

To handle the nonlinear equations (2.10) and (2.11), we introduce the Banach spaces $\mathcal{K}_r$, $1 \le r \le \infty$, given by the completion of

$$\{g\colon [0,\infty)\times[0,\infty) \to \mathbb{C} \text{ of class } C^\infty \ ; \ \|g\|_{\mathcal{K}_r} \equiv |||g|||_2 + |||g|||_r < \infty\},$$

where

$$|||g|||_r^2 = \sum_{a,b=0}^{p} \|(2\partial_+)^a(2\partial_-)^b g(\varphi_+^2,\varphi_-^2)\|^2_{L^r(\mathbb{R}^{2p}\times\mathbb{R}^{2p}, d^{2p}\varphi_+ d^{2p}\varphi_-)}.$$

We define linear operators

$$(\mathcal{T}g)(\varphi_+^2,\varphi_-^2) = \tag{3.1}$$
$$\frac{1}{\pi^{2p}} \int_{\mathbb{R}^{2p}\times\mathbb{R}^{2p}} e^{-i(\varphi_+\cdot\varphi'_+ - i\varphi_-\cdot\varphi'_-)} \partial_+^p\partial_-^p \{g({\varphi'_+}^2,{\varphi'_-}^2)\} d^{2p}\varphi'_+ d^{2p}\varphi'_-$$

and

$$\mathcal{B}(\lambda,z) = M(e^{iE(\varphi_+^2-\varphi_-^2)-\eta(\varphi_+^2+\varphi_-^2)} h(\lambda(\varphi_+^2-\varphi_-^2))), \tag{3.2}$$

where $M(g(\varphi_+^2,\varphi_-^2))$ denotes multiplication by the function $g(\varphi_+^2,\varphi_-^2)$. It turns out that $\mathcal{T}$ is unitary on $\mathcal{K}_2$ [**7**] and is a bounded linear operator from $\mathcal{K}_1$ to $\mathcal{K}_\infty$, $\mathcal{B}_{\lambda,z}$ is a bounded linear operator on all $\mathcal{K}_r$, and $g \to g^n$ is a continuous map from $\mathcal{K}_\infty$ to $\mathcal{K}_1$ for any $n = 2,3,\ldots$. We can prove the following lemma.

LEMMA 3.1. (I) *$\xi_{\lambda,z} \in \mathcal{K}_\infty$ for all $\lambda \in \mathbb{R}$ and $z = E + i\eta$ with $\eta > 0$. The map $(\lambda, E, \eta) \to \xi_{\lambda,E+i\eta}$ is continuous from $\mathbb{R} \times \mathbb{R} \times (0, \infty)$ to $\mathcal{K}_\infty$.*

(II) *If $|E| < \sqrt{K}$ we have $\xi_{0,E} \in \mathcal{K}_\infty$ and*

$$\lim_{\eta \downarrow 0} \xi_{0,E+i\eta} = \xi_{0,E} \quad \text{in } \mathcal{K}_\infty . \tag{3.3}$$

(III) *The integral equation(2.11) can be rewritten as a fixed point equation in $\mathcal{K}_\infty$:*

$$\xi_{\lambda,z} = \mathcal{T}\mathcal{B}_{\lambda,z}\xi^K_{\lambda,z} , \tag{3.4}$$

valid for all $\lambda \in \mathbb{R}$ and $z = E + i\eta$ with $\eta > 0$, and also valid for $\lambda = 0$ and $z = E$ with $|E| < \sqrt{K}$.

The next step is a fixed point analysis.

LEMMA 3.2. *The map $Q : \mathbb{R} \times \mathbb{R} \times [0, \infty) \times \mathcal{K}_\infty \to \mathcal{K}_\infty$, defined by*

$$Q(\lambda, E, \eta, g) = \mathcal{T}\mathcal{B}_{\lambda,E+i\eta}g^K - g , \tag{3.5}$$

is continuous. Q is continuously Frechet differentiable with respect to g, the partial derivative being

$$Q_g(\lambda, E, \eta, g) = K\mathcal{T}\mathcal{B}_{\lambda,E+i\eta}M(g^{K-1}) - I . \tag{3.6}$$

Moreover, for any E such that $|E| < \sqrt{K}$ we have $Q(0, E, 0, \xi_{0,E}) = 0$ and

$$0 \notin \sigma(Q_g(0, E, 0, \xi_{0,E})) . \tag{3.7}$$

PROOF. The proof of this lemma is routine except for (3.7). We have

$$Q_g(0, E, 0, \xi_{0,E}) = K\mathcal{A}_{0,E} - I$$

where $\mathcal{A}_{0,E} = \mathcal{T}\mathcal{B}_{0,E}M(\xi^{K-1}_{0,E})$. As in [**20**], following Acosta and Klein [**3**] (see their Propositions 3.2 and 3.3 and Theorem 3.5), we can show that $\mathcal{A}^2_{0,E}$ is a compact operator on $\mathcal{K}_\infty$ and

$$\sigma(\mathcal{A}_{0,E}) = \{\mathcal{E}_{i,j} = E_i\bar{E}_j \,;\; i, j = 0, 1, 2, \dots\} \cup \{0\} , \tag{3.8}$$

with

$$E_n = \left(\frac{-E + i\sqrt{K - E^2}}{K}\right)^{2n} \quad \text{for } n = 0, 1, 2, \dots . \tag{3.9}$$

Since $\mathcal{E}_{i,j} \neq \frac{1}{K}$ for any $i, j = 0, 1, 2, \dots$, (3.7) follows. □

Lemma 3.2 tells us that the hypotheses of the Implicit Function Theorem (see [**29**, 2.7.2]) are satisfied by the function $Q(\lambda, E, \eta, g)$ at $(0, E, 0, \xi_{0,E})$, if $|E| < \sqrt{K}$. It follows that for each E such that $|E| < \sqrt{K}$ there exist $\lambda_E > 0$, $\varepsilon_E > 0$, $\eta_E > 0$ and $\delta_E > 0$, such that for each

$$(\lambda, E', \eta) \in (-\lambda_E, \lambda_E) \times (E - \varepsilon_E, E + \varepsilon_E) \times [0, \eta_E)$$

there is a unique $\omega_{\lambda,E',\eta} \in \mathcal{K}_\infty$ with $\|\omega_{\lambda,E',\eta} - \xi_{0,E}\|_{\mathcal{K}_\infty} < \delta_E$, such that

$$Q(\lambda, E', \eta, \omega_{\lambda,E',\eta}) = 0.$$

Moreover, the map

$$(\lambda, E', \eta) \in (-\lambda_E, \lambda_E) \times (E - \varepsilon_E, E + \varepsilon_E) \times [0, \eta_E) \longrightarrow \omega_{\lambda,E',\eta} \in \mathcal{K}_\infty$$

is continuous. Combining with Lemma 3.1, and using the uniqueness of $\omega_{\lambda,E',\eta}$ as above for each (λ, E', η), we get

THEOREM 3.3. *For any E such that $|E| < \sqrt{K}$ there exist $\lambda_E > 0$, $\varepsilon_E > 0$ and $\delta_E > 0$, such that the map*

$$(\lambda, E', \eta) \in (-\lambda_E, \lambda_E) \times (E - \varepsilon_E, E + \varepsilon_E) \times (0, \infty) \longrightarrow \xi_{\lambda, E'+i\eta} \in \mathcal{K}_\infty$$

has a continuous extension to $(-\lambda_E, \lambda_E) \times (E - \varepsilon_E, E + \varepsilon_E) \times [0, \infty)$ satisfying (3.4).

Theorem 1.2 now follows from (2.10), Theorem 3.3, the translation invariance of expectations, and a simple compactness argument.

Full details for the $p = 1$ case can be found in Klein [**20**].

References

1. Abou-Chacra, R., Anderson, P., Thouless, D. J.: A selfconsistent theory of localization. J. Phys. C: Solid State Phys. **6**, 1734-1752 (1973)
2. Abou-Chacra, P., Thouless, D. J.: Selfconsistent theory of localization: II. Localization near the band edges. J. Phys. C: Solid State Phys. **7**, 65-75 (1974)
3. Acosta, V., Klein, A.: Analyticity of the density of states in the Anderson model in the Bethe lattice. J. Stat. Phys. **69**, 277-305 (1992)
4. Aizenman, M. : Localization at weak disorder: some elementary bounds. Rev. Math. Phys., to appear.
5. Aizenman, M., Molchanov, S.: Localization at large disorder and extreme energies: an elementary derivation. Commun. Math. Phys. **157**, 245-278 (1993)
6. Anderson, P.: Absence of diffusion in certain random lattices. Phys. Rev. **109**, 1492-1505 (1958)
7. Campanino, M., Klein, A.: A supersymmetric transfer matrix and differentiability of the density of states in the one-dimensional Anderson model. Comm. Math. Phys. **104**, 227-241 (1986)
8. Carmona, R., Klein, A., Martinelli, F.: Anderson localization for Bernoulli and other singular potentials. Commun. Math. Phys. **108**, 41-66 (1987)
9. Carmona, R., Lacroix, J.: Spectral Theory of Random Schrodinger Operators. Boston, MA: Birkhauser, 1990
10. Delyon, F., Levy, Y., Souillard, B.: Anderson localization for multidimensional systems at large disorder or low energy. Commun. Math. Phys. **100**, 463-470 (1985)
11. von Dreifus, H., Klein, A.: A new proof of localization in the Anderson tight binding model. Commun. Math. Phys. **124**, 285-299 (1989)
12. Figotin, A., Klein, A.: Localization phenomenon in gaps of the spectrum of random lattice operators. J. Stat. Phys. **75**, 997-1021 (1994)
13. Fröhlich, J., Martinelli, F., Scoppola, E., Spencer, T.: Constructive proof of localization in the Anderson tight binding model. Commun. Math. Phys. **101**, 21-46 (1985)
14. Fröhlich, J., Spencer, T.: Absence of diffusion in the Anderson tight binding model for large disorder or low energy. Commun. Math. Phys. **88**, 151-184 (1983)
15. Gol'dsheid, Ya., Molchanov, S., Pastur, L.: Pure point spectrum of stochastic one dimensional Schrödinger operators. Funct. Anal. Appl. **11**, 1-10 (1977)
16. Graf, G. M.: Anderson localization and the space-time characteristic of continuum states. J. Stat. Phys. **75**, 337-346 (1994)
17. Klein, A.: The supersymmetric replica trick and smoothness of the density of states for random Schrodinger operators. Proc. Symposia in Pure Mathematics **51**, 315-331 (1990)
18. Klein, A.: Localization in the Anderson model with long range hopping. Braz. J. Phys. **23**, 363-371 (1993)
19. Klein, A.: Absolutely continuous spectrum in the Anderson model on the Bethe lattice. Mathematical Research Letters **1**, 399-407 (1994)

20. Klein, A.: Extended states in the Anderson model on the Bethe lattice. Advances in Math., to appear.
21. Klein, A.: Spreading of wave packets in the Anderson model on the Bethe lattice. Commun. Math. Phys., to appear.
22. A. Klein, F. Martinelli and J. F. Perez, A rigorous replica trick approach to Anderson localization in one dimension. Commun. Math. Phys. **106**, 623-633 (1986)
23. Kunz, H., Souillard, B.: Sur le spectre des operateurs aux differences finies aleatoires. Commun. Math. Phys. **78**, 201-246 (1980)
24. Kunz, H., Souillard, B.: The localization transition on the Bethe lattice. J. Phys. (Paris) Lett. **44** , 411-414 (1983)
25. McKane, A. J., Stone, M.: Localization as an alternative to Goldstone's theorem. Ann. Phys. **131**, 36-55 (1981).
26. Miller, J., Derrida, B.: Weak disorder expansion for the Anderson model on a tree. J. Stat. Phys. **75**, 357-389 (1994)
27. Mirlin, A. D., Fyodorov, Y. V.: Localization transition in the Anderson model on the Bethe lattice: spontaneous symmetry breaking and correlation functions. Nucl. Phys. B **366**, 507-532 (1991)
28. Mott, N., Twose, W.: The theory of impurity conduction. Adv. Phys. **10**, 107-163 (1961)
29. Nirenberg, L.: Topics in Nonlinear Functional Analysis. New York: Courant Institute of Mathematical Sciences, 1974
30. Pastur, L.: Spectra of random selfadjoint operators. Russ. Math. Surv. **28**, 1-67 (1973)
31. Simon, B.: L^p norms of the Borel transform and the decomposition of measures. Proc. Amer. Math. Soc., to appear
32. Simon, B., Wolff, T.: Singular continuum spectrum under rank one perturbations and localization for random Hamiltonians. Commun. Pure. Appl. Math. **39**, 75-90 (1986)
33. Spencer, T.: Localization for random and quasiperiodic potentials. J. Stat. Phys. **51**, 1009-1019 (1988)
34. Spencer, T.: Lifshitz tails and localization. Preprint
35. Thouless, D.: Electrons in disordered systems and the theory of localization. Phys. Rev. **13**, 93-106 (1974)

DEPARTMENT OF MATHEMATICS, UNIVERSITY OF CALIFORNIA, IRVINE, CA 92717-3875
E-mail address: aklein@math.uci.edu

Proceedings of Symposia in Pure Mathematics
Volume **59**, 1996

QUANTIZATION BY DEFORMATION AND STATISTICAL MECHANICS

A. LICHNEROWICZ

ABSTRACT. This short summary of developments of the notion of star-product is divided into two parts. In the first one, we recall this notion as deformation and give a generalization based on conformal symplectic structure and well adapted to the formulation of Statistical Mechanics. The second part is relative to the famous Kubo-Martin-Schwinger (KMS) condition concerning the Gibbs states and we obtain simple and geometric expression for the (KMS)-conditions (both classical and quantum).

From 1973, Flato, Sternheimer and myself have studied, with increasing generality, deformations of the Poisson bracket structure of classical mechanics and related associative algebra structures [1]. The idea that quantum theories are deformations of classical theories was presumably in the back of the mind of many scientists, even before the mathematical notion of deformation was formalized by Gerstenhaber [2] for algebraic structures.

In all these approachs, people were always considering that in the end, quantum theories have to be formulated in operator language. The geometrical approach to quantization obtained by deformation can be developed in a complete autonomous manner on the algebra of classical observables by deformation of the algebraic structures. The connection with operatorial formulation, whenever possible, comes only afterwards and is not necessary. That is true for quantum mechanics and quantum field theories. The corresponding instrument is often referred as *star-products.*

This talk is divided in 2 parts. In the first part, we recall the essential of the theory of the star-products, in the second, we will sketch how a further deformation, necessarily based on *conformal symplectic structure*, with a second parameter, is adapted to the formulation of Statistical Mechanics, in particular to the famous Kubo-Martin-Schwinger (KMS) condition concerning the Gibbs states.

1. - Notion of star-product.

Let W be a Banach differentiable manifold (of finite or infinite dimension) on which there exists a partition of the unity. Denote by $N = C^\infty(W)$ the space of the real-valued infinitely differentiable functions on W, space which can be eventually restricted. (plus possibly a distribution). In the following, we use as technical

---

1991 *Mathematics Subject Classification.* Primary 58F06 ; Secondary 81S10.

instrument the *Schouten bracket* on the skew-symmetrical contravariant tensors or *p-tensors.*

a) *A Poisson manifold* (W, Λ) is a manifold endowed with a skewsymmetrical contravariant 2-tensor (or 2-tensor) satisfying in the sense of the Schouten bracket

$$(1.1) \qquad [\Lambda, \Lambda] = 0$$

Λ defines on W a *Poisson bracket* satisfying the Jacobi identity

$$(1.2) \qquad \{u, v\} = i(\Lambda)(du \wedge dv) = P(u, v) \qquad (u, v \in N)$$

where P is the Poisson bidifferential operator. If $u \in N$, the corresponding Hamiltonian vector field is $X_u = [\Lambda, u]$; (W, Λ) is symplectic if Λ is *non degenerate,* that is if $[\Lambda, u] = 0$ implies $du = 0$ for any $u \in N$.

Let $N[[\nu]]$ be the space of the formal functions in $\nu \in C$ with coefficients in N ; ν is the deformations parameter (for quantum Mechanics $\nu = \frac{\hbar}{2i}$). We consider the associative (and commutative) algebra $(N, .)$ given by the pointwise multiplication and a formal associative deformation of $(N, .)$. Such a deformation is strictly connected with the Hochschild cohomology of the composition law. More precisely, we say that we have a *star product* $*_\nu$ on the Poisson manifold (W, Λ) if we have [3]

$$(1.3) \qquad u *_\nu v = \sum_{r=0}^{\infty} v^r C_r(u, v) = u\,v + \sum_{r=1}^{\infty} v^r C_r(u, v) \qquad (u, v \in N)$$

where the C_r are bioperators or 2-cochains with values in N satisfying *the condition of associativity*

$$(u *_\nu v)) *_\nu w) = u *_\nu (v *_\nu w) \qquad (for\, u, v, w \in N\; and\, so\; \in N[[\,v\,]]$$

and in addition :

$$(1.4) \qquad C_1(u, v) - C_1(v, u) = 2P(u, v)$$

We assume often that the C_r are bidifferential operators (differential star-products) or that the C_r are null on the constants (*n.c case*), so that

$$(1.5) \qquad 1 *_\nu u = u *_\nu 1 = u$$

Moreover, we may suppose that the C_r are even in u, v if r is even, odd if r is odd *(parity condition)* that

$$u *_{-\nu} v = v *_\nu u \qquad (u, v \in N)$$

From (1.4), it follows that a star-product defines a deformation of the Poisson Lie algebra (N, P)

$$(1.6) \quad [u, v]_v = (2v)^{-1}(u *_\nu v - v *_\nu u) = P(u, v) + \sum_{r=2}^{\infty} (\frac{1}{2})\, v^{r-1}(C_r(u, v) - C_r(v, u))$$

In the differentiable case, this allows to use instead of the infinite-dimensional Hochschild cohomology , the finite-dimensional Chevalley cohomology space. In the d.n.c case Chevalley 2-cohomology has the dimension $b_2(W)$, where $b_2(W)$ is the second Betti number of W, which permits [3] to show that at each step there are only $1 + b_2(W)$ choices.

Under the hypothesis $b_3(W) = 0$. J. Vey [4] had obtained the existence of star-brackets for all symplectic manifolds and Neroslavsky-Vlasov the existence of star-products which are d.n.c and satisfy the parity condition. In 1983 P. Lecomte and M. De Wilde have proved that such a star-product exists always on a finite dimensional symplectic manifold [5] or regular Poisson manifold. In 1991 Omori and al have given another and more geometric proof of existence of non-products under the same hypothesis [6].

If (W, Λ) is a finite-dimensional or infinite-dimensional Poisson manifold, we suppose that there exists on (W, Λ) a star-product defined on a suitable algebra of functions. We have then a *starred Poisson manifold* $(W, \Lambda, *_\nu)$

2. - Star-products and quantization.

a) Consider $W = R^{2n} = R^n \times R^n$ with its usual symplectic structure Λ and its closed 2-forms F. In 1927, H. Weyl [7] gave a rule for passing from a classical observable $a \in N$ to an operator in $L^2(R^n)$ which represents a quantization of this observable. It can be written

$$a \in N \to \Omega_w(a) = \int \bar{a}(\xi, \eta)\ exp(i(P\xi + Q\eta)/\hbar)\ w(\xi, \eta) d\xi d\eta \tag{2.1}$$

where $\bar{a}$ is the Fourier transform of a, w is a weight function ($= 1$ in the case of Weyl), P and Q satisfy the canonical commutation relations and the integral is taken in the weak operator topology. The trace can be given by

$$Tr(\Omega_1(a)) = (2\pi\hbar)^{-n} \int_{R^{2n}} aF^n \tag{2.2}$$

In the end of the 40's, Moyal and Groenewold [8] found that the product and commutator of quantum observables correspond in the Weyl rule to exponential $exp((\frac{\hbar}{2i})P)(u, v)$ and sine of the Poisson bracket. Thus $\Omega_1(a)\Omega_1(b) = \Omega_1(a * b)$ with $C_r(a, b) = P^r(a, b)/r! \quad (r \geq 1)$ where P^r is the rth−power of P :

b) In 1975, inspired by our earlier works [1] on deformations of the Lie algebra (N, P), J. Vey [4] obtained the interpretation of the Moyal bracket as an example of differentiable n.c deformation. We then seen that quantization should in fact be considered as a deformation of a classical theory, with the same algebra of observables and a star-product [3]. In (2.1) the choice of w corresponds to a choice of the ordering (normal or standard ordering). It is then possible to develop quantum theories in terms of the star-product only (in general differentiable n.c), even if a Weyl-Wigner-type of quantization rule does not appear. In particular a spectral theory can be developed using the star-products.

In the case of a star-product (in the differentiable n.c case, with the parity condition), we can define the pth star-power of $a \in N$, for example of the classical

Hamiltonian H ; we denote this power by $H^{(*)p}$. It is easy to prove by induction on p that $H^{(*)p}$ is always an even function of ν. We can set formally

$$(2.3) \qquad Exp_* \left(\frac{i}{\hbar}\, t\, H\right) = \sum_0^{\infty} \left(\frac{i}{\hbar} t\right)^p \frac{H^{(*)p}}{p!}$$

For usual star-products and Hamiltonians, the right member of (2.3) defines a distribution on W (analogue of the evolution operator). The spectrum of H is the spectrum of $Exp\left(\frac{i}{\hbar}\, t\, H\right)$ in the sense of Schwartz.

c) Suppose $dim\ W = 2n$, and (W, Λ) symplectic. The manifold admits a measure $\tilde{\eta} = (2\pi\hbar)^{-n}\eta$ where η is the symplectic volume element of (W, Λ) ($\tilde{\eta}$ normalized Liouville measure). Let N^c be the complexified space of N ; the star-product can be extended to N^c in a natural way.

Suppose that our star-product (differentiable n.c) satisfies the following requirements

1) The star-product is non degenerate : if $u \in N^c$, $\bar{u} * u = 0$ implies $u = 0$.

2) If u, v are functions with compact supports or suitable asymptotic conditions

$$(2.4) \qquad \int_W C_r(u, v)\eta = \int_W C_r(v, u)\eta \qquad r = 1, 2...$$

Such a star-product is sayed to be *strongly closed.* Lecomte and De Wilde [5] have proved that all *differentiable, null on constants, star-products are strongly closed.*

The expectation value associated with an observable $a \in N$ and a state $\rho \in N$ which is a quasi projector $(\rho * \rho = \frac{1}{k}\, \rho)$ is defined by

$$(2.5) \qquad \int (a * \rho)\tilde{\eta} = \int (\rho * a)\tilde{\eta}$$

This formula is an extension of a classical formula of Wigner[7].

For the harmonic oscillator, one gets the spectrum $(m + \frac{1}{2}n)$ with Moyal ordering $(m \in N)$. Many other examples can be treated. Consider $E = (R^n - 0) \times R^n$. The 2-dimensional solvable group K acts on E in the following way

$$(x, y) \in E \rightarrow (x' = e^a x, y' = e^{-a}(u + b\, x) \qquad (a, b \in R)$$

K preserves the Moyal P^r. The space of the orbits of E by the group K is isomorphic to $T^*\, S^{n-1}$. We deduce from the Moyal product of E a star-product on $T^*\, S^{n-1}$ invariant under $So(n)$. The regular Kepler problem admits as phase space the so-called manifold of Moser-Sourian $T_0^* S^3$ (without null section). With the corresponding star-product, we obtain the quantization of the hydrogen atom.

A star-product is said *closed* if (2.4) is supposed only for $r \leq n$, i.e. if the coefficient of ν^n in $(u * v - v * u)$ for all u, v has vanishing integral. An interesting feature of closed star-products is that they are classified by cyclic cohomology of A.

Connes instead of only Hochschild cohomology. This suggests to define a notion of *character* of a closed star-product and leads to an index theorem [9].

3) Deformations of local associative algebras. Extensible infinitesimal deformations.

Let W be a smooth finite-dimensional manifold and $N = N(W)$ the space of real-valued infinitely differentiable functions on W.

a) *A local associative algebra* $(N, \square)$ is an associative algebra on N such that

$$Supp(u \square v) \subset Supp\, u \cap Supp\, v \qquad (u,\, v \in N)$$

It is known (Rubio [10]) that such an algebra is necessarily given by

$$u \square v = f\, u\, v \tag{3.1}$$

where f is a fixed element of N ; the algebra is necessarily commutative. If $(N, \square)$ admits a unit element e, the function f is $\neq 0$ everywhere and $e = f^{-1}$. In the following $(N, \square)$ *is the local associative algebra given by (3.1), where* f *is* $\neq 0$ *everywhere.*

Introduce the Hochschild cohomology of $(N, \square)$ with values in N. A p-cochain C is a mapping of $N \times N \times \ldots \times N$ (p factors) into N and the corresponding standard coboundary operator is denoted by $\tilde{\partial}$. In particular if C is a 2-cochain, we have

$$\begin{aligned} \tilde{\partial} C(u,v,w) = u \square C(v,w) - C(u \square v, w) + C(u, v \square w) \\ - C(u,v) \square w \ (u,v,w \in N) \end{aligned} \tag{3.2}$$

Consider a formal associative deformation of $(N, \square)$ defined by the map of N in $N[[\nu]]$ given by

$$u *_\nu^f v = \sum_{r=0}^{\infty} \nu^r C_r^f(u,\nu) = f\, u\, \nu + \sum_{r=0}^{\infty} \nu^r\, C_r^f(u,\nu) \tag{3.3}$$

where the $C_r^f (r \geq 1)$ are local 2-cochains satisfying the parity condition so that $u *_{-\nu}^f v = v *_\nu^f u$.

The notion of deformation of order q in ν is clear. In particular a deformation of order 1 is called an *infinitesimal deformation.*

b) Consider a deformation of order 2 of $(n, \square)$

$$u *_\nu^s v = f\, u\, \nu + \nu C_1^f(u,v) + \nu^2 C_2^f(u,v) \tag{3.4}$$

where the 2-cochains C_1^f, C_2^f are local, C_1^f being odd and C_2^f even. It is easy to verify that we have

$$\tilde{\partial}\, C_1^f = 0 \tag{3.5}$$

and

$$\tilde{\partial}\, C_2^f = \tilde{E}_2 \tag{3.6}$$

where $\tilde{E}_2$ is the 3-chain given by

$$\tilde{E}_2\,(u,v,w) = C_1^f\,(C_1^f\,(u,v,w) - C_1^f\,(u, C_1^f(v,w))$$

For an arbitrary 3-cochain B , we set

$$(\hat{\Sigma}\, B)(u,v,w) = B(u,v,w) - B(v,u,w) - B(u,w,v)$$

It is easy to see that if C is an even 2-cochain, we have $(\hat{\Sigma}\hat{\partial}\, C = 0$. It follows from (3.6) that $(\hat{\Sigma}\hat{E}_2 = 0$ that is

$$S\, C_1^f\,(C_1^f\,(u,v),w) = 0 \tag{3.7}$$

where S is the summation over cyclic permutations. Equation (3.7) expresses that C_1^f defines on W the brack of a *local Lie algebra.* It is known then (Kirillov [11]) that there exist on W a 2-tensor Λ^f and a vector E^f such that $C_1^f = P^f$, where

$$P^f\,(u,v) = i(\Lambda^f)\,(du \wedge dv) + i(E^f)(u\,dv - v\,du) \tag{3.8}$$

P^f satisfying the Jacobi identity, we have in terms of Schouten brackets

$$[\Lambda^f\,,\Lambda^f] = 2\,E^f \wedge \Lambda^f \qquad [E^f\,,\Lambda^f] = \mathcal{L}(E^f)\Lambda^f = 0 \tag{3.9}$$

where $\mathcal{L}$ is the operator of Lie derivative. We see that if $f\,u\,v + \nu\, C_1^f\,(u,v)$ is an infinitesimal deformation of $(N,\square)$ extensible to the order 2, $C_1^f = P^f$ where P^f is the above bracket of a Lie algebra. It is easy to see that if $\tilde{\partial}P^f = 0$ according to (3.5), we have $i(E^f)\,d\,f$ and so $E^f = f^{-1}[\Lambda^f, f]$. If we set $\Lambda^f = f\Lambda$, (3.9) is translated by $[\Lambda,\Lambda] = 0$, $E^f = [\Lambda, f]$; Λ defines on W a Poisson structure and E^f is the Hamiltonian vector field associated with f. We obtain

Proposition. - *Each extensible infinitesimal deformation of* $(N,\square)$ *is given by*

$$u *_\nu^f v = f\,u\,v + \nu P^f(u,v)$$

where P^f *is the bracket of the conformal Poisson structure deduced from a Poisson structure by the conformal factor* f.

c) Let W be a finite or infinite-dimensional Banach manifold. We consider on $N = N(W)$ an associative algebra $(N,\square)$ given by $u,v \in N \to f\,u\,v \in N$, where $f \in N$ is $\neq 0$ everywhere. Consider an associative deformation of $(N,\square)$ given by

$$u *_\nu^f\, v = f\,u\,v + \nu\, P^f(u,v) + \sum_{r=2}^{\infty} \nu\, C_r^f\,(u,v) \tag{3.10}$$

where the 2-cochains C_r^f are differentiable, satisfy the parity condition and where P^f is the bracket of the conformal Poisson structure (Λ^f, E^f) deduced from a Poisson structure Λ by the conformal factor f. *If (3.10) admits a unit element e_ν (necessarily even in ν) we say that (3.10) is a f-star-product on W.* We have proved [12].

Theorem. - *If (3.10) is a f-star-product, there is a differentiable n.c star-product $*_\nu$ of (W, Λ) and an even element f_ν of $N[[\nu]]$ satisfying $f_0 = f$ such that (3.10) is given by*

$$u *_\nu^f v = u *_\nu f_\nu *_\nu v \tag{3.11}$$

Conservely, for an arbitrary choice of f_ν and $*_\nu$, (3.11) defines a f-star-product admitting the unit element $f_\nu^{(*)-1}$.

The existence or equivalence problems concerning the f-star-products can be reduced to the same problems concerning the star-products.

4. - **Statistical Mechanics.**

Let $(W, \Lambda, *)$ be a starred symplectic Banach manifold, which is the phase space of the considered physical system. We introduce a new deformation parameter $\beta = 1/kT \geq 0$ (the inverse temperature, where k is Boltzmann's constant and T the temperature) adapted to the formulation of Statistical Mechanics. On the manifold, let $(A, *)$ be an algebra of observables contained in N (the algebra of the so-called quasi-local observables) and containing the function 1 (which is the unit element of the algebra). A state ω is described by a continuous linear functional on A and for $a \in A, \omega(a)$ is the estimation value corresponding to the observable and to the state ω. [13], [14].

Let H be the Hamiltonian of our physical system ; H does not belong necessarily to A. The quantum dynamics of the system is given by the one-parameter group of automorphisms of A

$$\alpha_t(u) = Exp_*((i/\hbar)tH) * u * Exp_*(-(i/\hbar)tH) \tag{4.1}$$

If $\beta = 1/kT$, we suppose that for every real τ $(0 \leq \tau \leq 1)$

$$\alpha_{t+i\hbar\tau\beta}(u) = Exp_*((i/\hbar)(t + i\hbar\tau\beta) H) * u * Exp_* (-(i/\hbar)(t + i\hbar\tau\beta) H) \tag{4.2}$$

is an automorphism of A. We note that

$$\alpha_{i\hbar\tau\beta} u = Exp_*(-\tau\beta H) * u * Exp_*(\tau\beta H) \tag{4.3}$$

and that $Exp_*(-\tau\beta H)$ satisfy the parity assumption in $\nu = \hbar/2i.Exp_*(-\tau\beta H)$ for example described the so-called *canonical ensemble*. We set

$$f_\beta = Exp_*(\tau\beta H) \tag{4.4}$$

and introduce eventually, in the following, the associative product

$$u \tilde{*}_\beta v = u * f_\beta * v \tag{4.5}$$

5. - **Classical KMS-condition.**

a) The (KMS) condition is an important and nice condition characterizing in Statistical Mechanics the Gibbs states. Haag-Kastler and Araki have studied, under strong analytical assumptions, the physical origin of this condition ([15] and [16]).

Haag and Kastler have proposed to consider the following as the defining properties of an equilibrium state or Gibbs state ω

1) *The stationarity of the state* $\omega(\alpha_t(a)) = \omega(a)$ for $a \in A$

2) *The stability under local perturbations of the Dynamics* (that is of H).

Consider a stationary state ω with relative purity : ω cannot be split into a convex combination of different stationary states. Under suitable analytic asumptions, Haag-Kastler et al have proved that, in the quantum case, the stability condition for a stationary state is translated by the (KMS)-condition. This in turn implies that it is a Gibbs state. In 1976 Aizenmann et al [17] have proved the same for classical Statistical Mechanics.

We see that the stability implies very strong constraints on ω. The quantum (KMS)-condition can be written in terms of star-products by translation of the usual operator condition : ω satisfies the (KMS)-condition if, for any observables $a, b \in A$

$$\omega(a * b) = \omega(b * \alpha_{i\hbar\beta}(a)) \tag{5.1}$$

b) Form the classical limit of Equation (5.1). We note that for $\nu = \hbar/2i$, we have up to terms of order ≥ 2 in ν

$$\alpha_{i\hbar\beta}(a) \simeq a - (\hbar/i)\,\beta P(H, a)$$

It follows from (5.1) that

$$\omega((i/\hbar)(a * b - b * a) \simeq -\beta\omega(b\,P(H, a))$$

We obtain thus for the classical (KMS)-condition

$$\omega(P(a, b)) = -\beta\omega(b\,P(H\,a)) \tag{5.2}$$

that is the form given by Aizenmann et al up to the notations [17].

c) It is easy to obtain an interesting geometric equivalent form. By skew symetrization of (5.2) in a, b, (5.2) implies

$$\omega(P(a, b)) = (\beta(2)\omega\,(a\,P(H, b) - b\,P(H, a) \tag{5.3}$$

Conversely assume (5.3). If we take $b = 1$, we have $\omega(P(H, a)) = 0$ for any a, relation which expresses that ω is stationary. For arbitrary observables a, b, we have then $\omega(P(H, a\,b) = 0$ and (5.3) can be written

$$\omega(P(a, b)) = (\beta(2)\omega\,(a\,P(H, b) - b\,P(H, a) - P(H, a\,b))$$

which, after simplifications, is nothing other but (5.2). Therefore (5.3) is equivalent with (5.2).

Take $f = f_\beta = e^{-(\beta/2)H}$ in (3.8). We have $\Lambda_\beta = e^{(\beta(2)H}\Lambda$ and $E_\beta = -(\beta/2)\, e^{-(\beta/2)H}[\Lambda, H]$. The bracket (3.8) takes the form

$$P^{f\beta}(u,v) = [u,v]_\beta = e^{-(\beta/2)H}(P(u,v) - (\beta/2)\, u\, P(H,v) + (\beta/2)\, v\, P(H,u))$$

It follows that (5.3) can be written

$$\omega[e^{(\beta/2)H}[a,b]_\beta] = 0 \qquad for\ f_\beta = e^{(-\beta/2)H} \tag{5.4}$$

we have [18].

Theorem. - *The classical (KMS)-condition is equivalent to (5.4), where* $[a,b]_\beta$ *is the bracket associated with the conformal symplectic structure corresponding to the factor* $f_\beta = e^{-(\beta/2)H}$

6. - **Quantum (KMS-condition**

a) Come back to the quantum (KMS)-condition (5.1). It is easy to prove the following Lemma

Lemma. - *If* ω *is a stationary state, we have for* $0 \leq \tau \leq 1$

$$\omega(\alpha_{i\hbar\tau\beta}(a)) = \omega(a) \tag{6.1}$$

Take $\tau = 1/2$; we set $f_\beta = Exp_*(-(\beta/2)\,H)$ and introduce the automorphism σ of Λ given by

$$\sigma a = \alpha_{i(\hbar/2)\beta}(a) = f_\beta * a * f_\beta^{(*)-1}$$

(6.1) implies $\omega(\sigma\, a) = \omega(a)$ for all $a \in A$; (5.1) can be written

$$\omega(a * b) = \omega(b * \sigma^2 a) \qquad forall\ a,\, b \in A \tag{6.2}$$

In (6.2) subtitute to the observable $\sigma\, b$. We have, according to (6.2)

$$\omega(a * \sigma\, b) = \omega(\sigma\,(b * \sigma a)) = \omega(b * \sigma\, a)$$

that is

$$\omega(a * \sigma\, b - b * \sigma a) = 0 \tag{6.3}$$

This relation can be written

$$\omega(Exp_*((\beta/2)\,H) * [a,b]_{\tilde{*}\,\beta}) = 0$$

where $[a,b]_{\tilde{*}\,\beta}$ is the bracket deduce from the associative product $\tilde{*}_\eta = *Exp_* - (\beta/2\,H)\,*$. The converse is evident. We have

Theorem. - *On a starred symplectic manifold* $(W, \Lambda, *)$, *the quantum (KMS)-condition is equivalent to*

$$\omega[Exp_*((\beta/2)\,H) \, * \, [a,b]_{\tilde{*}\,\beta}] = 0 \tag{6.4}$$

which corresponds to the classical case. In terms of operators (6.4) takes the form

$$\omega(\rho^{(\beta/2)\,H}\,A\,e^{-(\beta/2)\,H}\,B - e^{(\beta/2)\,H}Be^{-(\beta/2)\,H}A) = 0 \tag{6.5}$$

and it is possible to prove directly this equivalence.

Note that the deformation point of view leads us in a natural way to introduce conformal symplectic geometry. We obtain thus simple and geometric expressions for the (KMS)-conditions (both classical and quantum) ; it just tells us that, up to a temperature -dependant conformal factor, the KMS- states should be those which see a Lie algebra of observables as *abelian.*

Note also that in the second part of this talk, the point of view is geometrical and physical. We do not consider here the corresponding problems of Mathematical Analysis.

REFERENCE

[1] **M. Flato, A. Lichnerowicz, D. Sterheimer**, Composito Mathematica 31, (1975), 47-82. C.R. Acad Sci Paris 279, (1974), 887-881.

[2] **M. Gerstenhaber**, Ann Math 79 (1964), 59-103.

[3] **F. Bayen, M. Flato, C. Fronsdal, A. Lichnerowicz, D. Sternheimer**, Lett Math Phys 1 (1977), 521-530. Ann Phys 111 (1978), 61-151.

[4] **J. Vey**, Comm Math Helv 50, (1975), 421-454.

[5] **P. Lecomte and M. De Wilde**, Lett Math Phys 7 (1983), 487-496.

[6] **H. Omori, Y. Maeda and A. Yoshioka**, Adv in Math 85 (1991), 225-255 ; Lett Math Phys 26 (1992), 285-294.

[7] **H. Weyl**, Z Physik 46 (1927), 1-46 ; **E.P. Wigner**, Phys Rev 40 (1932) 749-759.

[8] **J.E. Moyal**, Proc Cambridge Phil Soc 45 (1949), 99-124 ; **A. Groenewold**, Physica 12 (1946) 405-460.

[9] **A. Connes, M. Flato, D. Sternheimer**, Lett Math Phys 24 (1992), 1-12.

[10] **R. Rubio**, C.R. Acad Sc Paris 299 I (1984), 370-375.

[11] **A.A. Kirillov**, Russian Math Surv 31 (1976), 55-75.

[12] **H. Basard and A. Lichnerowicz**, Lett Math Phys 10 (1985), 167-172.

[13] **R. Kubo**,J Phys Soc Japan 12 (1957), 570-586.

[14] **P.C. Martin and J. Schwinger**, Phys Rev 115 (1959), 1342-1373.

[15] **R. Haag, D. Kastler, E. Trich-Pohlmayer**,Comm Math Phys 38 (1974), 173-193.

[16] **H. Araki**, Lecture Notes in Math 650, Springer Berlin (1978), 66-84.

[17] **M. Aizenmann, G. Gallavotti, S. Goldstein and J. Lebowitz**, Comm Math Phys 48 (1976), 1-14.

[18] **H. Basard, M. Flato, A. Lichnerowicz and D. Sternheimer**, Lett Math Phys 8 (1984), 483-494.

COLLEGE DE FRANCE

Physique Mathématique

3, rue d'Ulm

75005 PARIS - FRANCE

Proceedings of Symposia in Pure Mathematics
Volume **59**, 1996

POSSIBLE CLASSIFICATION OF CONTINUOUS SPATIAL SEMIGROUPS OF $*$-ENDOMORPHISMS OF $\mathfrak{B}(\mathfrak{H})$

ROBERT T. POWERS

ABSTRACT. This talk concerns the structure of continuous spatial semigroups of $*$-endomorphisms of $\mathfrak{B}(\mathfrak{H})$. These are semigroups of $*$-endomorphisms of $\mathfrak{B}(\mathfrak{H})$ for which there exist a strongly continuous one parameter semigroup of intertwining isometries. New examples of such semigroups are discussed. Unlike all previous examples these semigroups are not cocycle conjugate to semigroups constructed using the CAR or CCR algebras. A scheme for the possible classification of all continuous spatial semigroups of $*$-endomorphisms is discussed.

I. INTRODUCTION.

This talk is concerned with the structure of spatial E_o-semigroups of $\mathfrak{B}(\mathfrak{H})$. An E_o-semigroup $\{\alpha_t\}$ of $\mathfrak{B}(\mathfrak{H})$ is a continuous unit preserving semigroup of $*$-endomorphisms of $\mathfrak{B}(\mathfrak{H})$. An E_o-semigroup $\{\alpha_t\}$ is said to be spatial if there exists a strongly continuous one parameter semigroup of isometries $\{U(t)\}$ which intertwine α_t in that $U(t)A = \alpha_t(A)U(t)$ for all $A \in \mathfrak{B}(\mathfrak{H})$ and $t \geq 0$.

At present there is only one known example of an E_o-semigroup of $\mathfrak{B}(\mathfrak{H})$ which is not spatial (see [P2]). We believe that there are uncountably many non spatial E_o-semigroups of $\mathfrak{B}(\mathfrak{H})$ which are mutually non cocycle conjugate (see next section for a the definition of cocycle conjugacy).

The spatial E_o-semigroups have been constructed in [P1] with aid of the CAR algebra and in [A1] with the CCR algebra. It was shown in [PR] that these E_o-semigroups are the same. The known examples of spatial E_o-semigroups are CAR-CCR flows of Arveson index $n = 1, 2, \cdots$ and $n = \infty$. These examples are what is known as completely spatial (a complete definition is given the next section) in that there are enough intertwining semigroups of isometries to uniquely determine the E_o-semigroup.

In this talk we discuss new examples of spatial E_o-semigroups constructed in [P4] and discuss a possible scheme for classifying all spatial E_o-semigroups.

II. DEFINITIONS, NOTATION AND BACKGROUND.

All Hilbert spaces which will be denoted by the characters such as $\mathfrak{H}, \mathfrak{K}, \mathfrak{J}$ and $\mathfrak{M}$ are assumed to be separable unless otherwise stated. On Hilbert spaces we use the physicist's inner product (f, g) which is linear in g and conjugate linear in f. If $\mathfrak{H}$

1991 *Mathematics Subject Classification.* Primary 46L40; Secondary 46L55, 46L30, 47B25.
Key words and phrases. $*$-endomophisms, E_o-semigroups.
Supported in part by a National Science Foundation Grant

is a Hilbert space we denote by $\mathfrak{B}(\mathfrak{H})$ the set of all bounded linear operators on $\mathfrak{H}$ and by $\mathfrak{B}(\mathfrak{H})_*$ the predual of $\mathfrak{B}(\mathfrak{H})$. Every element $\rho \in \mathfrak{B}(\mathfrak{H})_*$ can be represented in the form $\rho(A) = \Sigma_{i=1}^{\infty}(f_i, Ag_i)$ where $\Sigma_{i=1}^{\infty}\|f_i\|\,\|g_i\| < \infty$.

Definition 2.1. We say $\{\alpha_t; t \geq 0\}$ is an E_o-semigroup of a von Neumann algebra M with unit I if the following conditions are satisfied.

(i) α_t is a $*$-endomorphism of M for each $t \geq 0$.
(ii) α_o is the identity endomorphism and $\alpha_t \circ \alpha_s = \alpha_{t+s}$ for all $t, s \geq 0$.
(iii) For each $f \in M_*$ (the predual of M) and $A \in M$ the function $f(\alpha_t(A))$ is a continuous function of t.
(iv) $\alpha_t(I) = I$ for each $t \geq 0(\alpha_t$ preserves the unit of $M)$.

Occasionally we will want to drop condition (iv) in the above definition. A semigroup satisfying conditions (i) through (iii) of the above definition will be called an e_o-semigroup of M.

Definition 2.2. Suppose $\{\alpha_t : t \geq 0\}$ and $\{\beta_t : t \geq 0\}$ are E_o-semigroups $\mathfrak{B}(\mathfrak{H}_1)$ and $\mathfrak{B}(\mathfrak{H}_2)$. We say $\{\alpha_t\}$ and $\{\beta_t\}$ are conjugate denoted $\alpha_t \approx \beta_t$ if there is $*$-isomorphism ϕ of $\mathfrak{B}(\mathfrak{H}_1)$ onto $\mathfrak{B}(\mathfrak{H}_2)$ so that $\phi \circ \alpha_t = \beta_t \circ \phi$ for all $t \geq 0$. We say $\{\alpha_t\}$ and $\{\beta_t\}$ are cocycle conjugate denoted $\alpha_t \sim \beta_t$ if $\{\alpha_t'\}$ and $\{\beta_t\}$ are conjugate where α_t and α_t' differ by a unitary cocycle (i.e., there is a strongly continuous one parameter family of unitaries $U(t)$ on $\mathfrak{B}(\mathfrak{H}_1)$ for $t \geq 0$ satisfying the cocycle condition $U(t)\alpha_t(U(s)) = U(t+s)$ for all $t, s \geq 0$ so that $\alpha_t'(A) = U(t)\alpha_t(A)U(t)^{-1}$ for all $A \in \mathfrak{B}(\mathfrak{H}_1)$ and $t \geq 0$).

Definition 2.3. Suppose $\{\alpha_t : t \geq 0\}$ is an E_o-semigroup of $\mathfrak{B}(\mathfrak{H})$. We say $\{\alpha_t\}$ is spatial if there exists a strongly continuous one parameter semigroup of isometries $U(t) \in \mathfrak{B}(\mathfrak{H})$ which intertwine α_t, *i.e.*, $U(t)A = \alpha_t(A)U(t)$ for all $A \epsilon \mathfrak{B}(\mathfrak{H})$ and $t \geq 0$.

We show the property of being spatial is a cocycle conjugacy invariant. Suppose $\{\alpha_t : t \geq 0\}$ is a spatial E_o-semigroup of $\mathfrak{B}(\mathfrak{H})$ with a strongly continuous one parameter semigroup of intertwining isometries $U(t)$. Suppose $\{\beta_t\}$ is an E_o-semigroup of $\mathfrak{B}(\mathfrak{K})$ which is cocycle conjugate with $\{\alpha_t\}$. Then there is a unitary cocycle $S(t)$ for $\{\alpha_t\}$ so that $S(t)\alpha_t(S(s)) = S(t+s)$ for all $t, s \geq 0$ and an isometry W of $\mathfrak{H}$ onto $\mathfrak{K}$ so that $\beta_t(A) = WS(t)\alpha_t(W^{-1}AW)S(t)^{-1}W^{-1}$ for all $A \in \mathfrak{B}(\mathfrak{K})$ and $t \geq 0$. Let $V(t) = WS(t)U(t)W^{-1}$ for $t \geq 0$. Note

$$\begin{aligned} V(t)V(s) &= WS(t)U(t)S(s)U(s)W^{-1} = WS(t)\alpha_t(S(s))U(t)U(s)W^{-1} \\ &= WS(t+s)U(t+s)W^{-1} = V(t+s) \end{aligned}$$

so the $V(t)$ form a strongly continuous one parameter semigroup of isometries. Note the $V(t)$ intertwine β_t since

$$\begin{aligned} V(t)A &= WS(t)U(t)W^{-1}AWW^{-1} = WS(t)\alpha_t(W^{-1}AW)U(t)W^{-1} \\ &= \beta_t(A)WS(t)U(t)W^{-1} = \beta_t(A)V(t) \end{aligned}$$

for all $A \in \mathfrak{B}(\mathfrak{K})$ and $t \geq 0$. Hence, if $\alpha_t \sim \beta_t$ and one is spatial then so is the other. Furthermore, if $\alpha_t \sim \beta_t$ and there is a family $\mathcal{U}_\alpha$ of intertwining semigroups for α then there is a corresponding family $\mathcal{U}_\beta$ of intertwining semigroups for β since the mapping $U(t) \leftrightarrow V(t) = WS(t)U(t)W^{-1}$ is a bijection of the intertwining semigroups for $\{\alpha_t\}$ and the intertwining semigroups for $\{\beta_t\}$.

In an earlier paper [P2] we showed that there exists an E_o-semigroup of $\mathfrak{B}(\mathfrak{H})$ which is not spatial. It seems that there may be uncountably many mutually non conjugate non spatial E_o-semigroups of $\mathfrak{B}(\mathfrak{H})$. However, at present there is only one known example.

In [A1] Arveson introduced the notion of a spatial E_o-semigroup of $\mathfrak{B}(\mathfrak{H})$. Since Arveson's definition is considerably stronger than the notion of spatial semigroups just introduced we will call Arveson's notion "completely spatial" semigroups.

Definition 2.4. Suppose $\{\alpha_t : t \geq 0\}$ is an E_o-semigroup of $\mathfrak{B}(\mathfrak{H})$. Let $\mathcal{U}_\alpha$ be the set of all strongly continuous semigroups of isometries that intertwine α_t, i.e., $U(t)A = \alpha_t(A)U(t)$ for all $A \in \mathfrak{B}(\mathfrak{H})$ and $t \geq 0$ for $\{U(t)\} \in \mathcal{U}_\alpha$. Let $\mathfrak{H}_t = span\{U_1(t_1)U_2(t_2)\cdots U_n(t_n)f : f \in \mathfrak{H},\, U_i \in \mathcal{U}_\alpha,\, t_i > 0, \text{ and } t_1 + t_2 + \cdots + t_n = t\}$. Then $\{\alpha_t\}$ is completely spatial if $\mathfrak{H}_t = \mathfrak{H}$ for all $t > 0$.

In [A1] Arveson classified the completely spatial E_o-semigroups of $\mathfrak{B}(\mathfrak{H})$. He showed that each completely spatial E_o-semigroup is cocycle conjugate to a CAR flow of rank n for $n = 1, 2, \cdots$ and $n = \infty$. The CAR flows are E_o-semigroups of $\mathfrak{B}(\mathfrak{H})$ constructed using representations of the CAR algebra.

The completely spatial E_o-semigroups of $\mathfrak{B}(\mathfrak{H})$ are those spatial E_o-semigroups with sufficiently many intertwining semigroups of isometries to completely characterize them. If α_t is a completely spatial E_o-semigroup of $\mathfrak{B}(\mathfrak{H})$ then for each $t > 0$, α_t is determined by the set $\mathcal{U}_\alpha$ of strongly continuous one parameter semigroups of intertwining isometries. Following Arveson we define E_t to be the set of operators $S \in \mathfrak{B}(\mathfrak{H})$ so that $SA = \alpha_t(A)S$. If α_t is completely spatial then E_t is the closed linear span of operators of the form $U_1(t_1)U_2(t_2)\cdots U_n(t_n)$ with $U_i \in \mathcal{U}_\alpha$, $t_i > 0$ and $t_1 + t_2 + \cdots + t_n = t$.

The framework we will adopt for studying spatial E_o-semigroups is the following. Suppose $\{\alpha_t : t \geq 0\}$ is a spatial E_o-semigroup of $\mathfrak{B}(\mathfrak{H})$ and $\{U(t) : t \geq 0\}$ is a strongly continuous one parameter semigroup of intertwining isometries. Let $-d$ be the generator of $U(t)$ so

$$df = \lim_{t \to 0^+} (f - U(t)f)/t$$

and the domain $\mathfrak{D}(d)$ is set of all $f \in \mathfrak{H}$ so that the limit exists. In fact, if the sequence of numbers $\|f - U(t_n)f\|/t_n$ is bounded for any sequence $t_n \to 0^+$ as $n \to \infty$ then $f \in \mathfrak{D}(d)$. Let d^* be the hermitian adjoint of d. Since the $U(t)$ are isometries we have d is skew hermitian so $d^* \supset -d$ (i.e., $\mathfrak{D}(d^*) \supset \mathfrak{D}(d)$ and $d^*f = -df$ for all $f \in \mathfrak{D}(d)$).

Let δ be the generator of $\{\alpha_t\}$ so

$$\delta(A) = \lim_{t \to 0^+} (\alpha_t(A) - A)/t$$

and the domain of δ consists of all A so that the limit exists in the sense of strong convergence. In fact, one can show that if $\|(\alpha_{t_n}(A) - A)\|/t_n$ remains bounded for a sequence $t_n \to 0^+$ then $A \in \mathfrak{D}(\delta)$. We have that δ is a σ-weakly closed σ-weakly densely defined $*$-derivation of the $*$-algebra $\mathfrak{D}(\delta)$ into $\mathfrak{B}(\mathfrak{H})$ (see Chapter 3 of [BR] for details). Since $U(t)A = \alpha_t(A)U(t)$ we have by differentiation that if $A \in \mathfrak{D}(\delta)$ then $A\mathfrak{D}(d) \subset \mathfrak{D}(d)$ and

$$dAf = Adf - \delta(A)f \tag{2.1}$$

for all $f \in \mathfrak{D}(d)$. Replacing $A \in \mathfrak{D}(\delta)$ by $A^* \in \mathfrak{D}(\delta)$ and taking adjoints we find that $A\mathfrak{D}(d^*) \subset \mathfrak{D}(d^*)$ and

$$d^*Af = Ad^*f + \delta(A)f \tag{2.2}$$

for all $f \in \mathfrak{D}(d^*)$.

We define a non negative inner product on the domain of d^* via the relation

$$\langle f, g\rangle = \frac{1}{2}(d^*f, g) + \frac{1}{2}(f, d^*g) \quad \text{for } f, g \in \mathfrak{D}(d^*). \tag{2.3}$$

As is well known from the theory of the extensions of unbounded hermitian operators (see [DS] Lemma 10 *p.* 1227) since d is a closed skew hermitian operator every vector $f \in \mathfrak{D}(d^*)$ has a unique decomposition $f = f_o + f_+ + f_-$ where $f_o \in \mathfrak{D}(d)$, $f_+, f_- \in \mathfrak{D}(d^*)$ and $d^*f_+ = f_+$ and $d^*f_- = -f_-$. Since $-d$ is the generator of a semigroup of isometries it follows that the equation $d^*f = -f$ has no solutions other than $f = 0$. We denote by $\mathfrak{D}_+ = \{f \in \mathfrak{D}(d^*) : d^*f = f\}$. Then each vector $f \in \mathfrak{D}(d^*)$ has a unique decomposition $f = f_o + f_+$ with $f_o \in \mathfrak{D}(d)$ and $f_+ \in \mathfrak{D}_+$. Let E_+ denote the projection of $\mathfrak{H}$ onto $\mathfrak{D}_+$. In the decomposition $f = f_o + f_+$ the vector f_+ can be obtained from the formula $f_+ = \frac{1}{2}E_+(f + d^*f)$. If $f, g \in \mathfrak{D}(d^*)$ and $f = f_o + f_+$ and $g = g_o + g_+$ are the unique decompositions of f and g then one finds $\langle f, g\rangle = (f_+, g_+)$. It follows that if $f \in \mathfrak{D}(d^*)$ then $f \in \mathfrak{D}(d)$ if and only if $\langle f, f\rangle = 0$.

Let $\mathfrak{K}$ be the quotient of $\mathfrak{D}(d^*)$ mod $\mathfrak{D}(d)$. For $f \in \mathfrak{D}(d^*)$ we denote by $[f]$ the equivalence class in $\mathfrak{K}$ containing f. For $f, g \in \mathfrak{D}(d^*)$ we have $[f] = [g]$ if and only if $f - g \in \mathfrak{D}(d)$. Since the inner product $\langle f, g\rangle$ is zero if either f or g lies in $\mathfrak{D}(d)$ it follows that the inner product $([f], [g]) = \langle f, g\rangle$ is well defined on $\mathfrak{K}$. Note the mapping $\Theta[f] = \frac{1}{2}E_+(f + d^*f)$ is an isometry of $\mathfrak{K}$ onto $\mathfrak{D}_+$. Note also that $\Theta[\Theta[f]] = \Theta[f]$ for all $f \in \mathfrak{D}(d^*)$. Since the range of Θ is closed it follows that $\mathfrak{K}$ is complete. Hence, $\mathfrak{K}$ is a Hilbert space with its inner product $([f], [g]) = \langle f, g\rangle$.

We define a $*$-representation of $\mathfrak{D}(\delta)$ on $\mathfrak{K}$ as follows. For $A \in \mathfrak{D}(\delta)$ and $f \in \mathfrak{D}(d^*)$ we define $\pi(A)[f] = [Af]$. This gives us a well defined operator on $\mathfrak{K}$ since $A\mathfrak{D}(d) \subset \mathfrak{D}(d)$ and $A\mathfrak{D}(d^*) \subset \mathfrak{D}(d^*)$. Since $d^*Af = Ad^*f + \delta(A)f$ for $A \in \mathfrak{D}(\delta)$ and $f \in \mathfrak{D}(d^*)$ and since δ is a $*$-derivation one checks that $([f], \pi(A)[g]) = \langle f, Ag\rangle = \langle A^*f, g\rangle = (\pi(A^*)[f], [g])$ for $A \in \mathfrak{D}(\delta)$ and $f, g \in \mathfrak{D}(d^*)$ so π is a $*$-representation. One checks that $\pi(A)$ is bounded (see [P1] for details). We denote this representation by $\pi_{(\alpha,U)}$ where we include the symbols α and U to remind us of the dependence of π on α and U.

Definition 2.5. Suppose $\{\alpha_t : t \geq 0\}$ is an E_o-semigroup of $\mathfrak{B}(\mathfrak{H})$ and $\{U(t) : t \geq 0\}$ is a strongly continuous one parameter semigroup of intertwining isometries. Let $\pi_{(\alpha,U)}$ be the $*$-representation of $\mathfrak{D}(\delta)$ constructed above. We call this representation the boundary representation associated with α and U.

In [P1] we defined the index of α to be the multiplicity of $\pi_{(\alpha,U)}$. The difficulty with this definition is that the definition of the index seems to depend on the intertwining semigroup U. To overcome this difficulty D. Robinson and I found another definition of the index in [PR].

A better definition of the index was found by W. Arveson. We describe his construction briefly as follows (see [A1], [A2] for details). For an E_o-semigroup α on $\mathfrak{B}(\mathfrak{H})$, let $\mathcal{U}_\alpha$ denote the set of all intertwining semigroups $\{U(t) : t \geq 0\}$ of bounded operators on $\mathfrak{B}(\mathfrak{H})$. If $\{U(t)\}$ and $\{S(t)\}$ are a pair of intertwining semigroups in $\mathcal{U}_\alpha$, then is a complex number $c(S,U)$ which Arveson calls the covariance function such that

$$S^*(t)U(t) = e^{tc(S,U)}I. \tag{2.4}$$

Let $\tilde{\mathfrak{H}}(\alpha)$ be the vector space of all functions $f : \mathcal{U}_\alpha \to \mathbb{C}$ which are finitely non-zero and satisfy $\Sigma_{x\in U_\alpha} f(x) = 0$. Then $\tilde{\mathfrak{H}}(\alpha)$ becomes a pre-Hilbert space under the positive semi-definite inner product $(f,g) = \Sigma_{x,y}\overline{f(x)}g(y)c(x,y)$. Let d_α be the dimension of the Hilbert space completion $\tilde{\mathfrak{H}}(\alpha)$ of $\mathfrak{H}(\alpha)$. This is Arveson's index for E_o-semigroups of $\mathfrak{B}(\mathfrak{H})$.

As we remarked earlier if $\{\alpha_t\}$ and $\{\beta_t\}$ are cocycle conjugate E_o-semigroups then there is a natural isomorphism between $\mathcal{U}_\alpha$ and $\mathcal{U}_\beta$ the intertwining semigroups for $\{\alpha_t\}$ and $\{\beta_t\}$. Then it is clear that the Arveson index is a cocycle conjugacy invariant.

In [PP] G. Price and I showed the connection between Arveson's index d_α and the representation $\pi_{(\alpha,U)}$. It was shown that d_α is precisely the number of times the identity representation of $\mathfrak{D}(\delta)$ occurs in the representation $\pi_{(\alpha,U)}$. This means that if $\{\alpha_t\}$ is an E_o-semigroup of $\mathfrak{B}(\mathfrak{H})$ with a strongly continuous one parameter semigroup of intertwining isometries $\{U(t)\}$ then d_α is the maximum number of isometries $\{V_i : i = 1, \cdots n\}$ of $\mathfrak{H}$ into the representation space $\mathfrak{K}$ of $\pi_{(\alpha,U)}$ so that

$$\langle V_i f, AV_j g\rangle = \delta_{ij}(f, Ag)$$

for all $f, g \in \mathfrak{H}$ and $A \in \mathfrak{D}(\delta)$

The key to this result was the understanding of the pairs of strongly continuous semigroups of isometries satisfying (2.4). To describe these results we need to introduce the Wold decomposition. Suppose $\{U(t) : t \geq 0\}$ is a strongly continuous one parameter group of isometries of a Hilbert space $\mathfrak{H}$. Let $P(t) = U(t)U(t)^*$ for $t \geq 0$. We have that the $P(t)$ form a strongly continuous decreasing family of commuting projections. Let P_∞ the strong limit of $P(t)$ as $t \to \infty$ and let $\mathfrak{H}_o = P_\infty\mathfrak{H}$. One easily checks that $U(t)\mathfrak{H}_o \subset \mathfrak{H}_o$, $U(t)^*\mathfrak{H}_o \subset \mathfrak{H}_o$ and $U(t)$ is unitary on $\mathfrak{H}_o$. Let $\mathfrak{H}_1$ be the orthogonal complement of $\mathfrak{H}_o$ in $\mathfrak{H}$. The action of $U(t)$ on $\mathfrak{H}_1$ is unitarily equivalent to a number of copies of the shift operator. Specifically, we can express $\mathfrak{H}_1$ in the form $\mathfrak{H}_1 = \mathfrak{K} \otimes L^2(0,\infty)$ so a vector $F \in \mathfrak{H}_1$ is a $\mathfrak{K}$ valued measurable function defined almost everywhere on $(0,\infty)$ and is square integrable. The action of $U(t)$ on such a vector is given by $(U(t)F)(x) = F(x-t)$ for $x > t$ and $(U(t)F)(x) = 0$ for $0 \leq x \leq t$.

In this talk we will denote by $\{\mathfrak{H} = \mathfrak{H}_o \oplus \mathfrak{K} \otimes L^2(0,\infty)\}$ the Wold decomposition of $U(t)$. If F is a vector in $\mathfrak{H}$ we denote by $F(x)$ the $\mathfrak{K}$ valued function on $(0,\infty)$

representing the component of F in $\mathfrak{H}_1$. For example if $F \in \mathfrak{H}$ and $U(t)^*F = e^{-t}F$ the one can show that $F(x) = F(0)e^{-x}$ and $P_\infty F = 0$ (i.e., F has no component in $\mathfrak{H}_o$).

Suppose $-d$ is the generator of $U(t)$. If $F \in \mathfrak{D}(d)$ then $F(x)$ is of the form

$$F(x) = \int_0^x G(t)\, dt$$

with $G \in \mathfrak{H}_1$ and $(dF)(x) = G(x)$. The vector F may have a component in $\mathfrak{H}_o$ but here we concentrate on the component of F in $\mathfrak{H}_1$. We denote by d^* the adjoint operator of d. If $F \in \mathfrak{D}(d^*)$ then $F(x)$ is of the form

$$F(x) = F(0) + \int_0^x G(t)\, dt$$

with $G \in \mathfrak{H}_1$ and $(d^*F)(x) = -G(x)$. Strictly speaking for $F \in \mathfrak{H}$ the function $F(x)$ is only defined up to sets of measure zero. For vectors $F \in \mathfrak{D}(d^*)$ the representing function $F(x)$ may be chosen to be continuous and we will always assume the representing functions $F(x)$ have been so chosen. In this sense, it makes sense to speak of the value of F at x. In particular $F(0)$ is well defined.

In summary d act on $\mathfrak{H}_1$ by differentiation with the boundary condition $F(0) = 0$ and d^* acts on $\mathfrak{H}_1$ by minus differentiation with no boundary condition at $x = 0$. The connection between the Wold decomposition and the bilinear form $\langle \cdot, \cdot \rangle$ is given as follows. For $F, G \in \mathfrak{D}(d^*)$ we have $\langle F, G \rangle = \frac{1}{2}(F(0), G(0))$.

In [PP] Price and I showed that if $\{U(t) : t \geq 0\}$ and $\{S(t) : t \geq 0\}$ are strongly continuous semigroups of isometries on $\mathfrak{H}$ satisfying the condition $S(t)^*U(t) = e^{-\lambda t}I$ with $\lambda > 0$ and if $\mathfrak{H} = \mathfrak{H}_o \oplus \mathfrak{H}_1 = \mathfrak{H}_o \oplus \mathfrak{K} \otimes L^2(0, \infty)$ is the Wold decomposition of $\{U(t)\}$ then there is an isometry V from $\mathfrak{H}$ to $\mathfrak{K}$ so that the generator $-D$ of $\{S(t)\}$ is given by $DF = -d^*F + \lambda F$ and the domain $\mathfrak{D}(D) = \{F \in \mathfrak{D}(d^*) : F(0) = \sqrt{2\lambda} VF\}$. Conversely, if $\lambda > 0$ and V is an isometry of $\mathfrak{H}$ into $\mathfrak{K}$ then the operator D defined by the relation $DF = -d^*F + \lambda F$ for all $F \in \mathfrak{D}(D) = \{F \in \mathfrak{D}(d^*) : F(0) = \sqrt{2\lambda} VF\}$ is the generator of a strongly continuous one parameter semigroup of isometries $S(t)$ and $S(t)^*U(t) = e^{-\lambda t}I$ for all $t \geq 0$.

In the case where there are two strongly continuous one parameter semigroups $\{S_1(t)\}$ and $\{S_2(t)\}$ so that $S_1(t)^*U(t) = e^{-\lambda_1 t}I$ and $S_2(t)^*U(t) = e^{-\lambda_2 t}I$ and $S_1(t)^*S_2(t) = e^{-ct}I$ with $\lambda_1, \lambda_2 > 0$ and c a complex number we have the following situation. If D_i is the generator of $S_i(t)$ then $D_iF = -d^*F + \lambda_i F$ for all $F \in \mathfrak{D}(D_i) = \{F \in \mathfrak{D}(d^*) : F(0) = \sqrt{2\lambda_i} V_i F\}$ where V_i is an isometry from $\mathfrak{H}$ to $\mathfrak{K}$. And the isometries V_i satisfy $V_1^*V_2 = (\lambda_1\lambda_2)^{-\frac{1}{2}}(\lambda_1 + \lambda_2 - c)I$.

The connection between the Arveson index d_α and the representation $\pi_{(\alpha,U)}$ is the following. Suppose $\{\alpha_t : t \geq 0\}$ is an E_o-semigroup of $\mathfrak{B}(\mathfrak{H})$ with a strongly continuous one parameter semigroup of intertwining isometries. Let δ be the generator of α_t and let $-d$ be the generator of $U(t)$. Let $\langle \cdot, \cdot \rangle$ be the inner product associated with the operator d. Let $\mathfrak{H} = \mathfrak{H}_o \oplus \mathfrak{K} \otimes L^2(0, \infty)$ be the Wold decomposition of $U(t)$. Finally, let W be the mapping of $\mathfrak{K}$ into $\mathfrak{H}$ given by $(\mathrm{Wf})(\mathrm{x}) = \sqrt{2}e^{-x}f$ and the component of Wf in $\mathfrak{H}_o$ is zero. Note W is an isometry of $\mathfrak{K}$ onto $\mathfrak{D}_+$ the deficiency space of d. Suppose there is an intertwining operator between the $*$-representation

$\pi_{(\alpha,U)}$ and the identity representation of $\mathfrak{D}(\delta)$. This means there is an operator S from $\mathfrak{H}$ into $\mathfrak{D}_+$ so that $\langle F, ASG\rangle = \langle F, SAG\rangle$ for all $F \in \mathfrak{D}(d^*)$, $G \in \mathfrak{H}$ and $A \in \mathfrak{D}(\delta)$. We have for $F, G \in \mathfrak{H}$ and $A \in \mathfrak{D}(\delta)$

$$\begin{aligned}(F, S^*SAG) &= (SF, SAG) = \langle SF, SAG\rangle = \langle SF, ASG\rangle = \langle A^*SF, SG\rangle \\ &= \langle SA^*F, SG\rangle = (SA^*F, SG) = (F, AS^*SG).\end{aligned}$$

Hence, S^*S is a multiple of the unit. Multiplying S by a positive real number we may assume S is an isometry. Let $V = W^*S$. Then V is an isometry from $\mathfrak{H}$ to $\mathfrak{K}$. In [PP] it is shown that if $\lambda > 0$ and D is the operator defined by $DF = -d^*F + \lambda F$ for all $F \in \mathfrak{D}(D) = \{F \in \mathfrak{D}(d^*) : F(0) = \sqrt{2\lambda}VF\}$ then $-D$ is the generator of a strongly continuous semigroup of isometries $Q(t)$ which intertwine α_t. Conversely, suppose $Q(t)$ is a strongly continuous one parameter semigroup of isometries which intertwine α_t. We have $Q(t)^*U(t) = e^{-\lambda t}I$ for $t \geq 0$. Replacing the semigroup $Q(t)$ by $e^{ist}Q(t)$ we may with no loss of generality assume $\lambda > 0$. Then we have the generator $-D$ of $Q(t)$ is given by $DF = -d^*F + \lambda F$ for $F \in \mathfrak{D}(D) = \{F \in \mathfrak{D}(d^*) : F(0) = \sqrt{2\lambda}VF\}$ with V an isometry of $\mathfrak{H}$ into $\mathfrak{K}$. Then as shown in [PP] the isometry $S = WV$ from $\mathfrak{H}$ to $\mathfrak{D}_+$ intertwines the identity representation of $\mathfrak{D}(\delta)$ with $\pi_{(\alpha,U)}$ so $\langle f, ASg\rangle = \langle f, SAg\rangle$ for all $f \in \mathfrak{D}(d^*)$, $g \in \mathfrak{H}$ and $A \in \mathfrak{D}(\delta)$. We summaries these result in the following theorem.

Theorem 2.6. *Suppose $\{\alpha_t : t \geq 0\}$ is a spatial E_o-semigroup with a strongly continuous one parameter semigroup of intertwining isometries $U(t)$. Let $-d$ be the generator of $U(t)$, $\langle\cdot,\cdot\rangle$ be the bilinear form on $\mathfrak{D}(d^*) \times \mathfrak{D}(d^*)$ given by $\langle f, g\rangle = \frac{1}{2}(d^*f, g) + \frac{1}{2}(f, d^*g)$ and let $\pi_{(\alpha,U)}$ be the $*$-representation of $\mathfrak{D}(\delta)$ on $\mathfrak{D}(d^*)$ mod $\mathfrak{D}(d)$ given by $A[F] = [AF]$ where $[F]$ is the equivalence class in $\mathfrak{D}(d^*\}$ mod $\mathfrak{D}(d)$ containing $F \in \mathfrak{D}(d^*)$ of all $F' \in \mathfrak{D}(d^*)$ so that $F - F' \in \mathfrak{D}(d)$. Let $\mathfrak{H} = \mathfrak{H}_o \oplus \mathfrak{K} \otimes L^2(0,\infty)$ be the Wold decomposition of $\{U(t)\}$ and let W be the isometry of $\mathfrak{K}$ onto $\mathfrak{D}_+ = \{F \in \mathfrak{D}(d^*) : d^*F = F\}$ given by $(Wf)(x) = \sqrt{2}e^{-x}f$ and the component of Wf in $\mathfrak{H}_o$ is zero. Then there exist isometries $\{V_i : i = 1, \cdots, n\}$ of $\mathfrak{H}$ into $\mathfrak{K}$ with orthogonal ranges (i.e., $V_i^*V_j = 0$ for $i \neq j$) with the following two properties.*

(i) *If $\{S(t) : t \geq 0\}$ is a strongly continuous one parameter semigroup of isometries which intertwine $\{\alpha_t\}$ and such that $S(t)^*U(t) = e^{-\lambda t}I$ with $\lambda \geq 0$ (making $\lambda \geq 0$ can always be achieved by a trivial replacement of $S(t)$ with $e^{ist}S(t)$ with s real) and if $-D$ is the generator of $\{S(t)\}$ then $DF = -d^*F + \lambda F$ for all $F \in \mathfrak{D}(D) = \{F \in \mathfrak{D}(d^*) : F(0) = \sqrt{2\lambda}VF\}$ where V is an isometry of $\mathfrak{H}$ into $\mathfrak{K}$ of the form $V = \Sigma_{i=1}^n s_iV_i$ with the s_i complex numbers so that $\Sigma_{i=1}^n|s_i|^2 = 1$. And conversely, if if D is an operator of the form just given then $-D$ is the generator of a strongly continuous one parameter semigroup of isometries which intertwine $\{\alpha_t\}$.*

(ii) *If S is an operator from $\mathfrak{H}$ to the deficiency space $\mathfrak{D}_+$ of of d which intertwines the identity representation of $\mathfrak{D}(\delta)$ with $\pi_{(\alpha,U)}$ (i.e., $\langle F, ASG\rangle = \langle F, SAG\rangle$ for all $F \in \mathfrak{D}(d^*)$, $G \in \mathfrak{H}$ and $A \in \mathfrak{D}(\delta)$) then S is of the form $S = \Sigma_{i=1}^n s_iWV_i$ where the s_i are complex numbers so that $\Sigma_{i=1}^n|s_i|^2 < \infty$. And conversely, if S is of the form just given then S intertwines the identity representation of $\mathfrak{D}(\delta)$ with $\pi_{(\alpha,U)}$.*

If $\{V_j' : j = 1, \cdots, m\}$ is a second family of isometries from $\mathfrak{H}$ into $\mathfrak{K}$ with orthogonal ranges and properties (i) and (ii) then $m = n$. The extended integer $n (n = 0, 1, 2, \cdots$ or $n = \infty)$ is the Arveson index of $\{\alpha_t\}$.

Definition 2.7. Suppose $\{\alpha_t : t \geq 0\}$ is a spatial E_o-semigroup of $\mathfrak{B}(\mathfrak{H})$ with a strongly continuous one parameter semigroup of intertwining isometries $U(t)$. Assume the notation of the previous theorem. Suppose $\{V_i : i = 1, \cdots, n\}$ are isometries satisfying the conditions of the previous theorem. Let $\phi = \phi_{(\alpha,U)}$ be the $*$-representation of $\mathfrak{B}(\mathfrak{H})$ on $\mathfrak{K}$ given by

$$\phi(A) = \sum_{i=1}^{n} V_i A V_i^*$$

for all $A \in \mathfrak{B}(\mathfrak{H})$. We call ϕ the normal boundary representation of $\mathfrak{B}(\mathfrak{H})$ on $\mathfrak{K}$ associated with $\{\alpha_t\}$ and $\{U(t)\}$.

One can show from theorem 2.6 that the representation ϕ does not depend on the choice of the $\{V_i\}$ in the same way the number of vectors in a basis does not depend on the choice of the basis vectors. Note the representation ϕ is not necessarily unital in that $\phi(I)$ is not necessarily the unit operator on $\mathfrak{H}$ (in fact, it has not been proved that $\phi(I)$ is not zero).

In [P3] we determined the necessary and sufficient conditions on the normal boundary representation for it to be the boundary representation of a completely spatial E_o-semigroup. We summarize these results in the following theorem.

Theorem 2.8. *Suppose $\{\alpha_t\}$ is an E_o-semigroup of $\mathfrak{B}(\mathfrak{H})$ with a strongly continuous one parameter semigroup of intertwining isometries $U(t)$. Assume the notation of theorem 2.6. And let $\phi = \phi_{(\alpha,U)}$ be the normal boundary representation of $\mathfrak{B}(\mathfrak{H})$ on $\mathfrak{K}$ associated with $\{\alpha_t\}$ and $\{U(t)\}$ as defined in definition 2.7. For $\lambda \geq 0$ let Λ_λ be the linear mapping of $\mathfrak{B}(\mathfrak{K})$ into $\mathfrak{B}(\mathfrak{H})$ given as follows. Given a vector $F \in \mathfrak{H}$ we can according to the Wold decomposition represent F by a pair consisting of a vector $F_o \in \mathfrak{H}_o$ and a $\mathfrak{K}$-valued function $F(x)$. For $A \in \mathfrak{B}(\mathfrak{K})$ and $F = \{F_o, F(x)\}$ we define $(\Lambda_\lambda(A)F)_o = 0$ and $(\Lambda_\lambda(A)F)(x) = e^{-\lambda x} AF(x)$. Suppose $\lambda > 0$. Then $\{\alpha_t\}$ is completely spatial if and only if the normal boundary representation ϕ is unital and the equation $\Lambda_\lambda(\phi(A)) = A$ has only the trivial solution $A = 0$. Furthermore, given a normal unital boundary representation ϕ so that the equation $\Lambda_\lambda(\phi(A)) = A$ has only the trivial solution $A = 0$, then there is a completely spatial E_o-semigroup of $\mathfrak{B}(\mathfrak{H})$ with a precisely this boundary representation.*

Finally suppose $\{\alpha_t\}$ is completely spatial and $\phi = \phi_{(\alpha,U)}$ is the normal boundary representation of $\mathfrak{B}(\mathfrak{H})$ on $\mathfrak{K}$ associated with $\{\alpha_t\}$ and $\{U(t)\}$. Suppose δ is the generator of $\{\alpha_t\}$ and $-d$ is the generator of $\{U(t)\}$. Then for $F \in \mathfrak{D}(d^*)$ and $A \in \mathfrak{D}(\delta)$ we have $(AF)(0) = \phi(A)F(0)$. Let $\mathfrak{S}$ be the subset of $\mathfrak{B}(\mathfrak{H})_*$ (the predual of $\mathfrak{B}(\mathfrak{H})$) of linear functionals of the form $\rho(A) = (F, AG)$ for $A \in \mathfrak{B}(\mathfrak{H})$ where $F, G \in \mathfrak{D}(d^*)$. Since vectors $F \in \mathfrak{D}(d^*)$ are differentiable it make sense to speak of the vector $F(x) \in \mathfrak{K}$ where $x \to F(x)$ is the $\mathfrak{K}$ valued function representing F in the Wold decomposition of $\mathfrak{H}$ and in particular it makes sense to speak of $F(0)$. Let Δ be the linear mapping of $\mathfrak{B}(\mathfrak{H})_*$ into $\mathfrak{B}(\mathfrak{H})_*$ whose domain is the linear span of $\mathfrak{S}$ and for $\rho \in \mathfrak{S}$ of the above form we have

$$(\Delta\rho)(A) = -(d^*F, AG) - (F, Ad^*G) + (F(0), \phi(A)G(0))$$

for all $A \epsilon \mathfrak{B}(\mathfrak{H})$. *Then* Δ *is closable and its closure is equal to* δ_* *the action of* δ *on the predual. Specifically, this means that* $A \in \mathfrak{D}(\delta)$ *and* $\delta(A) = B$ *if and only if* $(\Delta\rho)(A) = \rho(B)$ *for all* $\rho \in \mathfrak{S}$.

The E_o-semigroups of $\mathfrak{B}(\mathfrak{H})$ themselves form a semigroup and the appropriate group operation is tensoring. If $\{\alpha_t\}$ and $\{\beta_t\}$ are E_o-semigroups of $\mathfrak{B}(\mathfrak{H})$ and $\mathfrak{B}(\mathfrak{K})$, respectively, then one can form a form a new semigroup $\gamma_t = \alpha_t \otimes \beta_t$ which acts on the tensor product space $\mathfrak{H} \otimes \mathfrak{K}$. Specifically, we define $\gamma_t(A \otimes B) = \alpha_t(A) \otimes \beta_t(B)$. In [P1] we showed the index was subadditive and in [A2] Arveson showed his index was additive (i.e., the index of $\{\gamma_t\}$ is the sum of the index of $\{\alpha_t\}$ and the index of $\{\beta_t\}$). One of the important results of the theory of E_o-semigroups obtained by Arveson is that if $\{\sigma_t\}$ is a one parameter group of $*$-automorphisms of $\mathfrak{B}(\mathfrak{H})$ (i.e., $\sigma_t(A) = U(t)AU(t)^{-1}$ with $U(t)$ a strongly continuous one parameter unitary group) then $\{\sigma_t\}$ acts like the unit under tensoring. This means that if $\{\alpha_t\}$ is an E_o-semigroup and $\{\sigma_t\}$ is one parameter group of $*$-automorphisms then $\{\alpha_t\}$ is cocycle conjugate to $\{\alpha_t \otimes \sigma_t\}$. This result follows from a fairly elaborate series of proofs and constructions in Arveson's work. Because of the importance of this result we would like to give a short direct proof. First we need the following theorem.

Theorem 2.9. *Suppose* $\{\alpha_t\}$ *is a proper* E_o*-semigroup of* $\mathfrak{B}(\mathfrak{H})$ *(so* $\alpha_t(\mathfrak{B}(\mathfrak{H})) \neq \mathfrak{B}(\mathfrak{H})$ *for* $t > 0$*) and* $E \in \mathfrak{B}(\mathfrak{H})$ *is an hermitian projection which is invariant under* α_t *(i.e.,* $\alpha_t(E) = E$ *for all* $t \geq 0$*). Let* $\mathfrak{M}$ *be the range of* E *and let* Q_E *be the set of all operators* $A \in \mathfrak{B}(\mathfrak{H})$ *so that* $A = EAE$*. Note* Q_E *is* $*$*-isomorphic with* $\mathfrak{B}(\mathfrak{M})$ *the algebra of all bounded operators on* $\mathfrak{M}$ *and note if* $A \in Q_E$ *then* $\alpha_t(A) \in Q_E$ *for all* $t \geq 0$*. Let* β_t *be the restriction of* α_t *to* Q_E *so* $\beta_t(A) = \alpha_t(A)$ *for all* $A \in Q_E$*. Then* $\{\beta_t\}$ *is an* E_o*-semigroup of* $\mathfrak{B}(\mathfrak{M})$ *which is cocycle conjugate to* $\{\alpha_t\}$*.*

Proof. Let $\{\alpha_t\}, E, Q_E$ and $\{\beta_t\}$ satisfy the hypothesis and notation of the theorem. A routine computation shows that β_t is an E_o-semigroup of $\mathfrak{B}(\mathfrak{M})$. Since α_1 is a proper $*$-endomorphism we have $\alpha_1(\mathfrak{B}(\mathfrak{H})) \neq \mathfrak{B}(\mathfrak{H})$ and since $\alpha_1(\mathfrak{B}(\mathfrak{H}))$ is a type I factor we have $\alpha_1(\mathfrak{B}(\mathfrak{H}))'$ is a type I_p factor where $p = 2, 3, \cdots$ or $p = \infty$. Then if follows that if d is the dimension of E then $d = p \times d$. Hence, E is of infinite dimension. Since the range of E and $\mathfrak{H}$ have the same dimension there is a partial isometry U from $\mathfrak{H}$ onto $\mathfrak{M}$. Then U satisfies the equations $U^*U = I$ and $UU^* = E$. Let $V(t) = U^*\alpha_t(U)$. We show $V(t)$ is a unitary cocycle. To see $V(t)$ is unitary note that

$$\begin{aligned} V(t)^*V(t) &= \alpha_t(U)^*UU^*\alpha_t(U) = \alpha_t(U)^*E\alpha_t(U) = \alpha_t(U)^*\alpha_t(E)\alpha_t(U) \\ &= \alpha_t(U^*EU) = \alpha_t(I) = I \end{aligned}$$

and

$$V(t)V(t)^* = U^*\alpha_t(U)\alpha_t(U)^*U = U^*\alpha_t(E)U = U^*EU = I$$

And $\{V(t)\}$ is a unitary cocycle since

$$\begin{aligned} V(t)\alpha_t(V(s)) &= U^*\alpha_t(U)\alpha_t(U^*\alpha_s(U)) = U^*\alpha_t(E)\alpha_{t+s}(U) \\ &= U^*E\alpha_{t+s}(U) = U^*\alpha_{t+s}(U) = V(t+s) \end{aligned}$$

The mapping $A \to$ UAU* is a $*$-isomorphism of $\mathfrak{B}(\mathfrak{H})$ onto $\mathfrak{B}(\mathfrak{M})$. Now we have

$$\begin{aligned}\beta_t(UAU^*) &= \alpha_t(UAU^*) = \alpha_t(EUAU^*E) = E\alpha_t(UAU^*)E \\ &= UU^*\alpha_t(U)\alpha_t(A)\alpha_t(U)^*UU^* = UV(t)\alpha_t(A)V(t)^*U^*.\end{aligned}$$

This displays the cocycle equivalence of $\{\alpha_t\}$ and $\{\beta_t\}$. □

Before proving the next theorem we characterize the E_o-semigroups which are cocycle conjugate to the trivial semigroup. Let $\{\sigma_t\}$ be the trivial E_o-semigroup of $\mathfrak{B}(\mathfrak{H})$ given by $\sigma_t(A) = A$ for all $A \in \mathfrak{B}(\mathfrak{H})$ and $t \geq 0$. Notice that a E_o-semigroup $\{\alpha_t\}$ is cocycle conjugate with $\{\sigma_t\}$ if and only if $\{\alpha_t\}$ is a one parameter group of automorphisms so $\alpha_t(A) = U(t)AU(t)^{-1}$ for $A \in \mathfrak{B}(\mathfrak{H})$ and $t \geq 0$ with $\{U(t)\}$ a strongly continuous unitary group. So the cocycle equivalence class of trivial E_o-semigroup $\{\sigma_t\}$ is just the E_o-semigroups coming from automorphism groups.

Theorem 2.10. *Suppose $\{\alpha_t\}$ is an E_o-semigroup of $\mathfrak{B}(\mathfrak{H})$. Suppose $\{U(t) : -\infty < t < \infty\}$ is a strongly continuous one parameter unitary group acting on a Hilbert space $\mathfrak{K}$ and let $\sigma_t(A) = U(t)AU(t)^{-1}$ be the associated one parameter group of $*$-automorphisms of $\mathfrak{B}(\mathfrak{K})$. Then $\{\alpha_t\}$ is cocycle conjugate with $\{\alpha_t \otimes \sigma_t\}$.*

Proof. Assume the hypothesis and notation of the theorem and assume $\{\alpha_t\}$ is proper. Let $\beta_t = \alpha_t \otimes \sigma_t$ and let γ_t be the tensor product of α_t with the identity automorphism on $\mathfrak{B}(\mathfrak{K})$ so $\gamma_t(A \otimes B) = \alpha_t(A) \otimes B$ for $A \in \mathfrak{B}(\mathfrak{H})$ and $B \in \mathfrak{B}(\mathfrak{K})$. Let $W(t) = I \otimes U(t)$ acting on $\mathfrak{H} \otimes \mathfrak{K}$. Note $W(t)$ is an $\{\gamma_t\}$ unitary cocycle and $\beta_t(A \otimes B) = W(t)\gamma_t(A \otimes B)W(t)^{-1}$ for all $A \in \mathfrak{B}(\mathfrak{H})$ and $B \in \mathfrak{B}(\mathfrak{K})$ so $\{\beta_t\}$ and $\{\gamma_t\}$ are cocycle conjugate. Let e_o be a rank one projection in $\mathfrak{B}(\mathfrak{K})$ and let $E = I \otimes e_o$. We see that E is a γ_t invariant projection so by the previous theorem $\{\gamma_t\}$ is cocycle conjugate to γ_t acting on $E\mathfrak{B}(\mathfrak{H} \otimes \mathfrak{K})E$ and this is cut down E_o-semigroup is clearly conjugate to $\{\alpha_t\}$. This proves the theorem in the case when $\{\alpha_t\}$ is proper. If $\{\alpha_t\}$ is not proper then α_t is invertible and there is a unitary group $\{W(t)\}$ so that $\alpha_t(A) = W(t)AW(t)^{-1}$ for all $A \in \mathfrak{B}(\mathfrak{H})$ and $t \geq 0$. Hence, both $\{\alpha_t\}$ and $\{\gamma_t\}$ are cocycle conjugate to the trivial E_o-semigroup and so they are themselves cocycle conjugate. □

Another result which has been well known since the beginning of this subject is that cocycle conjugacy is stable under bounded perturbations. We state this in the form of a theorem for easy reference later. A proof can be found in [P3].

Theorem 2.11. *Suppose $\{\alpha_t : t \geq 0\}$ is an E_o-semigroup of $\mathfrak{B}(\mathfrak{H})$ and δ is the generator of α_t. Suppose H is a bounded skew hermitian operator (i.e. $H^* = -H$) and $\delta_1(A) = \delta(A) + HA - AH$ for all $A \in \mathfrak{D}(\delta)$. Then δ_1 is the generator of an E_o-semigroup $\{\beta_t : t \geq 0\}$ of $\mathfrak{B}(\mathfrak{H})$ which is cocycle conjugate to $\{\alpha_t : t \geq 0\}$. Furthermore, if $U(t) = \exp(-td)$ is a strongly continuous one parameter semigroup of isometries which intertwine α_t (i.e. $U(t)A = \alpha_t(A)U(t)$ for $A \in \mathfrak{B}(\mathfrak{H})$ and $t \geq 0$) then $V(t) = \exp(-t(d - H))$ intertwine β_t.*

We would like to introduce a standard form for spatial E_o-semigroups. We will see in the next section that two completely spatial E_o-semigroups in standard form are cocycle conjugate if and only if they are conjugate. We define the notion of an absorbing state.

Definition 2.12. Suppose $\{\alpha_t\}$ is an E_o-semigroup of $\mathfrak{B}(\mathfrak{H})$. We say a normal state ω is an absorbing state for $\{\alpha_t\}$ if for every normal state ρ of $\mathfrak{B}(\mathfrak{H})$ and every $A \in \mathfrak{B}(\mathfrak{H})$ we have $\rho(\alpha_t(A)) \to \omega(A)$ as $t \to \infty$. Suppose $\{\alpha_t\}$ is a spatial E_o-semigroup of $\mathfrak{B}(\mathfrak{H})$. We say $\{\alpha_t\}$ is in standard form if there is a normal pure state ω of $\mathfrak{B}(\mathfrak{H})$ which is absorbing for $\{\alpha_t\}$.

We note that if $\{\alpha_t\}$ has an absorbing state ω then ω is the unique absorbing state. An absorbing state ω is automatically α_t invariant. Finally, we note that if ω is an absorbing state for $\{\alpha_t\}$ and ρ is an arbitrary normal state of $\mathfrak{B}(\mathfrak{H})$ then the states $\rho_t(A) = \rho(\alpha_t(A))$ converge in norm to ω, i.e. $\|\rho_t - \omega\| \to 0$ as $t \to 0$. In [P4] it is shown that

Theorem 2.13. *Every spatial E_o-semigroup is cocycle conjugate to a spatial E_o-semigroup in standard form.*

We in [P4] it is shown that two completely spatial E_o-semigroups in standard form are cocycle conjugate if and only if they are conjugate. We conjecture this result is true of all spatial E_o-semigroups.

III. Description of New Examples of Spatial E_o-semigroups.

In [P4] we constructed new examples of spatial E_o-semigroups. The basic idea was to construct the boundary representation π_o which is the direct sum of a normal representation of $\mathfrak{B}(\mathfrak{H})$ and a singular representation. In [P3] it was shown that if $\{\alpha_t\}$ is a spatial E_o-semigroup of $\mathfrak{B}(\mathfrak{H})$ with an intertwining semigroup of isometries $\{U(t)\}$ and $\pi_o = \pi_{(\alpha,U)}$ is the boundary representation associated with $\{\alpha_t\}$ and $U(t)$ then π_o is instantly normal. This means that for every $t > 0$ the representation $A \to \pi_o(\alpha_t(A))$ is normal. It was also shown in [P3] that if π_o has a singular part then the part of π_o is concentrated at $x = 0$ in the Wold decomposition of $\mathfrak{H}$. In more detail this means if ω is vector state of π_o coming from the singular part then $\omega(U(t)U(t)^*) = 0$ for all $t > 0$. If $\mathfrak{H} = \mathfrak{H}_o \oplus \mathfrak{K} \otimes L^2(0,\infty)$ is the Wold decomposition of $\mathfrak{H}$ with respect to $\{U(t)\}$ and we think of $\mathfrak{K} \otimes L^2(0,\infty)$ as $\mathfrak{K}$-valued functions of $x \in [0,\infty)$ then the singular state ω corresponds to a state with support at $x = 0$.

The basic idea of the construction in [P4] is as follows. We take the Fock flow of rank one and restricted it to operators A so that $AF_o = 0$ and $A^*F_o = 0$ where $F_o \in \mathfrak{H}_o$ is the vacuum vector. This produces an e_o-semigroup which does not preserve the identity. The associated boundary representation ϕ for this e_o-semigroup has the property that $I - \phi(I)$ is a rank one projection. Since there are no one dimensional representations of $\mathfrak{B}(\mathfrak{H})$ we need to create more room. This is achieved by tensoring this e_o-semigroup with the identity on a second Hilbert space $\mathfrak{Y}$. The associated boundary representation ϕ_1 of the resulting e_o-semigroup is a non unital normal $*$-representation and now $I - \phi_1(I)$ is of infinite rank. Then on this infinite dimensional subspace we build a singular instantly normal $*$-representation following the method of the last section. Taking the direct sum of these two representations way we obtain a boundary representation of a spatial E_o-semigroup which is not completely spatial. The actual construction is long and complicated so we omit further details.

In view of the fact that we now know there exist spatial E_o-semigroups of $\mathfrak{B}(\mathfrak{H})$ which are not completely spatial it seems reasonable to divide E_o-semigroups of

$\mathfrak{B}(\mathfrak{H})$ into three types in analogy with factors.

Definition 3.1. An E_o-semigroup of $\mathfrak{B}(\mathfrak{H})$ is of type I if it is completely spatial, type II if it is spatial but not completely spatial and type III if it is not spatial. The E_o-semigroups of type I and II can be further subdivided into those of type I_n and type II_n where $n = 0, 1, 2, \ldots$ and ∞ is the index.

¿From the work of Arveson it follows that E_o-semigroups of type I are completely classified up to cocycle conjugacy by the index n. If $\{\alpha_t\}$ and $\{\beta_t\}$ are E_o-semigroups $\mathfrak{B}(\mathfrak{H})$ and $\mathfrak{B}(\mathfrak{K})$ of types I_n and I_m then $\{\alpha_t \otimes \beta_t\}$ is an E_o-semigroup of $\mathfrak{B}(\mathfrak{H} \otimes \mathfrak{K})$ of type I_{n+m}.

In [A2] Arveson proved that the index of E_o-semigroups of $\mathfrak{B}(\mathfrak{H})$ are additive under tensor products. Arveson defined the index of a non spatial E_o-semigroup of $\mathfrak{B}(\mathfrak{H})$ to be the order of the continuum and showed with this definition of index the additivity under tensor products remained. It follows that if $\{\alpha_t\}$ is a non spatial E_o-semigroup of $\mathfrak{B}(\mathfrak{H})$ and $\{\beta_t\}$ is any E_o-semigroup of $\mathfrak{B}(\mathfrak{K})$ then $\{\alpha_t \otimes \beta_t\}$ is non spatial so the tensor product of a type III with any E_o-semigroup is of type III.

One can check that that tensor product of an E_o-semigroup of $\mathfrak{B}(\mathfrak{H})$ of type II with and E_o-semigroup of type I or II yields an E_o-semigroup of type II and since Arveson has shown the index is additive we have the tensor product of a type I_n with a type II_m is type II_{n+m} and the tensor product of a type II_m with a type II_m is type II_{n+m}.

In summary we have if $\{\alpha_t\}$ is an E_o-semigroup of $\mathfrak{B}(\mathfrak{H})$ of type A_n and $\{\beta_t\}$ is an E_o-semigroup of $\mathfrak{B}(\mathfrak{K})$ of type B_m where A and B are I,II or III then $\{\alpha_t \otimes \beta_t\}$ is an E_o-semigroup of $\mathfrak{B}(\mathfrak{H} \otimes \mathfrak{K})$ of type C_{n+m} where C is the maximum of A and B and for type III the index is superfluous.

The basic result of [P4] is the existence of E_o-semigroups of $\mathfrak{B}(\mathfrak{H})$ of type II. The example constructed in [P4] was of type II_1 and by tensoring with an E_o-semigroup of type I_n we obtain E_o-semigroups of type II_n for $n = 1, 2, \cdots$. The obvious question is how many E_o-semigroups are there of type II_1. This is what we believe to be the situation. Suppose $\{\alpha_t\}$ and $\{\beta_t\}$ are E_o-semigroups as constructed in [P4]. Suppose π_α and π_β are the singular part of the boundary representations of $\{\alpha_t\}$ and $\{\beta_t\}$. It is clear that if π_α and π_β are unitarily equivalent then $\{\alpha_t\}$ and $\{\beta_t\}$ are cocycle conjugate. We believe the converse is true, namely we believe that if $\{\alpha_t\}$ and $\{\beta_t\}$ are cocycle conjugate then π_α and π_β are unitarily equivalent. This would immediately lead to the result that there are uncountably many E_o-semigroups of type II_1 since by varying the parameters λ_k in the construction in [P4] we can produce uncountably many inequivalent representations.

We would like to mention an interesting conjecture which follows from the discussion of the previous paragraph.

Conjecture 3.2. Let $\{\alpha_t\}$ be the E_o-semigroup of $\mathfrak{B}(\mathfrak{H})$ obtained by setting $\lambda_k = a^k$ for $k = 1, 2, \cdots$ and $a > 1$ in the construction of section VII of [P4]. Let $\sigma_t = \alpha_{st}$ for $t \geq 0$. Then $\{\alpha_t\}$ and $\{\sigma_t\}$ are cocycle conjugate if and only if $s = a^r$ for some integer r.

Finally, we would like to sketch a possible classification scheme for spatial E_o-semigroups of $\mathfrak{B}(\mathfrak{H})$. Suppose $\{\alpha_t\}$ is an E_o-semigroup of $\mathfrak{B}(\mathfrak{H})$ with strongly continuous one parameter semigroup of intertwining isometries $\{U(t)\}$. Let $\pi_{(\alpha,U)}$ the

the boundary representation associated with $\{\alpha_t\}$ and $\{U(t)\}$. Recently, Alexis Alevras has shown in [AL] that the representation $\pi_{(\alpha,U)}$ does not depend on U. That is if $\{V(t)\}$ is a second intertwining semigroup of isometries for $\{\alpha_t\}$ then $\pi_{(\alpha,V)}$ is unitarily equivalent to $\pi_{(\alpha,U)}$.

We believe that the boundary representation $\pi_{(\alpha,U)}$ is a complete cocycle conjugacy invariant *i.e.* if (α, U) and (β, V) are two E_o-semigroupf of $\mathfrak{B}(\mathfrak{H})$ with intertwining semigroups $\{U(t)\}$ and $\{V(t)\}$, respectively, then α and β are cocycle conjugate if and only if the boundary representation $\pi_{(\alpha,U)}$ and $\pi_{(\beta,V)}$ are unitarily equivalent. We should emphasize that this is only a conjecture and a great deal of work has to be done to establish this.

REFERENCES

[Al] Alexis Alevras, *A note on the boundary representation of a continuous spatial semigroup of $*$-endomorphisms of $\mathfrak{B}(\mathfrak{H})$*, Proceeding of the AMS (to appear).

[A1] W.B. Arveson, *Continuous analogues of Fock Space*, Memoirs A.M.S. **80** (1989), no. 409.

[A2] ______, *An addition formula for the index of semigroups of endomorphisms of $\mathfrak{B}(\mathfrak{H})$*, Pac. J. Math. **137** (1989), no. 1, 19–36.

[A3] ______, *Continuous analogues of Fock space II: the spectral C^*-algebra*, J. Funct. Anal. **90** (1990), no. 1, 165-205.

[A4] ______, *Continuous analogues of Fock space III: Singular states*, J. Oper. Th. **22** (1989), 165–205.

[AW] H. Araki and E.J. Woods, *Complete Boolean Algebras of Type I Factors*, Publ. R.I.M.S. Kyoto Univ. **II** (1966), 157–242.

[BR] O. Bratteli and D. Robinson, *Operator Algebras and Quantum Statistical Mechanics I*, Springer-Verlag, 1979.

[DS] N. Dunford and J.T. Schwartz, *Linear Operators*, Interscience, New York, 1958-1971.

[P1] R.T. Powers, *An index theory for semigroups of $*$-endomorphisms of $\mathfrak{B}(\mathfrak{H})$ and type II_1 factors*, Can. Jour. Math. **40** (1988), 86–114.

[P2] ______, *A non-spatial continuous semigroup of $*$-endomorphisms of $\mathfrak{B}(\mathfrak{H})$*, Publ. R.I.M.S. Kyoto Univ. **23** (1987), 1053–1069.

[P3] ______, *On the structure of continuous spatial semigroups of $*$-endomorphisms of $\mathfrak{B}(\mathfrak{H})$*, International Journal of Math. **3**, 323–360.

[P4] ______, *New examples of continuous spatial semigroups of $*$-endomorphisms of $\mathfrak{B}(\mathfrak{H})$*, J. Funct. Anal. (to appear).

[PP] R.T. Powers and G. Price, *Continuous spatial semigroups of *-endomorphisms of $\mathfrak{B}(\mathfrak{H})$*, Transactions A. M. S. **321** (1990), 347–361.

[PR] R.T. Powers and D.W. Robinson, *An index for continuous semigroups of *-endomorphisms of $\mathfrak{B}(\mathfrak{H})$*, Jour. of Funct. Anal **84**, 85–96.

[SW] R. F. Streeter and A. S. Wightman, *PCT, Spin and Statistics and All That*, W. A. Bengiman, Inc., New York, 1964.

[vN] J. von Neumann, *On Infinite Direct Products*, Composito Mathematica **6** (1938), 1.

DEPARTMENT OF MATHEMATICS, UNIVERSITY OF PENNSYLVANIA, PHILADELPHIA, PA. 19104

Proceedings of Symposia in Pure Mathematics
Volume **59**, 1996

Rigorous Covariant Form of the Correspondence Principle

Irving Segal

ABSTRACT. The correspondence principle mapping classical hamiltonians and lagrangians into quantized counterparts is deduced in a unique and rigorous form from a generalization of the Poincare lemma applicable to infinite-dimensional non-commutative differential forms. In this form the principle incorporates symmetry and stability constraints, and shows the selfadjointness of a general class of quantized action integrals in four (or more) space-time dimensions. Relations to symplectic quantization, Wick products, constructive quantum field theory, and alternative space-times are discussed.

1. Introduction

My subject represents the classic interaction between pure and applied mathematics. In this an issue external to mathematics provides a challenge that motivates a study from a new mathematical viewpoint. As this study progresses, the mathematical aspect interacts with the external motivation, producing an integrated activity that can not be fully understood from either the pure or the applied standpoint. When the study is successfully completed, it separates momentarily at least into pure and applied aspects. The evolution of pure mathematical progress and of the external interests then normally leads to new challenges and interactions.

1991 *Mathematics Subject Classification.* Primary 81S99; Secondary, 47B25, 81T05, 81T20

The work of John von Neumann is an important and exemplary instance of these cultural dynamics. He was the first to bring the modern algebraic standpoint and technique to bear on functional analysis. Other outstanding mathematicians were inspired to proceed in similar directions, but I think all would agree that his work was the most seminal and influential. His work in Hilbert space is recognized by physicists as well as mathematicians as the core mathematical foundation of modern quantum theory. But it connects also with the (presently) extremely pure subject of continuous projective geometry, a highly original and characteristic work.

The present work is motivated by an external issue to which I shall try to apply von Neumann's general scientific standpoint as well as functional analytic technique. I believe it would somewhat falsify such work to present it in a purely mathematical way, and render it less rather than more comprehensible from a broad perspective. On the other hand, it is certainly true that logical precision and clarity are enhanced by presentation in the format of core mathematics. Accordingly, I shall try to inter-relate the physical background and motivation with the core mathematical formalism and technique, with emphasis on the former to start with and on the latter at the end.

2. Background

The "Correspondence Principle" has been a crucial but intuitive guide to "Quantization" since the invention of quantum mechanics by Heisenberg in 1925. The general idea was that the quantum hamiltonian for a system was to be derived from the classical hamiltonian for the system by substitution of noncommuting canonical p's and q's for the classical, commuting p's and q's. The Heisenberg commutation relation, $pq-qp = -i$ (where we choose units so that Planck's constant equals 2π) , was algebraically perfectly well-defined, but even for polynomials in the p's and q's, there was no apparent map from the classical to the quantized form that satisfied even rudimentary algebraic constraints, as shown by McCoy soon after Heisenberg's work. The lack of unicity was even more severe for general functions of the canonical operators, that is, the p's and q's.

Taking the trace on both sides of the Heisenberg commutation relation showed that the p's and q's could not be represented as finite-dimensional matrices, and indeed in the Schrœdinger representation they appear as unbounded operators in Hilbert space. To avoid the (apparently) irrelevant pathology associated with unbounded operators, Weyl developed a formally equivalent relation between the one-parameter unitary groups generated by the p's and q's, or linear combinations thereof. This in turn gave rise to a variant of the correspondence principle, which treated functions f(p,q) that could be expressed in the form

$$(1) \qquad f(p,q) = \iint \exp[i(sp+tq)]F(s,t)dsdt,$$

where F(s,t) is integrable on $\mathbb{R} \times \mathbb{R}$ (over which the integral is taken). But apart from the strong restriction on f(p,q), the Weyl quantization had limited invariance properties. Using the Weyl relations, von Neumann showed that the Schrœdinger representation is essentially unique, within unitary equivalence, which is all (presumptively) that is physically significant. So it appeared useless to look for alternative representations of the Heisenberg relations. Attempts to find essentially new representations in Banach spaces were equally unsuccessful.

The problem was of course no simpler in n-dimensions, in which the Heisenberg relations took the form $[p_j, q_k] = -i\delta_{jk}$ $(j,k = 1,2,\ldots,n)$. The issue took a more serious turn when Dirac invented quantum field theory, which essentially let n tend to infinity. The idea was a very simple one. Since it paid off empirically to assume the Heisenberg relations for the momenta and coordinates of a particle, why not do the same for the coordinates of a field? For example, the Cauchy (initial value) data, $\phi(0,x)$ and $\partial_t\phi(0,x)$ at the fixed time 0 for a solution ϕ of the wave equation (Dirac treated the Maxwell equations, but the difference is of no consequence here) could be expanded in terms of a complete orthonormal set $\{f_n\}$ of functions in $L_2(\mathbb{R}^3)$:

$$(2) \qquad \phi(0,x) = \Sigma_n q_n f_n(x), \quad \partial_t\phi(0,x) = \Sigma_n p_n f_n(x).$$

For the *classical* wave equation, the p's and q's are ordinary numbers. Dirac asked, why not replace them by quantities satisfying the Heisenberg commutation relations?, and showed that this paid off by explaining the Einstein transition probability laws.

The mathematics involved was almost totally heuristic; the hamiltonian involved was a formal expression, in no clear sense an operator in Hilbert space, and only the corresponding first-order perturbation theory played a role. The results seemed to explain brilliantly the emission and absorption of light from atoms, but it was profoundly disquieting that all nontrivial higher order perturbative terms appeared to be divergent. Von Neumann's 1932 book [1] gave perhaps the clearest account of the Dirac light theory, but proceeded at a formal level, and did not attempt to deal with the fundamental divergences. The Dirac hamiltonian was the prototype for the the quantum field theory of particle production, which is completely outside the scope of the classical theory, and the raison d'etre of the quantum field concept. But Dirac himself was throughout his life highly insistent on the crucial need, in physics as well as mathematics, to resolve the divergences. Thus he wrote in retrospect [2]:

"The difficulty of the divergences proved to be a very bad one. No progress was made for twenty years. Then a development came, initiated by Lamb's discovery and explanation of the Lamb shift, which fundamentally changed the character of theoretical physics. It involved setting up rules for discarding the infinities, rules which are precise, so as to leave well-defined residues that can be compared with experiment. But still one is using working rules and not regular mathematics.

"Most theoretical physicists nowadays appear to be satisfied with this situation, but I am not. I believe that theoretical phyics has gone on the wrong track with such developments... .We must realize that there is something radically wrong when we have to discard infinities from our equations, and we must hang on to the basic ideas of logic at all costs. Worrying over this point may lead to an important advance."

The Dirac theory was a non-relativistic approximation, in which only light, as described by the Maxwell equations, was quantized as a field. The relativistic Dirac equation for the electron was found a year later. Its quantization presented new problems, notably that of the formulation of "fermion" fields such as that represented by the electron. With the aid of the quantization theory for such fields due to Jordan and Wigner, Heisenberg and Pauli set forth in 1929 what has come to be the quintessential prototype for relativistic nonlinear quantum field theory, i.e. relativistic Quantum Electrodynamics (QED). But like the non-relativistic Dirac theory, it appeared irremediably divergent.

The problem was not mathematically well-defined, but was rather one of formulation of the physically seemingly compelling essential ideas in a mathematically meaningful form. To bypass the inessential (apparently) complications of treating the interaction of fields satisfying two different statistics,- Bose-Einstein for the photon (a boson) and Fermi-Dirac for the electron (a fermion), and involving fields of many components rather than just one,- both the physics and the mathematics literature often begin with scalar (essentially single-component) fields. Such fields ϕ are defined e.g. by nonlinear equations of the form

(3) $$\Box\phi + m^2\phi + g\phi^j = 0$$

where $\Box$ is the wave operator $\partial_0{}^2 - \partial_1{}^2 - \partial_2{}^2 - \partial_3{}^2$ on Minkowski space $\mathfrak{R}^1 \times \mathfrak{R}^3$, with time coordinate x_0 and space coordinates x_1, x_2, x_3 ($\partial_\mu = \partial / \partial x_\mu$), j is an integer > 1 and g and m are constants. As a classical nonlinear wave equation, in which the values of $\phi(t,x)$ are real numbers, this equation is unexceptionable and has been studied with considerable success in modern times since the pioneering work of Jorgens.

As a quantum system, the commutation relations at a fixed time (say 0) are a simple variant of the Poisson bracket relations for the solution manifold as a symplectic variety, and take the form

(4) $$[\phi(0,x), \ \partial_t\phi(0,y)] = -i\delta(x-y); \quad [\phi(0,x), \phi(0,y)] = 0;$$
$$[\partial_t\phi(0,x), \partial_t\phi(0,y)] = 0,$$

where again units are properly chosen. In view of the delta-distribution that occurs here, the values of the field can not be bona fide operators in the usual sense. What then is the meaning of the nonlinear expression ϕ^j? Thus the fundamental partial differential equation of the quantized theory has no a priori mathematical meaning. Until a rigorous interpretation is given for the nonlinear term, the issue of the existence, unicity, regularity, and properties of solutions to the equation can hardly begin to be studied.

When g = 0 the equation is linear and its quantization is mathematically well defined, and essentially uniquely within unitary equivalence, e.g. [3]. Consistency with the fundamental physical constraints of relativistic invariance, positivity of the energy, and the above "canonical" commutation relations for the field values at a fixed time are attained. But even in this simplest nontrivial case, the power ϕ^j is without a priori mathematical meaning. We will later see that there is a natural way to develope and apply such meaning. But as of about 1950, quantum field theory was generally accepted as theoretically probably irremediably divergent, although no rigorous theorem to this effect could be established because the issue fell outside the realm of (rigorous) mathematics.

This was the era of the "renormalization' theory of Feynman, Schwinger, and Tomonoga, which succeeded in meeting the challenge to theoretical physics provided by the Lamb shift measurements. These and some other measurements were "predicted" (after the fact) with the aid of perturbative scattering theory, in conjunction with systematic procedures for discarding the apparent infinities that occurred in the higher-order terms in QED. The key idea of the renormalization program was to consolidate all the apparent infinities in just two expressions, the feasibility of which represented the remarkable property of the QED equations called "renormalizability". The two apparently infinite expressions were then identified with the measured physical quantities of mass and charge which they in theory symbolized.

To the extent that the renormalization program succeeded, the scattering matrix S, from which all observable quantities were derivable, could be represented by a power series in the coupling constant, whose coefficients were finite well-defined quantities. But it was then and remains thought that the series itself is merely asymptotic and converges for no nonzero value of the coupling constant. Unfortunately, the renormalization procedures are difficult to rationalize in purely mathematical terms, and none of the attempts to prove the existence of well-defined renormalized and relatistically invariant coefficients to the S-matrix perturbative series appears clearly successful. In the introduction to the collection of basic papers on the renormalization of QED [4], Schwinger wrote: "We conclude that a convergent theory can not be formulated consistently within the framework of present space-time concepts".

The simplest form of renormalization theory was that of Feynman, which by using the "interaction representation", an application of the method of variation of constants that is particularly effective in connection with quantum field theory, reduced the theory to aspects of the theory of free fields. The Feynman formalism remains in active current use (e.g., [5]). The basic nonlinearities were dealt with by applying the prescription of Wick [6] for the standardization of polynomials in the fields. Wick's theorem was perfectly rigorous as regards polynomials in averages of the field with respect to smooth test functions, which were bona fide operators, unlike the values of the fields at single points. The Feynman formalism assumed the conclusion of Wick's theorem could be applied directly to the fields themselves, without averaging. Such application often gave rise to apparently meaningless integrals, which then were suitably "renormalized".

After the initial empirical success of the renormalization theory, dissatisfaction had grown, partly on foundational grounds, and partly because the renormalization eliminated the possibility of prediction for the physical mass and coupling constant. Heisenberg and others had earlier suggested the introduction of a fundamental length, designed to curb the singularity of the quantized field at a point, but this idea was not to prove fruitful until some decades later. More broadly, Heisenberg gave relatively high priority to new theoretical initiatives, relative to the accumulation of more experimental data. One such initiative was the very general axiomatic approach of Kallen, Wightman and others, which was mathematically rigorous, and clarified the situation, but before long had led to extremely difficult computations. The axiomatic approach avoided analytic specification of a particular hamiltonian or lagrangian, which limited its applicability to concrete systems. Another approach, that of Lehman, Symanzik, and Zimmerman (LSZ), was partially axiomatic, but applied to fields described in the traditional way via a given lagrangian. It was usefully operational in spirit, but never became clearly divergence free. Both the axiomatic and the LSZ approaches seem to have left the fundamental issues of the meaning of the nonlinear interaction term, and the existence of appropriate solutions to the quantized differential equations basically intact. So what remains in the direction of mathematical construction of nonlinear quantum fields?

3. Varieties of quantization

At the present time, there are two main quantization approaches that appear constructive, in the sense of being directly applicable to given equations or lagrangians, and potentially viable. The more natural from a traditional mathematical standpoint is perhaps that of symplectic quantization. This was applied to the solution manifold of a nonlinear wave equation in [7,8]. This approach starts from the *classical* solution variety, a subject that has made much progress in the past three decades, and is of much intrinsic interest.

Probably mathematically less natural, having a somewhat opportunistic tinge, is that of nonlinear functions of quantized distributions. This grows out of the theory of Wick products, which in turn developed in part from "subtraction physics", a somewhat inauspicious origin for a coherent mathematical theory. Even less reassuring is the association of the multiplication of quantized distributions with renormalization theory.

Having worked on both approaches, I can hope to be impartial in my view of their current status. It appears to me that the approach via a theory of nonlinear functions of quantized distributions is now the more promising route to a general mathematical theory of the quantization of nonlinear wave equations.

Taking for specificity the relativistic nonlinear wave equation (3), the solution manifold M forms a relativistically invariant infinite dimensional symplectic manifold,- indeed, conformally invariant in the case m = 0. Now the basic ideas of finite-dimensional quantum mechanics in euclidean space apply equally to general finite-dimensional manifolds. The problem is that M is the counterpart not to geometrical space, or 'configuration' space, but rather to the phase space, or cotangent bundle of the configuration space. In elementary quantum mechanics, state vectors are represented by functions on the configuration space $\mathbb{R}^{3n}$, where n is the number of particles, e.g. in the Schrœdinger formulation. This description can be carried over to one in terms of phase space in the case of $\mathbb{R}^{3n}$, but apparently not in the case of a general finite-dimensional manifold.

More specifically, replacing $\mathbb{R}^3$ by $\mathbb{R}^1$ and taking n = 1 for simplicity, the canonical operators p and q act in the Schrœdinger representation as $-i\partial/\partial x$ and multiplication by x, in the space $L_2(\mathbb{R}^1)$. Useful equivalent operators can be defined in the phase space $\mathbb{R}^2$ only by constraining the state vectors on $\mathbb{R}^2$, most simply by requiring them to be complex analytic. In fact, $L^2(\mathbb{R}^1)$ is unitarily equivalent to the space $AL_2(\mathbb{R}^2)$ of all entire (complex analytic) finite-norm functions with the inner product

$$(5) \qquad \langle f,g\rangle = \int f(z)\, g(z)^* \exp(-|z|^2)\, dxdy \qquad (z = x+iy)$$

in such a way that p and q correspond to linear combinations of z and $\partial/\partial z$. This alternative representation for the canonical operators is quite convenient in certain connections, and extends to the infinite-dimensional case, and so applies to the solution manifolds for the wave and Maxwell equations, etc. However, it seems to have no entirely satisfactory counterpart in the case of a general configuration space. In the case of a compact Lie group, Brian Hall has recently developed an effective analog, whose ramifications are treated in the contribution of Leonard Gross to this symposium.

There is a natural almost complex structure in the cotangent bundle B of a Riemannian manifold, which together with the intrinsic symplectic structure in B is almost Kæhler. There is also a natural measure on B analogous to the Gaussian measure on $\mathbb{R}^2$, similarly derived from the heat equation. However, the Kæhler structure is integrable only when the manifold is flat, as shown by Kodaira and others. In general the space of almost analytic functions on an almost Kæhler manifold is trivial, and so finally there is no known analog to the space $AL_2(\mathbb{R}^2)$ on other than Hilbert spaces and compact Lie groups.

These problems in the finite-dimensional case do not rule out the infinite-dimensional case of a nonlinear wave equation, whose Poincare or conformal invariance may be helpful. But one needs two things beyond the invariant symplectic structure, neither one of which is yet substantiated, despite much much effort. The first is an invariant almost Kæhler structure, to which the given symplectic structure is subordinate. This has interesting relations to scattering theory and to the infinite dimensional stability theory of the Mark Krein school. First steps were taken by the late Stephen Paneitz and myself [9,10]. In particular, sufficiently close to 0 in M, there exists a canonical complex hermitian structure, characterized by invariance under the scattering transformation. This suggests that the solution manifold may have intrinsic nontrivial structures comparable to those of finite-dimensional algebraic varieties. However, global existence of this canonical hermitian structure remains uncertain. In addition, it has not been possible to determine whether the almost complex structure is integrable, because of the very complicated form of the torsion.

The second thing that is needed is a suitably invariant notion of integration over the solution manifold, i.e. a counterpart to the isotropic Gaussian measures in e.g. the solution manifolds of linear relativistic wave equations. Even the existence of a nontrivially temporally invariant weak measure (or expectation functional on a general class of field functionals) is unknown. 'Nontrivially' here alludes to the fact that a (trivially) invariant measure is obtainable by starting with an arbitrary measure at time 0, and letting it evolve temporally. A natural way to characterize a nontrivial measure is rather by invariance under the scattering transformation.

The need for both things has been recognized for two decades or more, and it does not look like they will turn up or their existence even be decided any time soon. The idea of nonlinearizing the infinite-dimensional analysis in the solution manifolds of linear wave equations, which form *complex* Hilbert spaces (even when only real solutions are in question), by providing nonlinear solution manifold M with a canonical Kæhler structure, seems a natural one, and has influenced heuristic physical formalism in the past decade. But no nontrivial example of an infinite-dimensional Kæhler manifold whose automorphism group is a noncompact finite-dimensional Lie group such as the Poincare group is known. However, in this direction it is possible on occasion to introduce canonical complex structures, and thereby creation and annihilation operators and a vacuum state vector (which will be discussed later) in linear solution manifolds admitting no one-parameter automorphism groups. An example is the case of the Klein-Gordon equation in Minkowski space as modified by a small perturbation of the Minkowski metric of compact support [9]. But as yet there is nothing like this in the nonlinear case.

In contrast, there has been increasing progress in the past quarter century in treating quantized wave equations involving nonlinear functions of quantized distributions, and relating them to the corresponding classical equations.

4. Constructive quantum field theory

A general class of relativistic nonlinear quantized wave equations in two dimensional Minkowski space has been treated as regards existence and regularity of solutions (e.g., [11,12]). As earlier noted, the nonlinear terms were devoid of a priori mathematical significance, being formally powers of operator-valued distributions. Moreover, they are powers not of the *free* field, but of the *interacting* field, i.e the field satisfying the nonlinear equation. But for exploratory purposes, one might just substitute the free field into the nonlinear term, as first done by Nelson [13]. The limiting form of the Wick power, when applied to test functions whose support converged to a point, turned out to exist in an effective mathematical sense, as an operator-valued distribution on space, at fixed times, in two space-time dimensions, as required for this purpose [14].

But in the case of the interacting field, the Wick definition was inapplicable. This definition proceeded by expressing the field as a linear combination of the 'creation' and 'annihilation' operators, which exist only for free fields. A generalization of the notion of Wick product to interacting fields was needed, and will be discussed below. This eventually ennabled a partial reduction to the case of the products of free fields.

Nelson essentially proceeded by just plugging in the *free* quantum field ϕ in the expression for the energy $E(\phi)$ of the classical field defined by the differential equation (*):

(6) $$E(\phi) = \int\{(\tfrac{1}{2}[(\text{grad } \phi)^2 + (\partial_t\phi)^2] + gk^{-1}\phi^k\}dx \quad (k = j+1).$$

That is, the nonlinear expressions are interpreted as Wick powers of the free field ϕ, which is is evaluated at a fixed time. But the interacting field ϕ_{int} differs at finite times from the free field, so that $E(\phi)$ is not the energy operator for the putative interacting field, but rather a presumed approximation thereto. This energy $E(\phi_{int})$ is conserved, unlike $E(\phi)$, in which only the quadratic part is conserved (it represents the free energy), while the higher power term is explicitly time-dependent.

If as hoped and assumed in practical quantum field theory, the putative interacting field is asymptotic to a free field at very early or very late times, then it would be physically appropriate to assume the equivalence of the interacting to the free field at time '$-\infty$', in the sense of scattering theory, as the so-called 'incoming field'. More explicitly, the S-matrix would be expected to be the limit, modulo the 'free' motion, of the unitary transformation from time $-T$ to $+T$ generated by $E(\phi)$, as $T \to \infty$. It is nontrivial however to develope the spectral properties of $E(\phi)$ at a fixed time, and it remains an open problem whether the limit defining the putative S-matrix actually exists. But fortunately it turns out that the spectral analysis of the physically somewhat dubious appearing hamiltonian $E(\phi)$, where ϕ is free , is the same as that of $E(\psi_{int})$ where ψ_{int} is an interacting quantum field whose interaction energy is the given one augmented by a lower order polynomial in the field [12].

The spectral theory of the operator $E(\phi)$ was developed both by real-time and imaginary-time methods. As noted in the preceding paragraph, the analysis in the real-time approach [11] could in part be construed as working in the 'interaction representation'. Besides this, first steps towards a pilot model for nonlinear quantum field theory was established in a rigorous and natural way in two space-time dimensions. More specifically, let us take for simplicity the case in which the noncompact space $\mathbf{R}$ is replaced by the compact S^1 (otherwise, a sequence of spatial cutoffs must be used, as is made possible by the hyperbolicity of the underlying equation, which is inherited in the quantum context). Then given a real polynomial F in one variable that is bounded below, one first established the basic spectral properties of the hamiltonian

$$H = H_0 + \int :F(\phi(0,x)): \quad dx, \tag{7}$$

where H_0 is the hamiltonian of the free field, and : ... : indicates the Wick interpretation of the products that are involved (see below). One could then define a field whose hamiltonian was H and whose Cauchy data at time 0 were those of the free field, which appears as an approximation to the putative interacting field, but whose temporal evolution was not entirely local. Thus this field did not satisfy the quantized form of the original partial differential equation, in which the nonlinear term F' is local. But one could transform the result into a solution for the suitably local nonlinear quantized equation

(8) $$\Box\psi + m^2\psi + ::G'(\psi)::= 0,$$

where G has the same leading term as the given polynomial F, but additional lower order terms , and the $::G'(\psi)::$ indicates a generalization of the Wick concept to interacting fields. Time evolution formed a well-defined one-parameter group, but of automorphisms of a C*-algebra, rather than a unitary group on a Hilbert space, as in elementary quantum mechanics.

But the question of whether by modifying the starting F one could solve equation (8) for any given semibounded polynomial G remains unsolved. The problem arose mainly from the major difference between classical and quantized wave equations that the vacuum, a positive linear functional on the algebra generated by the smoothed field values, is an essential part of the quantum construction. For classical equations a rough counterpart would be a measure in the solution manifold that is invariant under temporal evolution and the scattering operator, but little is known the existence of such a measure in the nonlinear case.

When the full two-dimensional Minkowski space is used, with S^1 replaced by $\mathbf{R}^1$, time evolution forms a well-defined one-parameter group, thanks to the essential hyperbolicity of the underlying equations. But it is a group of automorphisms of a C*-algebra constructed from the local fields, rather than a unitary group on Hilbert space. This would be no problem if there existed an appropriate vacuum state for this group. In that event the canonical representation associated with the state would provide a Hilbert space of state vectors for the interacting field, on which the total hamiltonian acts as a semibounded self-adjoint operator (although the free and interaction components of this hamiltonian would not exist except as derivations of the C*-algebra). No such interacting field structure appears to exist, despite considerable efforts in this direction. A further problem is that the C*-algebra in question is not clearly independent of the initial representation of the canonical commutation relations (e.g., the construction could be made with an arbitrary value for the assumed mass of the free field). The problem seemed to be with the noncompactness of space.

This suggested that spacetime should be modelled as an alternative whose space component is compact. This might appear to restrict the invariance of the theory, but there is a natural alternative spacetime that is invariant under the Poincare group, and in which the space component is compact. After confirmation of the physical rationale for this alternative spacetime by its application to physics on the largest scale, the development of constructive quantum field theory was resumed, on the basis of this alternative model.

5. Spacetime geometry

Space*time* is distinguished from *space* by the intervention of the notion of *causality*, or of temporal precedence. It is natural to assume that if a point p of space-time M precedes another point q, relative to a given notion of causality in M, then for any truly physical space-time transformation T, Tp will also precede Tq. Causality is one of the key principles of theoretical physics, and the causal automorphism group of the manifold brings in the symmetries that are a central feature of quantum physics. Thus for Minkowski space $\mathbf{M}_0$, a one-to-one causal transformation differs from a Poincare transformation at most by a positive scalar factor (a theorem of Alexandrov & Ovchinnikova, later rediscovered by Zeeman). It is clear that Minkowski space is spatially and temporally isotropic, in that any two spacelike or and two timelike directions at a point can be connected by a transformation in the Poincare group, and homogeneous, in that any two points can be connected by a Poincare transformation.

There is no a priori reason to assume that the spacetime arena of physics is necessarily $\mathbf{M}_0$, but it seems reasonable in the light of modern physical principles to assume that physical spacetime enjoys spatial and temporal isotropy, and homogeneity. The conservation laws for angular momentum and energy-momentum are equivalent to spatial isotropy and homogeneity, while relativistic invariance is closely connected with temporal isotropy. But as Tits has shown, there are very few four-dimensional manifolds endowed with causal structure that have these properties, especially if causality is constrained to be global, e.g. in the sense of Leray (in particular, there can be no closed time-like loops). Moreover, among such spaces, there is a unique maximal one, of which all others are essentially causal submanifolds: namely the universal cover $\tilde{M}$ of the conformal compactification $\bar{M}$ of $\mathbf{M}_0$. Just as $\mathbf{M}_0$ can be factored into time and space components as $\mathbb{R}^1 \times \mathbb{R}^3$, so can $\tilde{M}$ can be factored as $\mathbb{R}^1 \times S^3$. Normalizable solutions of wave equations such as Maxwell's extend uniquely and maximally from $\mathbf{M}_0$ all of $\tilde{M}$, in which $\mathbf{M}_0$ is Poincare-equivariantly imbeddable, etc.

There is a good a priori theoretical case for modelling space-time as $\tilde{M}$ rather than $\mathbf{M}_0$. The causal (automorphism) group of $\tilde{M}$ is the universal cover of the conformal group of $\mathbf{M}_0$, and the singularities in the action of the conformal group on $\mathbf{M}_0$ are eliminated when conformal transformations in $\mathbf{M}_0$ are regarded as mapping it into $\tilde{M}$. But the natural notion of time in $\tilde{M}$ differs from that in $\mathbf{M}_0$, and the corresponding notions of energy (i.e. the generators of temporal evolution, as they act on a given field (e.g., the Maxwell field) are inequivalent. It has been proposed [15] that the cosmic redshift represents the difference between the respective energies. This has theoretical advantages over the traditional Doppler theory of the redshift, including restoration of the law of conservation of energy, explaining why the shift is to the red rather than the blue, eliminating the the adjustable cosmological parameters of Friedman-Lemaitre theory, and nevertheless making much more accurate predictions of the magnitude redshift relations of both galaxies and quasars in objectively selected and observed samples. And as noted, space is compact in $\tilde{M}$.

In the following the precise form of $\tilde{M}$ will not be needed, but only features that result from the compactness, and in certain connections, the positive curvature of its space component. To derive the rigorous form of the correspondence principle that is treated below, it is convenient to assume that in the fields considered, the hamiltonian H is positive and has the property that exp(-tH) is Hilbert-Schmidt for all $t > 0$. (If this property holds for the single-particle hamiltonian, it also holds for the quantized field hamiltonian, and conversely.) Some variant of the present correspondence principle is probably valid without this assumption, and thus directly in $\mathbf{M}_0$. But since it looks as if physical space-time may well be more efficiently modelled as $\tilde{M}$ rather than $\mathbf{M}_0$, and since sharp results on conformally invariant wave equations in $\mathbf{M}_0$ are presently established only via the imbedding of $\mathbf{M}_0$ in $\tilde{M}$ (e.g. [16,17]), we will not seek maximum generality as regards the spectral properties of H.

6. Wick products

The vacuum expectation value functional on an operator algebra $\mathcal{A}$ is a positive normalized linear functional E that is characterized by its relation to the temporal evolution in $\mathcal{A}$. If for example $\mathcal{A}$ is the algebra B(H) of all bounded operators on the Hilbert space H, and E is the linear functional on $\mathcal{A}$ defined by the equation E(A) = <Av,v> where v is a unit vector in H, then v is a 'state vector' of $\mathcal{A}$, and it is a vacuum relative to the one-parameter group U(.) of unitary operators on H (or corresponding automorphism group $A \to U(t)^{-1}AU(t)$) provided v is a lowest eigenvector for the selfadjoint generator of U(.), i.e. for the energy operator. The Wick product in a quantum field is defined relative to the vacuum vector v in the quantum field, i.e. the lowest eigenvector of the field hamiltonian (normally unique within a phase factor). The remarkable thing about it is that it renders products that appear meaningless by normal mathematical standards perfectly well-defined, albeit with an apparent touch of artificiality.

To give an example in the familiar context of brownian motion, let x(t) denote the usual Wiener process. Its fractional derivatives $x^{(\varepsilon)}(t)$ exist with probability one for $\varepsilon < ½$, but fail to exist for $\varepsilon \geq ½$. In particular, it is easy to verify that while $x^{½}(t)$ exists as a random distribution, its square does not. Thus, $\int_a^b [\, x^{½}(t)\,]^2 \, dt = \infty$ with probability one (a < b), if e.g. the integral is defined as the limit of the corresponding integrals for the partial sums in the Fourier expansion of x(t). However, from the standpoint of subtraction physics, this integral is an immaterial constant, and in fact when the expectation value of $x^{½}(t)^2$ is subtracted from $x^{½}(t)^2$, the result is again a random distribution (again defined using the limit of partial sums of corresponding Fourier series). Aha, you may say, that's cute, but the higher powers probably can't be made into rigorous random distributions just by subtracting expectation values. That's quite right, but the Wick product procedure provides additional subtractions, of lower order terms, that in the end produce random distributions for all the powers. The relation to Wick products comes about because $x^{½}(t)$ is abstractly equivalent, as a process, to the wave equation quantized field in two space-time dimensions, in *space* at a fixed *time* [18].

But what is this procedure good for? In quantum field theory, it ennables formal powers $\phi(x)^k$ of the free quantum field $\phi(x)$ to be established as well-defined operator-valued distributions. This is a natural way to get started on a rigorously viable formulation of quantized nonlinear wave equations.

The algebraic part of Wick's work was rigorous, but its extension to products of fields requires limiting processes whose formal results were merely assumed in practical quantum field theory. Being algebraic, Wick's really deals with the enveloping algebra of the Lie algebra of the Heisenberg group in n dimensions, and the essential idea is clear from the case $n = 1$. Given a canonical pair p and q, it is convenient to form the complex combinations

$$c = (p+iq)/\sqrt{2}, \quad c^* = (p-iq)/\sqrt{2}, \tag{9}$$

known as creation and annihilation operators. In particular c^* annihilates the vacuum vector v when the underlying complex structure is appropriate. This makes it convenient to standardize monomials in p and q by rewriting them in terms of c and c^*, and using the commutation relation in the form

$$c^*c - cc^* = 1 \tag{10}$$

to place all powers of c^* to the right of all powers of c, while treating c and c^* as commuting quantities (i.e. dropping commutators wherever they occur). The result of applying this procedure to a given monomial m is denoted as :m:. Wick's 'theorem' evaluated the vacuum expectation value of a product of Wick products, and is one of the most fundamental tools of computational quantum field theory.

The formation of the Wick product has a somewhat arbitrary aspect, and as formulated it applies only to free fields. An equivalent formulation can however be given that is mathematically coherent and can be extended to interacting fields. To explain the general idea briefly, let $\phi(x)$ denote a free quantum field that satisfies relations of the form

$$[\phi(x),\phi(y)] = D(x,y), \tag{11}$$

suitably regularized, where $D(x,y)$ is a scalar operator.
Whatever "$\phi(x)^2$" may be, its commutator with $\phi(y)$ would be expected to satisfy the relation:

$$[\text{"}\phi(x)^2\text{"},\phi(y)] = 2i\phi(x)D(x,y). \tag{12}$$

One can now ask: does there actually exist a bona fide operator A such that

$$[A, \phi(y)] = 2i\phi(x)D(x,y) \tag{13}$$

for all y? The right side of equation (13) is mathematically effectively defined, and this question has a definite anwer modulo the specification of the regularity of the type of operator considered. The answer is affirmative with great generality, but the regularity of the operator is considerably reduced in space-time dimensions greater than two. Because of the irreducibility of the totality of the quantum field operators $\phi(y)$, the operator A is unique modulo scalars, and it follows that the Wick product $:\phi(x)^2:$ is uniquely defined by requiring that A have vanishing expectation value.

Having defined $:\phi(x)^2:$, one can now proceed to define $:\phi(x)^k:$ ($k = 3,4,..$) by recursion, with similar existence and unicity. As a distribution in space at a fixed time, in which capacity it serves to give meaning to the right-hand side of the underlying differential equation in the interaction representation, it is selfadjoint operator valued in two space-time dimensions, while in higher-dimensions it is a densely-defined sesquilinear form, whose nonzero domain as a Hilbert space operator is typically vacuous. From the standpoint of the physical motivation, it is important that $:\phi(x)^k:$ is a *local* function of $\phi(x)$, e.g. in the sense in two dimensions that $\int :\phi(x)^k: f(x)dx$ is affiliated with the ring of operators generated by the values of $\phi(x)$ in a neighborhood of the support of f(x) (an analogous but less straightforward statement is valid in higher dimensions). Relativistic invariance is not required for the "free" field in question, e.g. the quantization of the Klein-Gordon equation with a suitably regular potential can be treated in the same way. But when the field is Lorentz-invariant, the same is true of its Wick powers. It is natural to require the vacuum expectation values of the Wick powers to vanish since they are temporally invariant, and the interacting field is expected to be asymptotic to the free field (in which the nonlinear term is absent) as time becomes infinite.

As the finite-dimensional algebraic level, the problem of showing the existence of :m: for any monomial m in p's and q's reduces to issues regarding the cohomology of the Heisenberg enveloping algebra $\mathfrak{H}$. In its simplest form, the question is: Given a linear map K from the Lie algebra L of the Heisenberg group, when is there an element X of $\mathfrak{H}$ such that $[X,u] = K(u)$ for all u in L. A solution exists if and only if

$$(14) \qquad dK(u,w) \equiv [K(u),w] - [K(w),u] = 0, \qquad \text{for all } w.$$

The solution is of course unique only modulo scalars, but can be made unique relative to a given linear functional on E on $\mathfrak{H}$ such that $E(e) \neq 0$, by requiring $E(X) = 0$. When E is the vacuum expectation functional, the result is the corresponding Wick product :m:, or the 'normal form' of m. This theory leads in a natural way to the introduction of quantized differential forms [19], in which the p's and q's are treated on an equal footing, and thereby to one of the earliest forms of noncommutative differential geometry.

The linear functional E can just as easily be the vacuum state of an interacting system as of a free system. In this way the cohomological reformulation of the Wick product provides the algebraic part of the generalization required to treat interacting fields. The analysis required to establish the existence and properties of products of quantized fields, such as the power $:\phi(x)^k:$, is quite another matter. This is nontrivial even for free fields, and is perhaps most comprehensibly accomplished from the standpoint of cohomology of the infinite Heisenberg group, as will be treated below. In two-dimensional space-time, the extension to interacting fields is established with some restrictions.

The quantization of a given nonlinear wave equation acquires natural mathematical meaning in this way. Take for example the equation

$$\Box\phi + m^2\phi + g\phi^k = 0. \tag{15}$$

To solve this in canonical quantized form means to find an operator-valued distribution $\phi(t,.)$ in space for given time t, *together with* a 'physical vacuum' expectation functional E on the operator algebra generated by the smoothed values of the distribution, satisfying the desiderata indicated above. For each t, the canonical commutation relations are to be satisfied by $\phi(t,x)$ and $\partial_t\phi(t,y)$. The nonlinear expression ϕ^k is defined as the solution of the recursive equation $[\mathbf{Z}_q(x), \ \partial_t\phi(y)] = iD(x-y)\mathbf{Z}_{q-1}(x)$ for $q = 1,2,\ldots,k$, with $Z_0(x)$ defined as 1, and unicity obtained by the constraints $E(\mathbf{Z}_q(x)) = 0$.

There is an existence theorem for solutions of nonlinear quantized scalar wave equations in two space-time dimensions, with ϕ^k replaced by $F'(\phi)$ where F is a polynomial that is bounded below. Unicity has however not been established, and it is not known whether the polynomial p can be arbitrary subject to the semiboundedness constraint, which is necessary to insure classical positivity of the energy. In practice this is correlated with positivity of the energy in the quantum case, but not in any direct or clearcut way.

The presumptive solution of this equation is known as the Heisenberg field. With the advent of renormalization theory, the interaction representation came increasingly to the fore, and is now used almost universally, as exemplified by the wide use of the Feynman rules. While the Heisenberg and the interaction representation fields are formally equivalent, it is arguable that at most one of them is likely to have a rigorous mathematical counterpart. Theoreticians have traditionally given priority to the Heisenberg field, but practical computations have largely been made in the interaction representation, which has a more direct interpretation in terms of the experimentally observed process of particle scattering. Such practical considerations have a way of turning out to be more fundamental than they originally appeared. In any event, the mathematical context is such that it presently appears necessary to begin with the interaction representation field, and in the following that is all I will treat.

7. Ultrasingular operators on boson fields

In general, the products of operator-valued distributions with which quantum field theory deals exist at most as sesquilinear forms on a dense domain in Hilbert space. In a formal way, e.g. in a finite-dimensional Hilbert space, such forms correspond to operators: to the form F(x,y) corresponds an operator **F** such that F(x,y) = <**F**x,y>, but in the present context the associated operator **F** typically has only 0 in its domain as a bona fide operator in Hilbert space. For brevity I will call such a generalized operator, i.e. sesquilinear form on a dense domain, an 'operator'. A calculus and representation theory for 'operators' exists that plays an essential role in our treatment of the Correspondence Principle. The work reported in this section and the next is the product of a collaboration with Zhengfang Zhou and our joint doctoral student, Jan Pedersen, as well as earlier work with Stephen Paneitz. The contribution of Zhou to this symposium explains how rigorous results of unprecedented precision in four-dimensional space-time regarding nonlinear massless fields follow from the general theory presented here, thus obviating the need to present examples here.

The interaction representation takes place in the free field Hilbert space, and a major part of free field theory is a function of an underlying single-particle Hilbert space, and to a considerable extent independent of the geometry of the fields represented by the vectors of the space. It will simplify and clarify matters to proceed in this abstract way. The classical solutions of the wave, Klein-Gordon, Maxwell, etc. equations form Hilbert spaces **H** in natural Lorentz-invariant ways. For each equation there is a corresponding irreducible unitary positive-energy representation of the Poincare group in **H**. In these terms one can more efficiently set up a calculus of 'operators' on the quantized field Hilbert space **K**.

Let **H** be a given complex Hilbert space. The boson field over **H**, intuitively, the quantum field whose quanta are the vectors of **H**,- can be defined in a variety of equivalent ways. For present purposes, it is efficient to define it in a way that emphasizes complex features, and so avoids the Weyl relations, in which they are less manifest. The quantum field **Q** over **H** is then defined as the mathematical structure with the following features and properties.

1) A complex Hilbert space **K**. (**K** is the 'Hilbert space of the quantized field'.)

2) A mapping C from **H** to densely defined operators on **K** that is complex linear, relative to the strong operations (involving closure following linear operations), and satisfying the commutation relations

$$C(z)^*C(w) - C(w)C(z)^* \subset \langle w,z\rangle \tag{16}$$

for arbitrary vectors w and z in **H**. (C(z) is the 'creation operator' for a particle of wave function z. The field $\phi(z)$ is the closure of the hermitian operator

$$\phi_0(z) = 2^{-1/2}[C(z)+C(z)^*] \quad). \tag{17}$$

3) A continuous unitary representation Γ of the unitary group U(**H**) on **H** into U(**K**) having the property that

$$\Gamma(U)C(z)\Gamma(U)^{-1} = C(Uz) \tag{18}$$

for arbitrary U ε U(**H**) and z ε H. (Γ(U) is how the unitary transformation U, e.g. time evolution or Lorentz transformation, on the 'single-particle space' **H** lifts up to the entire field **K**. But also, if **L** is a closed linear subspace of **H** and P is the projection on **L**, then the selfadjoint generator of the one-parameter group Γ(exp(itP) represents the 'number of particles' whose wave functions are in **L**.)

4) A unit vector v in K such that $\Gamma(U)v = v$. (v is the 'vacuum state vector'.)

These conditions admit a continuum of unitarily inequivalent solutions. For unicity and correspondence with heuristic physics, a stability condition, in the nature of energy positivity, is required.

5) For some strictly positive selfadjoint operator A in **H**, the selfadjoint generator $\partial\Gamma(A)$ of the one-parameter group $\Gamma(\exp(itA))$ is non-negative. (If e.g. A is the single-particle energy operator, $\partial\Gamma(A)$ is the energy operator of the quantized field.)

Perhaps the simplest construction for the boson field over **H** is to form the free commutative linear *-algebra **P** over **H** + **H***, where **H*** denotes the dual space, with unit e. There is a positive normalized linear functional E on P such that E(FG) = E(F)E(G) if F and G are elements of **P** based on orthogonal submanifolds of H+H*,where positive means that $E(uu^*) \geq 0$ and normalized that E(e) = 1. This has the property that $E(zz^*) = c||z||^2$ for z in **H**, where c is a constant, and becomes unique by specifying c = 1. The completion of the subalgebra **P**(**H**) generated by **H** (without **H***) relative to the inner product <F,G> = E(G*F) for F, G ε **P**, is then **K**. C(z) is the operation of multiplication by z, on the domain **P**(H). $\Gamma(U)$ is the unique unitary on **K** that carries z into Uz, for all z ε **H**. Finally, v = e.

In this form, the boson field over **H** is realized in the "complex wave" representation [20], so called because it pseudo-diagonalizes the creation operators C(z) (they become multiplication operators). This representation is equivalent to one in which the completion **K** can be expressed quite explicitly. **K** is then a subspace of the space of all antientire functions on **H** (i.e., such as have restrictions to finite-dimensional subspaces **M** that are the complex conjugates of entire functions on **M**). To be a member of **K**, all such restrictions must be square-integrable with respect to isotropic Gaussian measure dg on **M**: $dg(z) = (\pi)^{-m} \exp[-<z,z>]\, dz$, where dz is Lebesgue measure in **M**, as a 2m-dimensional real space), and the integrals of the absolute square must be uniformly bounded as **M** varies. The inner product in **K** is then the limit of the inner products in finite-dimensional subspaces, as they converge to all of **H**. There is a unique isomorphism of P(**H**) into the algebra of antientire functions on **H** that carries z ε **H** into the antientire function $f_z(u) = <u,z>$. (If entire rather than antientire functions were used, there would be a problem with this isomorphism and/or with positivity of the energy.)

A useful feature of this representation is the possibility of representing ultrasingular operators on K as integral operators with kernels K(z,z') that are entire and antientire in z and z' [21]. The simplest form of this representation involves a given operator B in H such that e^{-tB} is Hilbert-Schmidt for all t > 0. (B plays the role of the single-particle hamiltonian. In practical terms, the Hilbert-Schmidt character of B is equivalent to the compactness of space.) *Entire kernel* will mean a function K(z,z') on **E**(B) x **E**(B), where **E**(B) is the space of entire vectors for B in its natural topology, that is complex analytic as a function of z and anti-analytic as a function of z', and such that $K(e^{-tB}z, e^{-tB}z')$ is square-integrable over **H**+**H** for all sufficiently large real t. The ultrasingular operators treated are those represented by continuous sesquilinear forms F on **E**(H), where $H = \partial\Gamma(B)$. The representation theorem then states:

THEOREM 1. *For any such form F, there exists a unique entire kernel K(z,z') on E(B) such that for arbitrary u,u' in E(H),*

$$F(u,u') = \iint K(z,z')u(z)\overline{u}(z')dzdz', \tag{19}$$

where the integral is interpreted as the limit of the integrals over finite-dimensional subspaces as they converge to all of ***E****(B) (alternatively, integration in the sense of noncommutative integration theory, as adapted to the present commutative context, is applicable), and the inequality*

$$|K(z,z')| \leq C \exp[\,||e^{sB}z||^2 + ||e^{sB}z'||^2\,] \tag{20}$$

is satisfied for some s > 0. Conversely, if K(z,z') is an entire kernel on ***E****(B) that satisfies this inequality, then there exists a continuous sesquilinear form F on E(H) such that (19) holds.*

The analogous theorem holds if 'entire' vector is changed to 'analytic vector', and 'all t > 0' is replaced by 'all sufficiently large t'. With this change, the theorem is effectively conformally invariant, in the case of a field that is invariant under the conformal group (i.e. there is given aunitary positive-energy representation of the conformal group on **H**), if B corresponds to the generator of the center of the maximal essentially compact subgroup. Analytic vectors for B (resp. H) are then analytic for the entire group on **H** (resp. **K**) .

8. Cohomology of the infinite Heisenberg group

With the aid of the representation theory for ultrasingular operators, a variant of the quantum cohomology described above for the finite-dimensional case can be established in boson fields [22]. This in turn gives a basis for a canonical map from classical functionals such as hamiltonians or lagrangians on **H** to corresponding quantized entities, i.e. operators in **K**. We denote the spaces (with the natural topologies) of analytic vectors for B in **H** and for H in **K** as $\mathbf{H_0}$ and $\mathbf{K_0}$. An n-cocycle is here defined as a continuous multilinear map F from $\mathbf{H_0}$ to the space $\mathcal{L}(\mathbf{K_0})$ of all 'operators' in (i.e. continuous sesqulinear forms on) $\mathbf{K_0}$, that is a symmetric function of its n arguments, and has the property that

$$(21) \qquad [F(z_1, z_2, \ldots z_n), \phi(w)] = [F(w, z_2, \ldots, z_n), \phi(z_1)]$$

for arbitrary $z_1, z_2, \ldots, z_n, w$ in $\mathbf{H_0}$.

If F is an arbitrary 'operator' in $\mathcal{L}(\mathbf{K_0})$, its n-derivative is defined as the form $F(z_1, \ldots, z_n) = \partial(z_1)\ldots\partial(z_n)F$, where $\partial(z)$ denotes the operator $F \to [F, \phi(z)]$. The n-derivative is a cocycle, which may be called *exact.* The converse is a kind of non-commutative infinite-dimensional Poincare lemma for ultrasingular operators.

THEOREM 2. *Any cocycle F is the n-derivative of an 'operator' in* $\mathcal{L}(\mathbf{K_0})$.

This operator may be normalized so that it and its successive derivatives up to order n-1 have vanishing vacuum expectation value, and is then unique.

The mapping Q (for 'quantization') from cocycles to this unique operator is continuous.

If F is hermitian and invariant under the one-parameter group e^{itB}, *then Q(F) is essentially self-adjoint on* $\mathcal{L}(\mathbf{H_0})$.

There are two immediate applications for this result. First, Q carries conventional integrals of products of classical fields into integrals of Wick products of the corresponding quantized fields, without any further ado. Second, Q directly maps classical hamiltonians or lagrangians into corresponding quantum ones.

Suppose e.g. that $\phi(x)$ denotes a classical (numerically-valued) scalar field. Given a power $\phi(x)^k$ of this field, the corresponding power $\phi(x)^k$ of the quantized field is obtained by applying Q to the classical power. To do this, the classical power must be identified with a cocycle on the Hilbert space of normalizable classical fields.

Consider for example the wave or Klein-Gordon equation on $\mathbb{R}^1 \times S^3$:

$$[(\partial/\partial t)^2 - \Delta + 1 + m^2]\,\phi = 0, \tag{22}$$

where Δ denotes the Laplace-Beltrami operator on S^3, and m takes non-negative integral values (the 1 occurs because of the curvature of S^3, whose radius is normalized to unity). The Hilbert-Schmidt condition on the time evolution generator A for this equation is then satisfied. To correlate $\phi(x)^k$ with the corresponding cocycle, take successive commutators of the putative 'operator' $\phi(x)^k$ with the $\Phi(u)$ where the u are in $\mathfrak{E}(B)$, e.g. $\Phi(u_j) = \int \phi(x)\, f_j(x)dx$. If D denotes the commutator distribution for the quantized field, then formally

$$[\ldots[\phi(x)^k, \phi(y_1)], \phi(y_2)]\ldots, \phi(y_k)] = i^k k!\,\Pi_j\{\int D(x-y_j)\, f(y_j)\}. \tag{23}$$

The right hand side is a symmetric multilinear form on **E**(B) whose continuity, after integration with respect to $dy_1 \ldots dy_k$, is easily verified. Theorem 2 is applicable and shows the existence of a unique 'operator' characterized by the defining commutation relations and constraints on the vacuum expectation values. The result is in fact identical to the Wick product $:\phi(x)^k:$ (as a sesquilinear form on **E**(H)), which no longer appears only as an expedient or mere standardization device.

If $\phi(x)^k$ is a given interaction lagrangian, the classical action integral $\frac{1}{2}\int \phi(x)^k\, dx$ similarly maps via Q into the corresponding quantum action integral $\frac{1}{2}\int :\phi(x)^k:\, dx$, where the integration is over $S^1 \times S^3$. (The factor $\frac{1}{2}$ is required because $S^1 \times S^3$ is the two-fold cover of the conformal compactification of Minkowski space, in order to produce the counterpart to the conventional action integral in $\mathbf{M}_0$.) Because the solutions of equation (22) are periodic in time with period 2π, the quantum action integral is invariant under temporal evolution, and hence essentially self-adjoint.

In the interesting conformally invariant case p = 4 and m = 0, the action integral on $S^1 \times S^3$ is unitarily equivalent to the action integral $\int :\phi_0(x)^4:\, dx$, where the ϕ_0 denotes the wave equation free field in $\mathbf{M}_0$ and the integration is over all of $\mathbf{M}_0$. The nonvanishing character of the integral is implied e.g. by the nonvanishing of the corresponding cocycle, or by the analytically notably explicit analysis in the contribution of Zhengfang Zhou to this symposium. From the standpoint of conventional relativistic theory, it was quite unexpected that any nontrivial relativistic quantized action integral would be rigorously convergent, still less that it quantized action integral would be rigorously convergent, still less that it would be an essentially selfadjoint operator. There is no apparent way to obtain this result by analysis entirely within $\mathbf{M}_0$.

In a similar way, it can be shown that the wave equation quantized action integrals over $\mathbf{M}_0$, $\int :\phi_0(x)^k: dx$, where $k \geq 4$, are hermitian operators that have self-adjoint extensions. This result extends to Minkowski spaces of even higher dimension. It is not known if the action integrals are essentially self-adjoint, but their formulation as well-defined operators in Hilbert space makes them accessible by established mathematical techniques, and no longer possibly illusory. The action integrals are important in practical quantum field theory because they provide the leading nontrivial term in the perturbative S-matrix expansion, and in many particle contexts, the only term that is presently capable of being correlated with experiment.

Interaction hamiltonians are more singular than the corresponding interaction lagrangians. Theorem 2 can be used to establish interaction hamiltonians as continuous sesquilinear forms, in the cases just described, and in particular the integral over $\mathbf{R}^3$ defining the total interaction hamiltonian $\int :f(x)^k: d_3x$ is convergent as such. Whether the total hamiltonian is bounded below as a sesquilinear form is presently unknown, and appears dubious in $\mathbf{M}_0$, but has a better chance in $\tilde{M}$, in which the extra terms arising from the difference in the time evolution groups increases the hamiltonian.

The Klein-Gordon operator is not conformally covariant, and massive wave equations thereby show greater differences between their formulations in $\mathbf{M}_0$ and $\tilde{M}$. Massive nonlinear quantized scalar equations have convergent hamiltonians in $\tilde{M}$ for special group-theoretic reasons related to the positive curvature of space, which are beyond the scope of this article (see e.g. [23]). The results in $\tilde{M}$ suggest that convergence in $\mathbf{M}_0$ may be valid if the free particle Hilbert space is modified to represent the finite mass widths that are in fact physically observed for most particles. From the standpoint of practical quantum field theory, there would appear to be no essential difference between $\tilde{M}$ and $\mathbf{M}_0$ as space-time models. Indeed, $\mathbf{M}_0$ is often modified in practice by the imposition of periodic boundary conditions in time and space, arguably a more significant difference (e.g., Lorentz invariance is lost at the fundamental level, but retained in $\tilde{M}$).

These developments appear to provide a basis for convergent treatment of a significant part of nonlinear quantum field theory as modelled in the interaction representation. Many open questions remain, e.g. the analysis of the Heisenberg representations, the treatment of fermion fields and their interactions with bosons, and the asymptotics of physics in $\tilde{M}$ as its space curvature tends to zero. The Heisenberg fields are extremely interesting mathematically as the most direct quantum counterparts of the solutions of classical nonlinear wave equations, notwithstanding that the observable features of particles appear more practically and simply modelled in the interaction representation. A rigorous correspondence principle for fermion fields and their interaction with bosons, as in the prototypical case of quantum electrodynamics, will be treated elsewhere. Further properties of the quantization map Q need development, e.g. to what extent is positivity of a classical lagrangian or hamiltonian inherited by its quantization? In two-dimensional space-time, it is inherited to a substantial, but not total extent; in four space-time dimensions this question is open.

For many decades, the quantization of nonlinear wave equations in the physical context of four-dimensional space-time appeared a very natural and compelling idea, with considerable empirical success, which however always had the potential to turn out to be a specious, high-tech, illusion. Perhaps the main point of the developments reported here and in the contribution of Zhou is that the intuitively compelling idea of quantization, and its formulation of the fundamental physical principles of causality, symmetry, and stability, have basic mathematical reality as well as heuristic physical utility. Only the very first steps have been taken, but they show that there is some solid ground.

MATHEMATICS DEPARTMENT, MASSACHUSETTS INSTITUTE OF TECHNOLOGY, CAMBRIDGE, MA 02139

References

1. J. von Neumann, *Mathematical foundations of quantum mechanics*, Princeton Univ. Press, 1955 (Translated from the 1932 German work).

2. P. A. M. Dirac, in *From a life of physics*, International Atomic Energy Agency, Vienna, 1968, p. 26.

3. I. E. Segal, *Local nonlinear functions of quantum fields*, in *Functional Analysis*, ed. F. E. Browder, Springer, Berlin, 1970, pp. 188-210.

4. J. Schwinger, *Selected papers on quantum electrodynamics*, Dover, New York, 1958.

5. O. Nachtmann, *Elementary particle physics*, Springer, Berlin,1990.

6. G. C. Wick, *The evaluation of the collision matrix*, Phys. Rev. (2) **80** (1950, 268-272.

7. I. E. Segal, *Quantization of nonlinear systems*, J. Math. Phys.1 (1960), 468-488.

8. I. E. Segal, *Foundations of the theory of dynamical systems of infinitely many degrees of freedom, I*, Math.-fys. Medd. K. Danske Vidensk. Selsk **3 1**, no. 12 (1959), 1-38.

9. S. M. Paneitz, *Essential unitarization of symplectics and applications to field quantization*, J. Funct. Anal. **4 8** (1982), 310-359.

10. S. M. Paneitz & I. E. Segal, *Quantization and hermitian structures in partial differential varieties,* Proc. Natl. Acd. Sci. USA, **77** (1980), 6943-6947.

11. I. E. Segal, *Construction of nonlinear quantum processes, I,* Ann. Math. **9 2** (1970), 462-481.

12. I. E. Segal, *Construction of nonlinear quantum processes, II,* Inv. Math. **14** (1971), 211-242.

13. E. Nelson, *A quartic interaction in two dimensions*, in *Mathematical theory of elementary particles*, ed. R. Goodman & I. E. Segal, M.I.T. Press, Cambridge, MA, 1966, pp. 69-73.

14. I. E. Segal, Nonlinear functions of weak processes, I, J. Funct. Anal. **4** (1969), 404-456; II, ibid. **6** (1970), 29-75.

15. I. E. Segal, *Mathematical cosmology and extragalactic astronomy*, Academic, New York, 1976.

16. Y. Choquet-Bruhat, S. M. Paneitz, & I. E. Segal, The Yang-Mills equations on the universal cosmos, J. Funct. Anal. **5 3** (1983), 112-150.

17. J. Pedersen, I. E. Segal & Z. Zhou, *Massless nonlinear quantum field theories and the nontriviality of the biquadratic self-interaction in four dimensions*, Nucl. Phys. **B376** (1992), 129-142.

18. I. E. Segal, *Transformations in Wiener space and squares of quantum fields*, Adv. Math. **4** (1970), 91-108.

19. I. E. Segal, *Quantized differential forms*, Topology **7** (1968), 147-171.

20. I. E. Segal, *The complex-wave representation of the free boson field*, in Suppl. Studies, vol. 3, Adv. Math., Academic, New York, 1978, pp. 321-343.

21. S. M. Paneitz, J. Pedersen, I. E. Segal & Z. Zhou, *Singular operators on boson fields and the implementation of symplectic transformations*, J. Funct. Anal. **100** (1991), 36-58.

22. I. E. Segal & Z. Zhou, *Convergence of nonlinear massive quantum field theory in the Einstein Universe*, Ann. Phys. **218** (1992), 279-292.

23. J. Pedersen, I. E. Segal & Z. Zhou, *Nonlinear quantum fields in ≥ 4 dimensions and the cohomology of the infinite Heisenberg group*, Trans. Amer. Math. Soc. 345 (1994), 73-95

Proceedings of Symposia in Pure Mathematics
Volume **59**, 1996

THE RELATIVISTIC BOLTZMANN EQUATION

WALTER A. STRAUSS
Brown University

There are three main ways to model a gas; by statistical mechanics, by a kinetic model, or by a fluids model. In this lecture I will discuss the kinetic theory, which goes back to Boltzmann around 1870. Here we are interested in a gas that is relativistic. We write the distribution of particles (molecules) in relativistic notation as F(X,V) where X and V are 4-vectors. We take the speed of light c=1 and the mass of a single particle m=1. The Boltzmann equation has the form

$$V \cdot \nabla_X F = \mathcal{C}(F) \tag{1}$$

where the left side is the streaming term, the right side the collision term, and the dot is the Lorentz inner product.

In coordinates, the dot has the signature $(-, +, +, +)$, a point in space-time is denoted by $X = (x_0, ..., x_3)$ and the 4-momentum is denoted by $V = (v_0, ..., v_3)$. For any 4-vector $Y = (y_0, ..., y_3)$, we denote $Y^2 = Y \cdot Y$ (which can be positive or negative) and $|Y| = \sqrt{|Y^2|}$. We assume the 4-momentum satisfies

$$V^2 = 1 \text{ and } v_0 > 0. \tag{2}$$

We alternatively denote $V = (v_0, v)$ and $X = (-t, x)$ where v and x are 3-vectors, v_0 is the energy and t is the time. Hence $v_0 = \sqrt{1 + v^2}$. In this notation, (1) takes

---

1991 *Mathematics Subject Classification.* Primary 82Cxx, 81D25, 35B35.

the explicit form

$$(3) \qquad (-\sqrt{1+v^2})\left(-\frac{\partial F}{\partial t}\right) + v \cdot \nabla_x F = \mathcal{C}(F).$$

What about the collision operator? The only rule is the conservation of 4-momentum: $U + V = U' + V'$. Here U and V are the 4-momenta of the incident particles and U' and V' are the 4-momenta of the scattered particles. Only binary collisions are considered, as triple collisions form a set of measure zero in phase space. Motivated by these considerations, the collision operator takes the form

$$(4) \qquad \mathcal{C}(F)(V) = \frac{1}{2}\int_{R^{12}} [F(U')F(V') - F(U)F(V)]\ \delta(U^2 - 1)\ \delta(U'^2 - 1)$$
$$\delta(V'^2 - 1)\ \delta^{(4)}(U + V - U' - V')\ s\sigma(s,\theta)\ dUdU'dV'.$$

Here $F(U')F(V')$ represents the gain of the scattered particles, $-F(U)F(V)$ represents the loss of the incident particles, $s\sigma(s,\theta)$ is the scattering cross-section (the probability of collision), δ is the delta function in one variable, and $\delta^{(4)}$ is the delta function in four variables. The variables are $s = (U + V)^2$, which is related to the momentum difference of the particles, and

$$cos\theta = \frac{(V - U)\cdot(V' - U')}{|V - U||V' - U'|},$$

the generalized scattering angle.

For the rest of this lecture, I will use the non-relativistic notation $F(t, x, v)$. Carrying out the delta functions and simplifying, we then can rewrite the relativistic Boltzmann equation (4) in the form

$$(5) \qquad \frac{\partial F}{\partial t} + \frac{v}{\sqrt{1+v^2}} \cdot \nabla_x F = \int_{S^2 \times R^3} \sigma\ v_M\ [F(u')F(v') - F(u)F(v)]\ d\omega\ du$$

where

$$u' + v' = u + v, \quad \sqrt{1+u'^2} = \sqrt{1+v'^2} = \sqrt{1+u^2} + \sqrt{1+v^2}$$

$$v_M = \text{Möller velocity} = \left(\frac{s(s-4)}{4(1+v^2)(1+u^2)}\right)^{1/2}.$$

The equation can be shown to satisfy the principle of causality.

In the classical limit $c \to \infty$, we have $\sqrt{1+v^2} \to 1 - v^2/2 + \ldots$, as well as $v_M \to |u-v|$ and, by conservation of energy, $u'^2 + v'^2 = u^2 + v^2$.

The conservation laws for RB are mass, momentum and energy:

$$\iint F \, dvdx, \quad \iint v \, F \, dvdx \quad \text{and} \quad \iint \sqrt{1+v^2} \, F \, dvdx.$$

Boltzmann's H-Theorem asserts that the entropy of the system increases:

$$\frac{d}{dt}\iint F \log F \, dvdx \leq 0. \tag{6}$$

(The integral is the negative of the physical entropy.) Thus it is a natural conjecture that the system is driven to a state of maximal entropy. The states of maximal entropy are found by minimizing the entropy subject to the constraints of constant mass, momentum and energy. We find $\log F == a + b \cdot v + c\sqrt{1+v^2}$. Normalizing, we get

$$\mu(v) = e^{-\sqrt{1+v^2}}, \tag{7}$$

which is called the (global) relativistic Maxwellian or the Jüttner equilibrium. This is the expected limiting distribution.

In the classical (non-relativistic) case, it reduces to the Gaussian $exp(-v^2)$. The classical Boltzmann equation goes back to Boltzmann (1872). Expansions in small mean-free-path were introduced by Hilbert (1912) and improved by Chapman and Enskog (1916). The x-independent case was studied analytically by Carleman (1933). Finally, in a series of papers around 1960, Grad began the analytic study of the full Boltzmann equation. The question of the existence of solutions is not a trivial issue, because the gas is expected to develop vortices that should lead to turbulence. The state of mathematical knowledge today is as follows.

(A) For initial data near the Maxwellian μ , solutions F exist globally, are unique and are smooth. Furthermore $F \to \mu$ as $t \to \infty$. These results are due to Ukai and Nishida and Imai in the classical case, and Glassey and Strauss in the relativistic case.

(B) For"arbitrary" initial data, weak solutions exist globally (i.e., for all time) but are not known to be unique. This was proved by DiPerna and Lions in the classical case, and Dudiński and Ekiel in the relativistic case.

(C) If a gas is sufficiently dense, the mean-free-path ϵ is small. (Its reciprocal ϵ^{-1} is a coefficent in front of the collision term.) The Hilbert-Chapman-Enskog expansion is in powers of ϵ. The limiting behavior as $\epsilon \to 0$ is believed to be governed by the classical equations of fluids like those of Euler or Navier-Stokes. Recent analytical results of this type are due to Bardos, Golse and Levermore, among others.

(D) Discrete models of the Boltzmann equation have been found to be much more tractable. The best known such model is Broadwell's.

From now on, we restrict ourselves to the relativistic case (5). First we state particular theorems alluded to in (A) and (B) above. In both of them we omit the precise assumptions on the scattering cross-section σ.

Theorem 1 [GS]. *Let μ be given by (7). Let the initial data have the form $F_0(x,v) = \mu(v) + \sqrt{\mu(v)} f_0(x,v)$. Assume*

$$\sup_v (1+v^2)\|f_0\|_{H_x^2} + \int (\int |f_0|\; dx)^2 dv$$

is sufficiently small. Then there exists a unique global solution $(0 \le t < \infty)$ *such that*

$$\frac{1+v^2}{\sqrt{\mu(v)}}\; \|F-\mu\|_{H_x^2} \;\le\; \frac{c}{(1+t)^{3/4}} \qquad (0 \le t < \infty).$$

Theorem2 ([DE],[A]). *Assume $F_0 \ge 0$ and merely*

$$\int_B \int_{R^3} (\sqrt{1+v^2} + |\log F_0|)\; F_0\; dvdx \; < \infty$$

for all bounded sets $B \subset R^3$.

(a) There exists a global solution $F \ge 0$ for which

$$\sup_{0 \le t < \infty} \int_B \int_{R^3} (\sqrt{1+v^2} + |\log F|)\; F\; dvdx \; < \infty.$$

(b) There exists a sequence $t_k \to +\infty$ such that $f(\cdot + t_k)$ converges strongly in L^1 to a relativistic global Maxwellian.

The deficiency of Theorem 1 is that it applies only to configurations that are close to equilibrium. The deficiency of Theorem 2 is that the solution in (a) might not be unique (nor smooth) and the limits in (b) might also not be unique (the constants a, b and c above (7) possibly being different for different sequences).

The idea of the proof of Theorem 2(a) is to approximate the equation, for instance by replacing $\mathcal{C}(F)$ by $\mathcal{C}(F)\; [1+\delta \int F dv]^{-1}$ for small δ, then prove the boundedness of the energy and entropy for the approximate equation, and finally pass to the weak limits. In the weak limit the energy might not be conserved but is only known to be bounded above by the initial energy.

The idea of the proof of Theorem 1 is as follows. We write $F(t,x,v) = \mu(v) + \sqrt{\mu(v)}f(t,x,v)$ to obtain

$$\frac{\partial f}{\partial t} + \frac{v}{\sqrt{1+v^2}} \cdot \nabla_x f + \nu f + Kf = \tilde{Q}(f), \tag{8}$$

where the last term is quadratic in the perturbation f, K is a certain (rather complicated) linear operator, and ν is a positive function. Two main ingredients are needed: estimates on the nonlinear term $\tilde{Q}$ and decay of the linearized equation. Here we sketch the proof of decay. The linearized equation is (8) with the term on the right side omitted. If the Kf term were also omitted, then the solutions decay exponentially because $\nu(v) \geq const > 0$. How do we handle K?

We begin with the L^2 estimate

$$\frac{d}{dt}\|f\|_{L^2}^2 + ((\nu + K)f, f) = 0.$$

By entropy, the last term is non-negative and vanishes only on the 5-dimensional nullspace $\mathcal{N}$ provided by the conservation laws. It is only $\mathcal{N}$ that impedes our proof of decay. In order to get around this difficulty, we use the streaming term following the method of Kawashima [K]. The nullspace $\mathcal{N}$ is generated by the five functions: $\sqrt{\mu(v)}$ times 1, v_1, v_2, v_3, and $\sqrt{1+v^2}$. The streaming term multiplies $\mathcal{N}$ by its coefficient $v_k/\sqrt{1+v^2}$ so that we are led to the 14-dimensional subspace $\mathcal{M}$ of pseudo-moments generated by the functions: $\sqrt{\mu(v)}$ times 1, v_j, $\sqrt{1+v^2}$, $v_j/\sqrt{1+v^2}$, and $v_jv_k/\sqrt{1+v^2}$. Next we use a "multiplier" or "energy" method within $\mathcal{M}$ to obtain the missing estimate on $\mathcal{N}$. This is accomplished on the

Fourier-transformed (linearized) solution $\hat{f}(t,\xi,v)$. We obtain an estimate of the type

$$\frac{dE}{dt}+\delta\frac{|\xi|^2}{1+|\xi|^2}E\leq \text{error terms},$$

where E is approximately $\|f\|_{L^2}^2$. For $|\xi|>1$ we deduce exponential decay, while for $|\xi|<1$ we deduce merely

$$\int_{R^3} e^{-\delta t|\xi|^2}d\xi = O(t^{-3/2}).$$

REFERENCES

[A] H. Andréasson, *A regularity property and strong L^1 convergence to equilibrium for the relativistic Boltzmann equation*, preprint.

[BGL] C. Bardos, F. Golse and D. Levermore, *Fluid dynamical limits of kinetic equations*, J. Stat. Phys. **63** (1991), 323-344; Comm. Pure Appl. Math. **46** (1993), 667-753.

[B] L. Boltzmann, *Weitere Studien über das Wärmegleichgewicht unter Gasmolekülen*, Sitzungsberichte Akad. Wiss. Wien **66** (1872), 275-370.

[C] T. Carleman, *Sur la théorie intégro-differentielle de Boltzmann*, Acta Math. **60** (1933), 91-146.

[DL] R. DiPerna and P-L. Lions, *On the Cauchy problem for Boltzmann equations*, Ann. Math. **130** (1989), 321-366.

[DE] M. Dudińskii and M. Ekiel-Jezewska, *Global existence proof for relativistic Boltzmann equation*, J. Stat. Phys. **66** (1992), 991-1001.

[GS] R. Glassey and W. Strauss, *Asymptotic stability of the relativistic maxwellian via fourteen moments*, Trans. Th. Stat. Phys. **24** (1995), 657-678; see also Publ. RIMS Kyoto Univ. **29** (1993), 301-347.

[G] H. Grad, *Asymptotic theory of the Boltzmann equation II*, Rarified Gas Dynamics (J. Laurmann, ed.), 1963, pp. 26-59.

[H] Hilbert, *Begründung der kinetischen Gastheorie*, Math. Ann. **72** (1916), 562-577.

[K] S. Kawashima, *The Boltzmann equation and thirteen moments*, Japan J. Appl. Math. **7** (1990), 301-320.

[NI] T. Nishida and K. Imai, *Global solutions to the initial-value problem for the nonlinear Boltzmann equation*, Publ. RIMS Kyoto Univ. **12** (1976), 229-239.

[U] S.Ukai, *On the existence of global solutions of a mixed problem for the nonlinear Boltzmann equation*, Proc. Japan. Acad. **50** (1974), 179-184; see also C. R. Acad. Sci. Paris **282A** (1976), 317-320.

Proceedings of Symposia in Pure Mathematics
Volume **59**, 1996

Microlocal Analysis and Nonlinear PDE

MICHAEL E. TAYLOR

University of North Carolina
Chapel Hill NC 27599

The process of localization, i.e., a map of the form $u \mapsto \varphi(x)u$ where $\varphi \in C_0^\infty(\mathbb{R}^n)$, is of frequent use in both linear and nonlinear PDE. In addition to localizing in space, one often localizes in frequency, i.e., one uses $u \mapsto \varphi(D)u$, where $\varphi(D)$ has the effect of multiplying the Fourier transform $\hat{u}(\xi)$ of u by $\varphi(\xi)$. One can combine these two types of operations, to produce 'microlocal analysis.' Thus, we consider operators of the form

$$\varphi(x, D)u = \int \varphi(x, \xi)\hat{u}(\xi)e^{ix\cdot\xi}\, d\xi. \tag{1}$$

If $\varphi(x, \xi) = \sum_{|\alpha|\le m} a_\alpha(x)\xi^\alpha$, then $\varphi(x, D)$ is just the linear differential operator $\sum a_\alpha(x)D^\alpha$. There are various 'symbol classes,' such as $S^m_{\rho,\delta}$, introduced in [H1], where, with $\langle\xi\rangle = (1 + |\xi|)^{\frac{1}{2}}$,

$$\varphi(x, \xi) \in S^m_{\rho,\delta} \iff |D_x^\beta D_\xi^\alpha \varphi(x, \xi)| \le C_{\alpha\beta}\langle\xi\rangle^{m-\rho|\alpha|+\delta|\beta|}. \tag{2}$$

We say $\varphi(x, D) \in OPS^m_{\rho,\delta}$. Typically, we require $0 \le \delta \le \rho \le 1$. For example, the differential operator of order m mentioned above belongs to $OPS^m_{1,0}$.

If $\delta < \rho$, there is a useful symbol calculus, arising from

$$p_1(x, D)p_2(x, D) = a(x, D), \tag{3}$$

where

$$p_j(x, \xi) \in S^{m_j}_{\rho,\delta} \Longrightarrow a(x, \xi) = p_1(x, \xi)p_2(x, \xi) \bmod S^{m_1+m_2-(\rho-\delta)}_{\rho,\delta}. \tag{4}$$

On the other hand, if $\delta \le \rho$ and also $\delta < 1$, one has the following boundedness on Sobolev spaces. If $p(x, \xi) \in S^m_{\rho,\delta}$,

$$p(x, D) : H^{s,p}(\mathbb{R}^n) \longrightarrow H^{s-m,p}(\mathbb{R}^n), \quad 1 < p < \infty, \tag{5}$$

1991 *Mathematics Subject Classification.* Primary 35S50.

for all $s \in \mathbb{R}$. There are also Hölder estimates:

$$p(x, D) : C^s(\mathbb{R}^n) \longrightarrow C^{s-m}(\mathbb{R}^n), \tag{6}$$

if $s, s-m \in (0, \infty) \setminus \mathbb{Z}^+$. Proofs of these results can be found in a number of places, such as [H2], [S2], and [T1].

These results have a well-known role in linear PDE. For the simplest application, suppose $p(x, D)$ is an elliptic differential operator (with smooth coefficients) of order m, i.e., $|p(x, \xi)| \geq C|\xi|^m$ for $|\xi| \geq B$. Then

$$q(x, \xi) = \psi(\xi) p(x, \xi)^{-1} \in S^{-m}_{1,0}, \tag{7}$$

if $\psi \in C^\infty(\mathbb{R}^n)$, $\psi(\xi) = 0$ for $|\xi| \leq B$, $\psi(\xi) = 1$ for $|\xi| \geq 2B$. It follows from (4) that

$$q(x, D) p(x, D) u = u + r(x, D) u, \quad r(x, \xi) \in S^{-1}_{1,0}. \tag{8}$$

Standard results on elliptic regularity follow easily from this.

There is a plethora of other applications to linear PDE, many of which are given in [H2], [T1], and [Tr]. Our aim here is to discuss some applications to nonlinear PDE, which have played a significant role since the foundational work of J.-M.Bony [B1] and Y.Meyer [M1], [M2] in about 1980.

We begin with an analysis of $F(u)$, for smooth F, given in [M1]. Take $\Psi_0 \in C^\infty_0(\mathbb{R}^n)$, $\Psi_0(\xi) = 1$ for $|\xi| \leq 1$, and set $\Psi_k(\xi) = \Psi_0(2^{-k}\xi)$, $u_k = \Psi_k(D)u$. Then

$$F(u) = M(x, D)u + F(u_0) \tag{9}$$

where the formula

$$M(x, D)u = \sum_{k \geq 0} \{F(u_{k+1}) - F(u_k)\} \tag{10}$$

yields

$$M(x, \xi) = \sum_k m_k(x) \psi_{k+1}(\xi), \quad m_k(x) = \int_0^1 F'(\Psi_k(\tau; D)u)\, d\tau, \tag{11}$$

with

$$\psi_{k+1}(\xi) = \Psi_{k+1}(\xi) - \Psi_k(\xi), \quad \Psi_k(\tau; D) = \Psi_k(D) + \tau \psi_{k+1}(D). \tag{12}$$

To estimate $M(x, \xi)$, given $u \in L^\infty(\mathbb{R}^n)$, we have, by the chain rule,

$$\|D^\ell_x m_k\|_{L^\infty} \leq C_\ell \sum_{1 \leq \nu \leq \ell} \|D^{\ell_1} u_{k+1}\|_{L^\infty} \cdots \|D^{\ell_\nu} u_{k+1}\|_{L^\infty} \cdot \|F''\|_{C^{\nu-1}}. \tag{13}$$

Also,

$$\|D^{\ell_j} u_{k+1}\|_{L^\infty} \le C_{\ell_j} 2^{k\ell_j} \|u\|_{L^\infty}. \tag{14}$$

Since $2^{k\ell} \sim \langle\xi\rangle^\ell$ on the support of $\psi_{k+1}(\xi)$, we see that

$$u \in L^\infty(\mathbb{R}^n) \Longrightarrow M(x,\xi) \in S^0_{1,1}. \tag{15}$$

Now $\delta = \rho = 1$ is the 'bad' case we avoided when writing down (4)-(6). However, as proved in [S1] (see also [Bour]), for $p(x,\xi) \in S^m_{1,1}$, (5) holds, provided $s - m > 0$, and so does (6).

One immediate consequence of this is that, for $s > 0$, $p \in (1,\infty)$,

$$\begin{aligned} \|F(u)\|_{H^{s,p}} &\le \|M(x,D)u\|_{H^{s,p}} + \|F(u_0)\|_{H^{s,p}} \\ &\le C(\|u\|_{L^\infty})\{\|u\|_{H^{s,p}} + 1\}, \end{aligned} \tag{16}$$

which is a 'Moser estimate,' established in [Mos] and of frequent use in nonlinear PDE. The fact that this rather subtle and powerful estimate follows so readily is a good preliminary indication of the power of (9)-(15) as a tool in nonlinear analysis.

In order to have a symbol calculus available, one splits such a symbol as $M(x,\xi)$ into two pieces:

$$M(x,\xi) = M^\#(x,\xi) + M^b(x,\xi), \tag{17}$$

where

$$M^\#(x,\xi) = \sum_k J_k m_k(x)\, \psi_{k+1}(\xi), \tag{18}$$

and J_k are smoothing operators (in the x variable), forming an approximate identity. Possible choices of J_k are

$$J_k = \Psi_0(2^{-k\delta}D), \quad \text{or} \quad J_k = \Psi_{k-5}(D), \tag{19}$$

where $\delta \in (0,1)$. Given $r > 0$, we have

$$u \in C^r \Longrightarrow M^\#(x,\xi) \in S^0_{1,\delta}, \quad M^b(x,\xi) \in S^{-r\delta}_{1,1}. \tag{20}$$

If we take $\delta < 1$, then the symbol calculus (4) applies. If instead we take $J_k = \Psi_{k-5}(D)$, then there is a replacement operator calculus, given by [M1], [B1]. We have $M^\#(x,\xi)$ in the symbol class $\mathcal{B}S^0_{1,1}$, where

$$p(x,\xi) \in \mathcal{B}S^m_{1,1} \Longleftrightarrow p(x,\xi) \in S^m_{1,1}, \text{ and } \operatorname{supp} \hat{p}(\eta,\xi) \subset \{|\eta| \le \rho|\xi|\}, \tag{21}$$

for some $\rho \in (0,1)$. Operators in $OP\mathcal{B}S^m_{1,1}$ satisfy (5), for all $s \in \mathbb{R}$, not just $s > m$. A more general operator calculus has been developed in [Bour] and [H3].

For a nonlinear differential operator, there is a similar construction. We can write

$$F(x, D^m u) = M(x, D)u + R(u), \tag{22}$$

where $R(u) = F\big(x, D^m \Psi_0(D)u\big)$ and

$$u \in C^m \Longrightarrow M(x,\xi) \in S^m_{1,1}, \tag{23}$$

and we can write $M(x,\xi) = M^{\#}(x,\xi) + M^b(x,\xi)$ where, given $r > 0$,

$$u \in C^{m+r} \Longrightarrow M(x,\xi) \in S^m_{1,\delta}, \quad M^b(x,\xi) \in S^{m-r\delta}_{1,1}. \tag{24}$$

As an application of the results stated above, we can establish a Schauder type elliptic regularity result for a solution to a completely nonlinear elliptic PDE. Suppose

$$F(x, D^m u) = g(x). \tag{25}$$

That this is elliptic implies that $M(x,\xi) \in S^m_{1,1}$ is elliptic, and that $M^{\#}(x,\xi) \in S^m_{1,\delta}$ is elliptic. Pick $\delta < 1$. Then the symbol calculus (4) yields a parametrix $E \in OPS^{-m}_{1,\delta}$ of $M^{\#}(x,D)$, such that $EM^{\#}(x,D) = I$ modulo a smoothing operator. Writing (25) as $M^{\#}(x,D)u = g - M^b(x,D)u$ and applying E, we have

$$u = Eg - EM^b(x,D)u, \quad \text{mod } C^\infty. \tag{26}$$

Suppose we assume initially that

$$u \in C^{m+\varepsilon}, \quad \varepsilon > 0. \tag{27}$$

Also assume that $g \in H^{s,p}(\mathbb{R}^n)$, $1 < p < \infty$, $s > 0$. The hypothesis (27) implies $M^b(x,D) \in OPS^{m-\varepsilon\delta}_{1,1}$. Hence the right side of (26) belongs to $H^{s+m,p}(\mathbb{R}^n) + C^{m+\varepsilon+\varepsilon\delta}(\mathbb{R}^n)$. This is contained in $H^{m+\gamma,p}(\mathbb{R}^n)$, for any $\gamma < \min(s, \varepsilon + \varepsilon\delta)$. Given this, we have $Eg - EM^b(x,D)u \in H^{s+m,p}(\mathbb{R}^n) + H^{m+\gamma+\varepsilon\delta,p}(\mathbb{R}^n)$. Iterating this argument, we obtain

$$g \in H^{s,p}(\mathbb{R}^n) \Longrightarrow u \in H^{s+m,p}(\mathbb{R}^n), \tag{28}$$

for a solution to (25) when this PDE is elliptic, given $1 < p < \infty$, $s > 0$, and (27). A similar argument yields

$$g \in C^s(\mathbb{R}^n) \Longrightarrow u \in C^{s+m}(\mathbb{R}^n), \tag{29}$$

for a solution u to (25), in the elliptic case, given $s \in \mathbb{R}^+ \setminus \mathbb{Z}^+$ and (27).

Note that the seminorms of $E(x,\xi)$ in $S^{-m}_{1,\delta}$ depend only on $||u||_{C^m}$ while those of $M^b(x,\xi)$ in $S^{m-\epsilon\delta}_{1,1}$ depend on $||u||_{C^{m+\epsilon}}$. Using this, we can establish the following Moser type estimate for a solution u to (25), when it is elliptic:

$$(30) \qquad ||u||_{C^{m+s}} \leq A_s(||u||_{C^m})||g||_{C^s} + B_s(||u||_{C^{m+\epsilon}}),$$

the significant part of this estimate being the linear dependence on $||g||_{C^s}$. Note that the $m = 0$ case of (30), which is an estimate on a solution to $F(u) = g$ when F is invertible, is just a little weaker than the estimate (16) (with F replaced by F^{-1} and $H^{s,p}$ replaced by C^s).

Another important Moser estimate is the commutator estimate

$$(31) \quad ||P(fu) - fPu||_{H^{s,p}} \leq C||f||_{\mathrm{Lip}^1}||u||_{H^{m-1+s,p}} + C||f||_{H^{m+s,p}}||u||_{L^\infty},$$

when $s \geq 0$, $p \in (1,\infty)$, and P is a differential operator of order $m \in \mathbb{Z}^+$. This was extended in [KP] to the case $P \in OPS^m_{1,0}$, $m > 0$. In [T2] it was shown how this result can be derived from the paradifferential operator calculus.

In the approach of [T2], to establish (31), one starts with the following representation of a product:

$$(32) \qquad fg = T_f g + T_g f + R(f,g),$$

where T_f is Bony's 'paraproduct,' defined by

$$T_f g = \sum_{k\geq 5} \Psi_{k-5}(D)f \cdot \psi_{k+1}(D)g.$$

This arises from the construction (9)-(19), applied to $F(f,g) = fg$, and with J_k given by the second formula in (19). Clearly $f \in L^\infty \Rightarrow T_f \in OPS^0_{1,1}$. In fact, it belongs to $OP\mathcal{B}S^0_{1,1}$, and hence is bounded on all the Sobolev spaces $H^{s,p}$, for $s \in \mathbb{R}, p \in (1,\infty)$. There are the following important estimates:

$$(33) \quad ||T_f g||_{L^p} \leq C_p||f||_{L^p}||g||_{\mathrm{BMO}}, \quad ||R(f,g)||_{L^p} \leq C_p||f||_{\mathrm{BMO}}||g||_{L^p},$$

for $p \in (1,\infty)$, which follow from work of [CM]; proofs are also given in [T2]. Another useful estimate, established in [T2], is

$$(34) \quad ||R(f,g)||_{H^{\sigma,p}} \leq C||f||_{\mathrm{Lip}^1}||g||_{H^{\sigma-1,p}}, \quad \sigma \in [0,\infty), \ p \in (1,\infty).$$

To apply the decomposition (32) to (31), we write

$$(35) \qquad \begin{aligned} f(Pu) &= T_f Pu + T_{Pu} f + R(f, Pu), \\ P(fu) &= PT_f u + PT_u f + PR(f,u). \end{aligned}$$

The operator calculus readily yields

$$\|T_f Pu - PT_f u\|_{H^{s,p}} \le C\|f\|_{\mathrm{Lip}^1}\|u\|_{H^{m-1+s,p}}, \tag{36}$$

for all $m, s \in \mathbb{R}$, $p \in (1,\infty)$. Meanwhile, (34) implies

$$\|R(f,Pu)\|_{H^{s,p}} + \|PR(f,u)\|_{H^{s,p}} \le C\|f\|_{\mathrm{Lip}^1}\|u\|_{H^{m-1+s,p}}, \tag{37}$$

provided $s \ge 0$, $m+s \ge 0$, and $p \in (1,\infty)$. To estimate $PT_u f$, we use the fact that $u \in L^\infty \Rightarrow T_u \in OP\mathcal{B}S^0_{1,1}$, so $T_u : H^{\sigma,p} \to H^{\sigma,p}$ for all $\sigma \in \mathbb{R}$, to get

$$\|PT_u f\|_{H^{s,p}} \le C\|u\|_{L^\infty}\|f\|_{H^{m+s,p}}. \tag{38}$$

Finally, if $u \in L^\infty$ and $m > 0$, Pu belongs to the 'Zygmund space' C^{-m}_* (which we define below; see (55)), and one can show that T_{Pu} belongs to $OP\mathcal{B}S^m_{1,1}$ and thus maps $H^{\sigma,p} \to H^{\sigma-m,p}$, for all $\sigma \in \mathbb{R}$. Hence

$$\|T_{Pu} f\|_{H^{s,p}} \le C\|u\|_{L^\infty}\|f\|_{H^{m+s,p}}, \tag{39}$$

provided $m > 0$, $1 < p < \infty$. Thus we have (31). For details of the arguments sketched above, see [T2].

We next sketch a method for establishing existence of solutions to first order quasilinear symmetric hyperbolic systems:

$$\frac{\partial u}{\partial t} = K(u,D)u, \quad u(0,x) = f(x), \tag{40}$$

where

$$K(u,D)u = \sum_{j=1}^{n} K_j(u)\partial_j u, \tag{41}$$

with $\partial_j u = \partial u/\partial x_j$. Here, f and u take values in $\mathbb{R}^\ell$, and each K_j is an $\ell \times \ell$ matrix valued function. The hypothesis that (40) is symmetric hyperbolic is that $K_j(u)^* = K_j(u)$. We could just as easily have $K_j = K_j(x,u)$.

We obtain a solution to (40) as a limit of solutions u_ε to

$$\frac{\partial u_\varepsilon}{\partial t} = J_\varepsilon K(J_\varepsilon u_\varepsilon, D)J_\varepsilon u_\varepsilon, \quad u_\varepsilon(0) = f, \tag{42}$$

where $J_\varepsilon = \Psi_0(\varepsilon D)$. For simplicity, write $K_\varepsilon = K(J_\varepsilon u_\varepsilon, D)$. For any $\varepsilon > 0$, (42) has a unique solution on some t-interval containing 0. We need to show that the size of the t-interval does not shrink to zero as $\varepsilon \to 0$, and to obtain appropriate bounds on u_ε. To do this, we estimate the rate of change of

$\|\Lambda^m u_\varepsilon(t)\|^2_{L^2}$, where $\Lambda^m \in OPS^m_{1,0}$ is Fourier multiplication by $\langle\xi\rangle^m$. Here m is some positive real number. We have

$$\begin{aligned}\frac{d}{dt}\|\Lambda^m u_\varepsilon\|^2_{L^2} &= 2(\Lambda^m J_\varepsilon K_\varepsilon J_\varepsilon u_\varepsilon, \Lambda^m u_\varepsilon)_{L^2} \\ &= 2(\Lambda^m K_\varepsilon J_\varepsilon u_\varepsilon, \Lambda^m J_\varepsilon u_\varepsilon)_{L^2} \\ &= A_1 + A_2,\end{aligned} \tag{43}$$

where

$$A_1 = 2(K_\varepsilon \Lambda^m J_\varepsilon u_\varepsilon, \Lambda^m J_\varepsilon u_\varepsilon)_{L^2}, \quad A_2 = 2\big([\Lambda^m, K_\varepsilon] J_\varepsilon u_\varepsilon, \Lambda^m J_\varepsilon u_\varepsilon\big)_{L^2}. \tag{44}$$

To estimate A_1, we have

$$\begin{aligned}(K_\varepsilon v, v)_{L^2} &= \sum \big(K_j(J_\varepsilon u_\varepsilon)\partial_j v, v\big)_{L^2} \\ &= -\sum \big(\partial_j[K_j(J_\varepsilon u_\varepsilon)v], v\big)_{L^2},\end{aligned} \tag{45}$$

upon integrating by parts and using $K_j = K_j^*$. Taking the sum of the two expressions on the right, we obtain

$$(K_\varepsilon v, v)_{L^2} = \frac{1}{2} \sum \big([\partial_j K_j(J_\varepsilon u_\varepsilon)]v, v\big)_{L^2}. \tag{46}$$

Hence

$$\begin{aligned}|A_1| &\le \sum \|\partial_j K_j(J_\varepsilon u_\varepsilon)\|_{L^\infty} \|\Lambda^m J_\varepsilon u_\varepsilon\|^2_{L^2} \\ &\le C(\|J_\varepsilon u_\varepsilon\|_{C^1}) \|\Lambda^m J_\varepsilon u_\varepsilon\|^2_{L^2}.\end{aligned} \tag{47}$$

To estimate A_2, we have

$$[\Lambda^m, K_\varepsilon]v = \sum_j [\Lambda^m, K_j(J_\varepsilon u_\varepsilon)]\partial_j v, \tag{48}$$

and applying the estimate (31), with $s = 0$, $p = 2$, gives

$$\begin{aligned}&\|[\Lambda^m, K_\varepsilon]v\|_{L^2} \le \\ &C\sum_j \big\{\|K_j(J_\varepsilon u_\varepsilon)\|_{H^m} \|\partial_j v\|_{L^\infty} + \|K_j(J_\varepsilon u_\varepsilon)\|_{\mathrm{Lip}^1} \|\partial_j v\|_{H^{m-1}}\big\}.\end{aligned} \tag{49}$$

Hence, using (16) to estimate $\|K_j(J_\varepsilon u_\varepsilon)\|_{H^m}$, we have

$$|A_2| \le C(\|J_\varepsilon u_\varepsilon\|_{C^1}) \|\Lambda^m J_\varepsilon u_\varepsilon\|^2_{L^2}. \tag{50}$$

Thus we have

$$\frac{d}{dt}\|\Lambda^m u_\varepsilon\|_{L^2}^2 \leq C(\|J_\varepsilon u_\varepsilon\|_{C^1})\|\Lambda^m u_\varepsilon\|_{L^2}^2. \tag{51}$$

If $f \in H^m(\mathbb{R}^n)$ and $m > \frac{n}{2}+1$, so $\|v\|_{C^1} \leq C\|v\|_{H^m}$, we have

$$\frac{d}{dt}\|u_\varepsilon\|_{H^m}^2 \leq B(\|u_\varepsilon\|_{H^m}), \tag{52}$$

where the form of $B(\lambda)$ is independent of ε. Gronwall's inequality then yields an estimate

$$\|u_\varepsilon(t)\|_{H^m} \leq C(\|f\|_{H^m}), \quad \text{for } t \in (-T, T), \tag{53}$$

with T independent of ε. The PDE (42) then yields a bound on $\|\partial_t u_\varepsilon\|_{H^{m-1}}$, and a standard argument yields a limit $u_{\varepsilon_\nu} \to u$, solving (40). A variant of (43)-(51) demonstrates that the solution $u(t)$ persists as long as there is a bound on $\|u(t)\|_{C^1}$.

A more general class of first order quasilinear hyperbolic system (40) is the class of *symmetrizable* hyperbolic systems. If we do not assume that $K_j(u)^* = K_j(u)$, but that there exists $R(u)$, positive definite, such that $R(u)K_j(u)$ is symmetric for $1 \leq j \leq n$, then $R(u)$ is called a symmetrizer for (40). Basic examples arise in the equations of compressible fluid flow. There is a more general notion of symmetrizer, introduced by P.Lax. Namely, we consider $R(u, x, \xi)$, smooth on $\mathbb{R}^\ell \times \mathbb{R}^n \times (\mathbb{R}^n \setminus 0)$, homogeneous of degree 0 in ξ, which is positive definite and satisfies the property that $R(u, x, \xi)\sum K_j(u)\xi_j$ is symmetric. One can modify the analysis in (42)-(53) to treat such symmetrizable hyperbolic systems; see [T2].

Higher order quasilinear hyperbolic systems can be reduced to systems of a more complicated form than (40):

$$\frac{\partial u}{\partial t} = K(Au, x, D)u + F(x, Au), \quad u(0, x) = f(x), \tag{54}$$

where $A \in OPS^0_{1,0}$, and $K(v, x, \xi) \in S^1_{1,0}$. Again a treatment parallel to (42)-(53) is effective, though there are some significant differences. For example, dependence on $\|u\|_{C^1}$ is replaced by dependence on $\|Bu\|_{C^1}$, for some $B \in OPS^0_{1,0}$, which might not be bounded on C^1. This appears to lead to a weaker sort of persistence result than the one mentioned above. In fact, in all these cases such a persistence result can be strengthened, using a technique from [BKM]. Namely, for a solution to (54) to persist, in the hyperbolic case, it suffices to have a bound on $\|u(t)\|_{C^1_*}$, where we use the *Zygmund norm*:

$$\|u\|_{C^r_*} = \sup_{k \geq 0} \langle k\rangle^r \|\psi_k(D)u\|_{L^\infty(\mathbb{R}^n)}, \tag{55}$$

with $r = 1$. This result is proved in [T2].

Solutions to hyperbolic equations involve the propagation of waves, and at this point let us mention that the first major application of paradifferential operator calculus, in [B1], was to the study of propagation of singularities of solutions to nonlinear PDE. At the same time, an alternative approach to propagation of singularities, in the semilinear case, was given in [RR]. Variants of Bony's proof are also given in [H4] and [T2], and in [Be] there is an extensive discussion of the semilinear case. Further important developments include [Del], [MSZ]; see also [BMR] for a number of survey papers. A related subject is the propagation of highly oscillatory solutions to nonlinear wave equations; papers on this include [HMR], [JMR].

We next discuss another commutator estimate, first established in [CRW], and then used in [CLMS]: given $P \in OPS^0_{1,0}$, $1 < p < \infty$,

$$\|fPu - P(fu)\|_{L^p} \le C_p \|f\|_{\mathrm{BMO}} \|u\|_{L^p}. \tag{56}$$

In [AT], there is a proof of this result (with the BMO-norm replaced by the slightly stronger bmo-norm) using paradifferential operator calculus, in particular the decomposition (35). In place of (37)-(39), we have

$$\begin{aligned} \|T_{Pu}f\|_{L^p} + \|R(f, Pu)\|_{L^p} + \|PT_u f\|_{L^p} + \|PR(f,u)\|_{L^p} \\ \le C_p \|f\|_{\mathrm{BMO}} \|u\|_{L^p}, \end{aligned} \tag{57}$$

as a consequence of (33). One gets (56) (with bmo) from this and the following result of [AT]:

$$\|[P, T_f]u\|_{L^p} \le C \|f\|_{C^0_*} \|u\|_{L^p}, \tag{58}$$

where C^0_* is a Zygmund norm, as in (55). Several proofs of this are given in [AT]. One approach used there is to give a refinement of the analysis of products of operators in [M1] and then show that

$$f \in C^0_*(\mathbb{R}^n) \Longrightarrow [P, T_f] \in OP\mathcal{B}S^0_{1,1}, \tag{59}$$

modulo a smoothing operator, given $P \in OPS^0_{1,0}$, or more generally, (essentially) $P \in OP\mathcal{B}S^0_{1,1}$.

One primary corollary of the commutator estimate (56) is a 'div-curl lemma,' as discussed in [CLMS]. We give here an abstract version of such a result, using a formulation of P.Auscher and the author. Consider a bilinear form

$$\mathcal{P}u \cdot \mathcal{Q}v = \sum_{j=1}^{N} (P_j u)(Q_j v), \tag{60}$$

where $\mathcal{P}$, $\mathcal{Q} \in OPS^0_{1,0}$ (more generally, we can take $\mathcal{P}, \mathcal{Q} \in OPBS^0_{1,1}$). Here, u and v can take values in $\mathbb{R}^k$ and $\mathbb{R}^\ell$, respectively, so $\mathcal{P}$ is a $k \times N$ matrix of operators and $\mathcal{Q}$ is an $\ell \times N$ matrix of operators. Take

$$f \in \text{bmo}, \quad u \in L^p(\mathbb{R}^n), \quad v \in L^{p'}(\mathbb{R}^n). \tag{61}$$

We have

$$\begin{aligned}\Bigl(f, \sum_j (P_j u)(Q_j v)\Bigr) &= \int f(\mathcal{P}u) \cdot (\mathcal{Q}v)\, dx = \int v\ \mathcal{Q}^t(f\mathcal{P}u)\, dx \\ &= (v, [\mathcal{Q}^t, M_f]\mathcal{P}u) + (fv, \mathcal{Q}^t\mathcal{P}u).\end{aligned} \tag{62}$$

Now, we make the hypothesis that

$$h = \mathcal{Q}^t\mathcal{P}u = \sum Q_j^t P_j u \in L^r(\mathbb{R}^n), \quad r > p. \tag{63}$$

Then, if f has support in a compact set $K \subset \mathbb{R}^n$, we have

$$|(fv, h)| \le C_K \|f\|_{\text{bmo}} \|v\|_{L^{p'}} \|h\|_{L^r}.$$

Since (56) implies that

$$\|[\mathcal{Q}^t, M_f]w\|_{L^p} \le C_p \|f\|_{\text{bmo}} \|w\|_{L^p}, \tag{64}$$

for $1 < p < \infty$, we have (when supp $f \subset K$)

$$\Bigl|\Bigl(f, \sum (P_j u)(Q_j v)\Bigr)\Bigr| \le C_{pK} \|f\|_{\text{bmo}} \Bigl(\|u\|_{L^p} + \|h\|_{L^r}\Bigr) \|v\|_{L^{p'}}. \tag{65}$$

In the standard div-curl lemma, $u = (u_1, \dots, u_n)$, $P_j u = u_j$, and $Q_j v = \partial_j \Lambda^{-1} v$, $1 \le j \le n$. Then (63) is the hypothesis that div $u \in H^{-1,r}(\mathbb{R}^n)$. One particularly successful application of this result is given in [Hel], on the regularity of harmonic maps of 2-dimensional Riemannian manifolds into spheres. See [Ev] for another proof of the div-curl lemma and application to results on partial regularity. We also mention the proof of [DM], which makes use of 'product renormalization.'

A number of variants of the div-curl lemma, discussed in [CLMS] and elsewhere, involve estimation of wedge products. We take the space here to mention an approach to such estimates, using the following estimate on a 'super-commutator.' Let M be a compact, oriented, n-dimensional Riemannian manifold. Let f be an ℓ-form, on M, set $W_f u = f \wedge u$, and define

$$\begin{aligned}[[\Lambda^{-1}d, W_f]] = [\Lambda^{-1}d, W_f] &\text{ if } \ell \text{ is even}, \\ \{\Lambda^{-1}d, W_f\} &\text{ if } \ell \text{ is odd},\end{aligned} \tag{66}$$

where $[A, B] = AB - BA$ and $\{A, B\} = AB + BA$. Here, d is the exterior derivative and $\Lambda = (I - \Delta)^{\frac{1}{2}}$. We prove the following estimate.

LEMMA. *For $1 < p < \infty$, we have*

$$\| [[\Lambda^{-1} d, W_f]]\beta\|_{L^p} \le C_p \|f\|_{\mathrm{bmo}} \|\beta\|_{L^p}. \tag{67}$$

PROOF: Write $W_f = \sum M_{f_i} W_{e_i}$ where e_i are smooth ℓ-forms and $\sum \|f_i\|_{\mathrm{bmo}} \sim \|f\|_{\mathrm{bmo}}$. Then

$$[[\Lambda^{-1} d, W_f]]\beta = \sum_i [\Lambda^{-1} d, M_{f_i}] W_{e_i}\beta + \sum_i M_{f_i}[[\Lambda^{-1} d, W_{e_i}]]\beta. \tag{68}$$

Now the estimate (56) applies to the first sum on the right. Since the principal symbol of $\Lambda^{-1} d$ is wedge by $i|\xi|^{-1}\xi$, we have

$$[[\Lambda^{-1} d, W_{e_i}]] \in OPS^{-1}_{1,0}, \tag{69}$$

so the estimate on the second term on the right side of (68) is elementary.

We apply this lemma, first to an estimate of $du \wedge dv$. Let u be a j-form and v a k-form on M, $j + k \le n - 2$. Let f be an ℓ-form, $\ell = n - j - k - 2$. We set $u = \Lambda^{-1}\tilde{u}$, $v = \Lambda^{-1}\tilde{v}$, and desire to estimate

$$\int f \wedge du \wedge dv = (W_f d\Lambda^{-1}\tilde{u}, \delta\Lambda^{-1} * \tilde{v}). \tag{70}$$

Here, δ is the adjoint of d, and $*$ is the Hodge star operator. Since $W_f \Lambda^{-1} dd\Lambda^{-1} = 0$, the right side of (70) is equal to

$$(\Lambda^{-1} dW_f d\Lambda^{-1}\tilde{u}, *\tilde{v}) = \big([[\Lambda^{-1} d, W_f]] d\Lambda^{-1}\tilde{u}, *\tilde{v}\big). \tag{71}$$

Applying the Lemma, we deduce that

$$\Big| \int f \wedge du \wedge dv \Big| \le C_p \|f\|_{\mathrm{bmo}} \|u\|_{H^{1,p}} \|v\|_{H^{1,p'}}. \tag{72}$$

Next, we estimate k-fold wedge products. Assume u_j are ℓ_j-forms, $\sum_{j=1}^{k} (\ell_j + 1) = m \le n$. Let f be an $(n - m)$-form. Then we will show that

$$\Big| \int f \wedge du_1 \wedge \cdots \wedge du_k \Big| \le C_p \|f\|_{\mathrm{bmo}} \|u_1\|_{H^{1,p_1}} \cdots \|u_k\|_{H^{1,p_k}}, \tag{73}$$

provided $p_j \in (1, \infty]$ and

$$\frac{1}{p_1} + \cdots + \frac{1}{p_k} = 1, \quad p_k \in (1, \infty). \tag{74}$$

To prove this, note that, since $du_1 \wedge \cdots \wedge du_{k-1}$ is closed, we can use Hodge theory to write

$$du_1 \wedge \cdots \wedge du_{k-1} = du + h, \tag{75}$$

where h is a harmonic form and

$$\begin{gathered} \|u\|_{H^{1,p}} + \|h\|_{L^\infty} \leq C\|u_1\|_{H^{1,p_1}} \cdots \|u_{k-1}\|_{H^{1,p_{k-1}}}, \\ \frac{1}{p_1} + \cdots + \frac{1}{p_{k-1}} = \frac{1}{p}, \quad p \in (1,\infty), \quad p_k = p'. \end{gathered} \tag{76}$$

Then, with $v = u_k$, we have

$$\int f \wedge du_1 \wedge \cdots \wedge du_k = \int f \wedge du \wedge dv + \int f \wedge h \wedge dv. \tag{77}$$

The last integral in (77) is easy to estimate, and the estimate (72) applies to the other integral on the right side of (77). This proves the desired estimate (73). The case $k = n$, $\ell_j = 0$ yields a Jacobian determinant estimate, which played a particularly significant role in [CLMS].

There are a number of other topics in nonlinear PDE in which microlocal analysis has been influential recently. We mention particularly the study of the Euler equation for incompressible fluid flow; see [Che]. Microlocal analysis in nonlinear PDE is still a young area, and one can expect a good deal of development to alter the landscape considerably over the next decade.

References

[AG] S.Ahlinac and P.Gerard, Operateurs Pseudo-differentiels et Theoreme de Nash-Moser, Editions du CNRS, Paris, 1991.

[AT] P.Auscher and M.Taylor, Paradifferential operators and commutator estimates, Comm. PDE, to appear.

[BKM] T.Beale, T.Kato, and A.Majda, Remarks on the breakdown of smooth solutions for the 3-d Euler equations, Commun. Math. Phys. 94(1984), 61-66.

[Be] M.Beals, Propagation and Interaction of Singularities in Nonlinear Hyperbolic Problems, Birkhauser, Boston, 1989.

[BMR] M.Beals, R.Melrose, and J.Rauch (eds.), Microlocal Analysis and Nonlinear Waves, IMA Vol.30, Springer, New York, 1991.

[B1] J.Bony, Calcul symbolique et propagation des singularities pour les equations aux derivees nonlineaires, Ann. Sci. Ecole Norm. Sup. 14(1981), 209-246.

[B2] J.Bony, Second microlocalization and propagation of singularities for semi-linear hyperbolic systems, Taniguchi Symp. Katata 1984, pp.11-49.

[Bour] G.Bourdaud, Une algebre maximale d'cperateurs pseudodifferentiels, Comm. PDE 13(1988), 1059-1083.

[Cal] A.P.Calderon, Singular integrals, Bull. AMS 72(1966), 427-465.

[Che] J.Chemin, Une facette mathematique de la mecanique des fluides, I, Publ. CNRS #1055, Paris, 1993.

[Chr] M.Christ, Lectures on Singular Integral Operators, CBMS Reg. Conf. Ser. in Math. #77, AMS, Providence, RI, 1990.

[CLMS] R.Coifman, P.Lions, Y.Meyer, and S.Semmes, Compensated compactness and Hardy spaces, J. Math. Pures Appl. 72(1993), 247-286.

[CM] R.Coifman and Y.Meyer, Au-dela des Operateurs Pseudo-differentiels. Asterisque, #57, Soc. Math. de France, 1978.

[CRW] R.Coifman, R.Rochberg, and G.Weiss, Factorization theorems for Hardy spaces in several variables, Ann. of Math. 103(1976), 611-635.

[Del] J.Delort, F.B.I. Transformation, Second Microlocalization, and Semilinear Caustics, LNM #1522, Springer, New York, 1992.

[DM] S.Dobyinsky and Y.Meyer, Lemme div-curl et renormalization, Sem. EDP, Centre de Mathematique, Ecole Polytechnique, 1991-92.

[Ev] L.Evans, Partial regularity for stationary harmonic maps into spheres, Arch. Rat. Mech. Anal. 116(1991), 101-113.

[FS] C.Fefferman and E.Stein, H^p spaces of several variables, Acta Math. 129(1972), 137-193.

[Hel] F.Helein, Regularite des applications faiblement harmonique entre une surface et une variete riemannienne, CRAS Paris 312(1991), 591-596.

[H1] L.Hörmander, Pseudo-differential operators and hypoelliptic equations, Proc. Symp. Pure Math. 10(1967), 138-183.

[H2] L.Hörmander, The Analysis of Linear Partial Differential Operators, vols.3-4, Springer-Verlag, 1985.

[H3] L.Hörmander, Pseudo-differential operators of type 1,1. Comm. PDE 13(1988), 1085-1111.

[H4] L.Hörmander, Non-linear Hyperbolic Differential Equations. Lecture Notes, Lund Univ., 1986-87.

[HMR] J.Hunter, A.Majda, and R.Rosales, Resonantly interacting weakly non-linear hyperbolic waves II: several space variables, Studies Appl. Math. 75(1986), 187-226.

[JMR] J.Joly, G.Metivier, and J.Rauch, Generic rigorous asymptotic expansions for weakly nonlinear multidimensional oscillatory waves, Duke Math. J. 70(1993), 373-404.

[KP] T.Kato and G.Ponce, Commutator estimates and the Euler and Navier-Stokes equations, CPAM 41(1988), 891-907.

[Mal] J.Marschall, Pseudo-differential operators with non regular symbols, Inaugural-Dissertation, Freien Universität Berlin, 1985.

[Ma2] J.Marschall, Pseudo-differential operators with coefficients in Sobolev spaces, Trans. AMS 307(1988), 335-361.

[MSZ] R.Melrose, A.SaBaretto, and M.Zworski, Diffraction of conormal waves, Preprint, 1993.

[M1] Y.Meyer, Regularite des solutions des equations aux derivees partielles non lineaires, Springer LNM #842(1980), 293-302.

[M2] Y.Meyer, Remarques sur un theoreme de J.M.Bony, Rend. del Circolo mat. di Palermo (suppl. II:1) (1981), 1-20.

[Mos] J.Moser, A rapidly convergent iteration method and nonlinear partial differential equations, I, Ann. Scuola Norm. Sup. Pisa 20(1966), 265-315.

[RR] J.Rauch and M.Reed, Propagation of singularities for semilinear hyperbolic equations in one space variable, Ann. Math. 111(1980), 531-552.

[Sem] S.Semmes, A primer on Hardy spaces, and some remarks on a theorem of Evans and Müller, Comm. PDE 19(1994), 277-319.

[S1] E.Stein, Singular Integrals and Pseudo-differential Operators, Graduate Lecture Notes, Princeton Univ., 1972

[S2] E.Stein, Harmonic Analysis, Princeton Univ. Press, 1993.

[T1] M.Taylor, Pseudodifferential Operators, Princeton Univ. Press, 1981.

[T2] M.Taylor, Pseudodifferential Operators and Nonlinear PDE, Birkhauser, Boston, 1991.

[T3] M.Taylor, Analysis on Morrey spaces and applications to Navier-Stokes and other evolution equations, Comm. PDE 17(1992), 1407-1456.

[T4] M.Taylor, Partial Differential Equations, Vols. 1-3, Springer, New York, to appear.

[Tr] F.Treves, Introduction to Pseudodifferential Operators and Fourier Integral Operators, Plenum, New York, 1980.

Research partially supported by NSF grant DMS 9400905.

Other Titles in This Series

(Continued from the front of this publication)

36 **Robert Osserman and Alan Weinstein, editors,** Geometry of the Laplace operator (University of Hawaii, Honolulu, March 1979)

35 **Guido Weiss and Stephen Wainger, editors,** Harmonic analysis in Euclidean spaces (Williams College, Williamstown, Massachusetts, July 1978)

34 **D. K. Ray-Chaudhuri, editor,** Relations between combinatorics and other parts of mathematics (Ohio State University, Columbus, March 1978)

33 **A Borel and W. Casselman, editors,** Automorphic forms, representations and L-functions (Oregon State University, Corvallis, July/August 1977)

32 **R. James Milgram, editor,** Algebraic and geometric topology (Stanford University, Stanford, California, August 1976)

31 **Joseph L. Doob, editor,** Probability (University of Illinois at Urbana-Champaign, Urbana, March 1976)

30 **R. O. Wells, Jr., editor,** Several complex variables (Williams College, Williamstown, Massachusetts, July/August 1975)

29 **Robin Hartshorne, editor,** Algebraic geometry – Arcata 1974 (Humboldt State University, Arcata, California, July/August 1974)

28 **Felix E. Browder, editor,** Mathematical developments arising from Hilbert problems (Northern Illinois University, Dekalb, May 1974)

27 **S. S. Chern and R. Osserman, editors,** Differential geometry (Stanford University, Stanford, California, July/August 1973)

26 **Calvin C. Moore, editor,** Harmonic analysis on homogeneous spaces (Williams College, Williamstown, Massachusetts, July/August 1972)

25 **Leon Henkin, John Addison, C. C. Chang, William Craig, Dana Scott, and Robert Vaught, editors,** Proceedings of the Tarski symposium (University of California, Berkeley, June 1971)

24 **Harold G. Diamond, editor,** Analytic number theory (St. Louis University, St. Louis, Missouri, March 1972)

23 **D. C. Spencer, editor,** Partial differential equations (University of California, Berkeley, August 1971)

22 **Arunas Liulevicius, editor,** Algebraic topology (University of Wisconsin, Madison, June/July 1970)

21 **Irving Reiner, editor,** Representation theory of finite groups and related topics (University of Wisconsin, Madison, April 1970)

20 **Donald J. Lewis, editor,** 1969 Number theory institute (State University of New York at Stony Brook, Stony Brook, July 1969)

19 **Theodore S. Motzkin, editor,** Combinatorics (University of California, Los Angeles, March 1968)

18 **Felix Browder, editor,** Nonlinear operators and nonlinear equations of evolution in Banach spaces (Chicago, April 1968)

17 **Alex Heller, editor,** Applications of categorical algebra (New York City, April 1968)

16 **Shiing-Shen Chern and Stephen Smale, editors,** Global analysis, Part III (University of California, Berkeley, July 1968)

15 **Shiing-Shen Chern and Stephen Smale, editors,** Global analysis, Part II (University of California, Berkeley, July 1968)

14 **Shiing-Shen Chern and Stephen Smale, editors,** Global analysis, Part I (University of California, Berkeley, July 1968)

13 **Dana S. Scott (Part 1) and Thomas J. Jech (Part 2), editors,** Axiomatic set theory (University of California, Los Angeles, July/August 1967)

(See the AMS catalog for earlier titles)

ISBN 0-8218-0381-6

9 780821 803813